Stainless Steel Wires-Engineered Multifunctional Ultra-High Performance Concrete

The emergence and application of stainless steel wires-engineered multifunctional ultra-high performance concrete advances the safety, durability, function/intelligence, resilience, and sustainability of infrastructure, thus prolonging the service life and reducing the maintenance to lower the life-cycle cost of infrastructure. This is the first reference work on this multifunctional concrete, which combines high performance with functional/smart properties, such as thermal, electrical, self-sensing, and electromagnetic properties, as well as a sustainable profile.

The book delivers both fundamentals and applications about multifunctional concrete, covering basic principles, properties, mechanisms, engineering application cases, and future development challenges and strategies.

Stainless Steel Wires-Engineered Multifunctional Ultra-High Performance Concrete opens up a new horizon for researchers and specialist technologists in the field of concrete materials and structures.

Stainless Steel Wires-Engineered Multifunctional Ultra-High Performance Concrete

Sufen Dong, Xinyue Wang
and Baoguo Han

CRC Press
Taylor & Francis Group
Boca Raton London New York

CRC Press is an imprint of the
Taylor & Francis Group, an **informa** business

MATLAB® is a trademark of The MathWorks, Inc. and is used with permission. The MathWorks does not warrant the accuracy of the text or exercises in this book. This book's use or discussion of MATLAB® software or related products does not constitute endorsement or sponsorship by The MathWorks of a particular pedagogical approach or particular use of the MATLAB® software.

First edition published 2024
by CRC Press
6000 Broken Sound Parkway NW, Suite 300, Boca Raton, FL 33487-2742

and by CRC Press
4 Park Square, Milton Park, Abingdon, Oxon, OX14 4RN

CRC Press is an imprint of Taylor & Francis Group, LLC

© 2024 Sufen Dong, Xinyue Wang and Baoguo Han

ISBN: 978-1-032-23236-2 (hbk)
ISBN: 978-1-032-23238-6 (pbk)
ISBN: 978-1-003-27635-7 (ebk)

DOI: 10.1201/9781003276357

Typeset in Sabon
by Deanta Global Publishing Services, Chennai, India

To the families!

Sufen Dong, Xinyue Wang, Baoguo Han

Contents

8 Wear resistance, damping, and electromagnetic
   properties of stainless steel wires-engineered
   ultra-high performance concrete   247

9 Stainless steel wires-engineered multifunctional ultra-high
   performance concrete incorporating nanofillers   265

About the authors

Sufen Dong, associate professor of transportation and logistics at Dalian University of Technology, Dalian, China, since 2021, earned her PhD in the field of structure engineering from the Dalian University of Technology in 2018. Her main research interests include multifunctional/smart concrete, nanotechnology, sensing technology, structural health monitoring, and traffic detection. She has published four book chapters and 50+ research papers, and holds five authorized national invention patents. Her papers have been published in journals including *Engineering, Cement and Concrete Composites, Materials and Design, Composite Part B: Engineering*, among others. Her research work has been supported by the National Natural Science Foundation of China and China Postdoctoral Science Foundation. She has served as an associate editor of the Chinese journal *Progress in Chinese Materials Science* and a reviewer for scientific research projects and for more than 20 journals. She was named in the world's top 2% scientists by Stanford University in 2022.

Xinyue Wang earned his PhD degree from Dalian University of Technology in 2021. He is now a postdoctoral fellow at the School of Civil Engineering, Dalian University of Technology. His main research interests include nano-engineered cementitious composites and interfaces in concrete materials and structures. He has published an authored book, one book chapter, and more than 30 technical papers in reputable journals, and holds two authorized national invention patents. He was honored with the Young Scientist Medal by the International Association of Advanced Materials in 2022.

Baoguo Han, professor of civil engineering at Dalian University of Technology since 2012, holds a BS and MS in Material Science and Engineering (Harbin Institute of Technology, Harbin, China, 1999, 2001) and a PhD in Engineering Mechanics (Harbin Institute of Technology 2005). He was invited to the University of Minnesota and has worked as a visiting research scholar there for three years. His main research interests include cement and concrete materials, nano-engineered cementitious composites, smart materials and structures, multifunctional concrete, fiber reinforced concrete, and ultra-high performance concrete. He has published five authored books, three edited books, 13 book chapters, and more than 200 technical papers. The citations and h-index/i10-index in Google are 10538 and 54/154, respectively. The research outcomes have been reported as research highlights more than 20 times. He has held more than 20 authorized national invention patents and was awarded the first prize of Natural Science by the Ministry of Education of China. Professor Han is a Fellow of the International Association of Advanced Materials, and has served as a member in more than 30 scientific professional societies (including an academic organization, as journal chief editor and editorial board member, and on a review panel for scientific research projects and awards) and as a reviewer for more than 150 books and journals. He has been awarded Top Peer Reviewer in the Global Peer Review Awards powered by Publons in both Materials Science and Cross-Field, listed in Highly Cited Chinese Researchers by Elsevier, and named in the world's top 2% scientists by Stanford University.

Foreword

In spite of being the most often used building material, concrete continues to encounter enormous challenges (e.g. a low fracture toughness, propensity to crack easily, poor deformability, inadequate durability, and an enormously negative impact on the environment). Clearly, concrete often fails to meet the requirements of safety, carbon-neutrality, longevity, functionality, and structural resilience. Multifunctional ultra-high performance concrete (UHPC) has the potential to address these lacunas of concrete by taking full advantage of our technological innovations of the last hundred years. Highly optimized for packing, using a micron-scale particle gradation, and toughened using stainless steel wires (SSWs), UHPC not only has reduced defect sizes but also has greater strength, enhanced toughness, and improved physical and electrical/thermal properties. SSWs-engineered UHPC thus simultaneously presents ultra-high properties (including excellent workability, mechanical performance, and durability) and functional/ smart properties (e.g. electrical, thermal, self-sensing, and electromagnetic abilities), thereby generating multifunctional UHPC indeed for use in our structures. Associate Professor Sufen Dong, Dr. Xinyue Wang, and Professor Baoguo Han from Dalian University of Technology have, based on their years of research experience in SSWs-engineered multifunctional UHPC, co-authored this excellent academic monograph.

The book consists of 12 chapters, covering the basic design principles, fabrication, properties, mechanisms, and application of SSWs-engineered multifunctional UHPC, three types of derived SSWs-engineered multifunctional UHPC, and future developmental challenges and suggested strategies of SSWs-engineered multifunctional UHPC. Clearly, compared with the other books in this field, the distinguishing feature of this book is the introduction of novel SSWs in multifunctional UHPC. The influence of SSWs on mechanical properties and durability as well as functional/smart properties such as electrical, thermal, self-sensing, electrothermal, and damping are elaborated and analyzed. The book further provides discussion on design, fabrications, life-cycle performance, and applications of SSWs-engineered multifunctional UHPC and three types of derived SSWs-engineered multifunctional UHPC. Finally, it elaborates on research and applications (codes and standards) challenges in the use of SSWs-engineered multifunctional UHPC.

I strongly recommend this book for both academics and professional engineers in the field of concrete materials and structures. I believe this book will promote the scientific research and engineering applications of multifunctional UHPC, promote further developments in this field, and lay the foundation for much-desired sustainable and carbon-neutral infrastructure.

Sincerely,

npBanthia

Nemkumar Banthia PhD PEng FACI FCAE FICI FCSCE FNAE FRSC
University Killam Professor and Canada Research Chair
The University of British Columbia, Vancouver, Canada

Preface

In its simplest form, concrete is a mixture of paste and aggregates. The paste, mainly composed of Portland cement and water, coats the surface of the fine or/and coarse aggregates. Through a chemical reaction called hydration, the paste hardens and gains strength to form the rock-like mass known as concrete. In this book, concrete is a collective term referring to conventional concrete (containing coarse and fine aggregates) and cement mortar (containing fine aggregates). Thanks to its excellent mechanical strength and resistance to water, being easily formed into various shapes and sizes, and being cheap and readily available everywhere, concrete has become the backbone of infrastructure worldwide – houses, schools, and hospitals as well as airports, bridges, highways, and rail systems. It has been reported that the anthropogenic mass on the earth will surpass bio-mass in the year 2020 ±6 years, and concrete will account for 40% of the anthropogenic mass. Therefore, concrete is the material most used to support human social civilization. Its production and application have a significant impact on resources and energy as well as the environment. For example, making 1 ton of cement requires about 2 tons of raw material (limestone and shale), consumes about 4 GJ of energy in electricity and process heat as well as transport (energy equivalent to 131 cubic meters of natural gas), and produces approximately 1 ton of CO_2 (creating concrete's substantial carbon footprint), about 3 kg of NO_X (an air contaminant that contributes to ground-level smog), and about 0.4 kg of PM10 (airborne particulate matter harmful to the respiratory tract when inhaled). Meanwhile, the requirement for natural materials in concrete production and application has been sharply increasing because of population growth, urbanization, and rapid industrial development in some countries or regions in the world, resulting in the depletion of natural aggregates and the shortage of fresh water. In addition, the extractive process of natural aggregates also has adverse impacts on the environment.

Although the production of concrete binder (i.e. cement) needs intensive resources and energy as well as generating a large environmental footprint (especially carbon footprint), concrete has a better ecological profile than other construction materials such as metal, glass, and polymers. Compared with other construction materials, the production of concrete consumes the lowest amount of materials and energy, produces the lowest amount of harmful by-products, and causes the least damage to the environment. In the foreseeable future, concrete will continue to play an important role in constructing infrastructure and promoting the development of human society. However, the development of concrete is encountering enormous problems and challenges. (1) Cement manufacturing has a direct and visible negative impact on the world's resources, energy consumption, and environmental footprint. Reducing the release of greenhouse gases (i.e. CO_2) associated with cement production and use must be a high priority for concrete development. (2) Many countries currently suffer from a shortage of the natural aggregates and fresh water that are essential for concrete production, leading to the high cost of concrete and other environmental impacts. (3) Increasing attention has been paid to security of infrastructure, since concrete is a brittle material, and it usually works even with cracks. Due to the large size and long life cycle of concrete infrastructure, enhancing the anti-cracking properties and toughness of concrete, and thus avoiding severe deterioration caused by uncontrolled cracking, can contribute significantly to green infrastructure and development of construction materials. (4) The durability of concrete infrastructure is becoming an increasingly important issue to ensure the predetermined service life and maintain community resilience, especially in extreme service environments. Improving the durability of concrete is therefore beneficial to keep the infrastructure sound and avoid unexpected failures, which can pose significant threats to public safety, and can decrease the life-cycle cost of infrastructure. However, due to the degeneration of concrete materials, the complex interaction between concrete materials and their service environment, and the absence of advanced design and condition assessment tools and timely maintenance, many concrete structures are in a state of utter disrepair. It is therefore necessary to render the failing infrastructures back to a serviceable and safe state. (5) Traditional concrete serving as a structural material cannot meet the upgrading requirements in terms of safety, longevity, resilience, and function of advanced engineering infrastructure. Multifunctional ultra-high performance concrete provides a suite of capabilities to address these unmet needs in the infrastructure field by making the material work to the maximum extent.

The introduction and development of ultra-high performance concrete (UHPC) can be traced back to 1981, when researchers proposed macro defect free cement. In 1993, Richard, a French scientific researcher, found that the absence of coarse aggregate and the application of pressure during the setting period can enhance the uniformity and density of concrete, while incorporating steel fibers can improve the toughness of concrete. This type of concrete with high strength, toughness, and density was named reactive powder concrete (RPC), one of the earlier research products of UHPC. In

1994, the concept of UHPC was formally proposed by De Larrard and Sedran in order to remove the commercial color of RPC and better express the excellent performance of concrete, which is gradually being widely accepted and adopted by researchers and engineers. Commonly, UHPC should have a compressive strength greater than 120 MPa and a sustained post-cracking tensile strength greater than 5 MPa; furthermore, it has high density and a discontinuous pore structure, significantly enhancing durability compared with conventional concrete. This leads to UHPC's ratio of compressive strength to density being about twice to four times that of normal concrete. To realize this excellent strength, toughness, and durability, the raw materials of UHPC mainly include cement, fly ash, silica fume, quartz powder, fine aggregate, and superplasticizer. Coarse aggregate was removed from UHPC to achieve high packing and reduce defects in the interface transition zone between the aggregate and the matrix. Meanwhile, the water–binder ratio of UHPC was usually controlled to between 0.11 and 0.25 to improve the density, silica-rich components were added to enhance the SiO_2 to CaO ratio of the concrete matrix, and steel fibers were added to reduce shrinkage, especially caused by the unevenly distributed hydration heat, and improve the toughness of UHPC. Based on the excellent performance of UHPC, it will definitely play an important role in fabricating large-scale infrastructure, represented by super-high and large-span structures. In addition, endowing UHPC with additional properties different from those of conventional concrete, such as self-sensing, self-healing, electrically conductive, thermal, electromagnetic, (super) hydrophobic, light-transmitting/emitting, photocatalytic, energy harvesting, and anti-bacterial/anti-viral properties, or the ability to react to an external stimulus, such as loading/deformation, temperature, and humidity, gives significant value to UHPC through the contributions of these properties to the durability, resilience, and sustainability of infrastructure. The multifunctionality of concrete is mainly achieved through material composition design, special processing, introduction of other functional components, or modification of microstructures. Among these, incorporating functional fillers is a commonly used method in existing research, and the traditional functional fillers mainly include carbon- and metal- based fillers with excellent thermal and electrical conductivity. Because it is easy for them to overlap into an electrically conductive or heat transfer network at the same dosage, the addition of functional fibers endows concrete with better functionality compared with that of particle fillers. Due to the uneven distribution and high price of carbon fibers, the performance of carbon fibers modified concrete is not easy to guarantee, and its large-scale application is limited. Because of their similar thermal expansion coefficient and good adaptability to the concrete matrix, steel fibers can not only improve the toughness of UHPC but can also be used as functional fillers. However, what is remarkable is that the diameter of steel fibers used in UHPC is usually larger than 0.16 mm, which is close to the particle size of fine aggregate. Therefore, although the addition of steel fibers improves the toughness of UHPC, it places a series of limitations on the performance optimization and application of UHPC. (1) The incorporation of steel fibers increases the heterogeneity of concrete and introduces a weak interface transition zone and macro defects into the concrete matrix, especially when the distribution of steel fibers is not uniform, and steel fibers sink and accumulate in high-fluidity UHPC. (2) The modification of the mechanical properties of UHPC by commonly used steel fibers usually occurs after cracking of the concrete matrix, endowing UHPC with strain-hardening and multi-cracking characteristics by performing a bridging function and being pulled out to improve the toughness of UHPC. However, it is worth noting that the multi-cracking state of UHPC during service accelerates the invasion of harmful media in extreme environments, thus accelerating the deterioration of infrastructure. (3) The high stiffness and large geometry of steel fibers limits the minimum size of UHPC elements, restricting the reduction of structure weight and the optimization of component design. (4) The high content of steel fiber has an adverse effect on the workability of UHPC, thus influencing the construction performance and application of UHPC. Although it has electrically and thermally conductive properties, steel fiber cannot realize percolation by forming an overlapping network at commonly used dosages; thus, UHPC with commonly used steel fibers is still mainly used as a structural material and cannot serve as multifunctional ultra-high performance concrete.

Refining the diameter of commonly used steel fiber to micron in order to prepare stainless steel wires (SSWs) is expected to endow UHPC with both excellent mechanical and multifunctional properties, providing feasibility for the development of multifunctional ultra-high performance concrete to meet the upgrading requirements of advanced engineering infrastructures. This results from the following series of advantages of SSWs. (1) The diameter of SSWs is close to that of cementitious material particles, which can eliminate the weak interface transition zone between the wires and the concrete matrix and reduce the influence on structural inhomogeneity caused by the large diameter of commonly used steel fibers. Meanwhile, due to the large specific surface area, silica fume and hydration products are adsorbed on the surface of wires to improve the dispersion of SSWs and enhance the structural uniformity and compactness of hardened UHPC. Furthermore, the high strength and modulus of SSWs are also conducive to improving the strength of UHPC. The improvement of matrix homogeneity, compactness, and strength of UHPC enables concrete infrastructure to remain crack free at high stress levels, enhancing the service capability and durability of infrastructure in extreme environments. (2) The high aspect ratio and flexibility prevent SSWs from settling in UHPC with high fluidity, which makes it possible to form a widely distributed overlapped network at low dosage, thus effectively hindering the initiation and propagation of microcracks to enhance the anti-cracking property and toughness of UHPC. The excellent strength and toughness of SSWs-engineered UHPC directly support infrastructure sustainability through reducing the section size and self-weight of elements. (3) Due to their good thermally and electrically conductive properties, the overlapped network of SSWs constitutes thermal and electrical conductive pathways in UHPC. The formation of a heat conduction network reduces shrinkage cracks caused by the uneven distribution of hydration heat to improve the durability of UHPC, and also endows UHPC with electrothermal and thermoelectric conversion performance. The existence of an electrically conductive network formed by SSWs has the potential to give UHPC additional functional and smart properties, such as self-sensing, electrically conductive, and electromagnetic properties. (4) The intrinsic stainless nature of SSWs is beneficial to maintain the long-term stability of the mechanical and multifunctional properties of UHPC and avoid the adverse effects of steel fiber corrosion on the long-term properties of concrete.

The application of SSWs-engineered multifunctional UHPC has huge potential to advance the safety, durability, multifunctionality/intelligence, resilience, and sustainability of civil infrastructure (e.g. low carbon footprint, low energy consumption, low resource consumption) and prolong the service life as well as reducing repair and maintenance to lower the life-cycle cost of infrastructure. Under the same bearing capacity, the employment of SSWs-engineered multifunctional UHPC can also decrease the section size of elements and optimize structural design to reduce the relative demand for concrete materials, and this is an effective approach to address the ecological issues (e.g. carbon dioxide emission, energy consumption, resource consumption) caused by the extensive use of concrete. Therefore, this book provides a summary report on the research results of SSWs-engineered multifunctional UHPC to help people working on this particular aspect to do their job better. This book covers basic principles, properties, mechanisms, engineering application cases, and future development challenges and strategies of SSWs-engineered multifunctional UHPC. The book is organized as shown in Figure 0.1. The first part provides a general introduction to the design principle and fabrication of SSWs-engineered multifunctional UHPC (Chapter 1). The second part presents the comprehensive properties of SSWs-engineered multifunctional UHPC, comprising the flexural and compressive properties (Chapter 2), bending and fracture properties (Chapter 3), impact properties (Chapter 4), fatigue properties (Chapter 5), electrical and self-sensing properties (Chapter 6), thermal and electrothermal properties (Chapter 7), as well as wear resistance, damping, and electromagnetic properties (Chapter 8). The third part features three types of derived SSWs-engineered multifunctional UHPC: SSWs-engineered multifunctional UHPC incorporating nanofillers (Chapter 9), SSWs-engineered

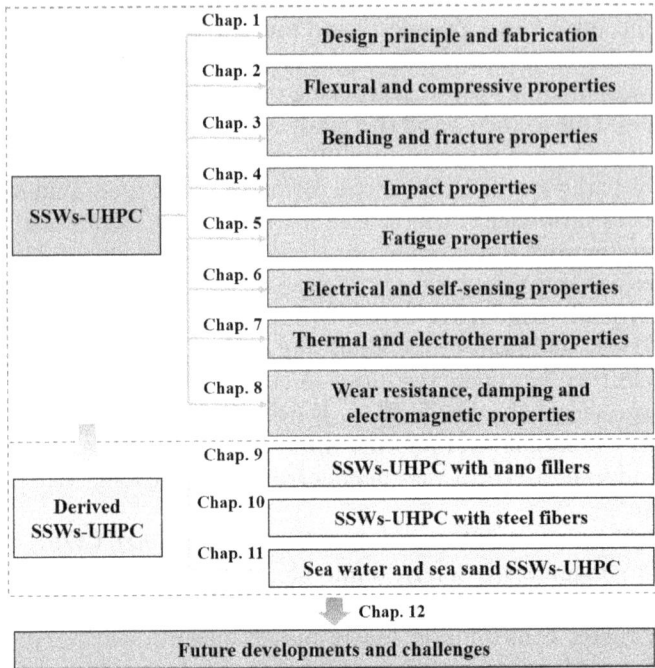

Figure 0.1 The main contents of this book

multifunctional UHPC incorporating steel fibers (Chapter 10), and SSWs-engineered multifunctional UHPC fabricated with seawater and sea sand (Chapter 11). Finally, the fourth part discusses the future developments and challenges of SSWs-engineered multifunctional UHPC (Chapter 12).

Acknowledgements

The authors are greatly indebted to Professor Jinping Ou (an academician of the Chinese Academy of Engineering) at Harbin Institute of Technology and Dalian University of Technology for giving us overall planning, detailed guidance, and great help with high-performance and multifunctional concrete. The authors are also deeply grateful to Professor Nemkumar Banthia (a Killam University Professor and Canada Research Chair, a fellow of the Royal Society of Canada [Canada's highest academic honor], an academician of the Canadian Academy of Engineering, and editor-in-chief of *Journal of Cement and Concrete Composite* at the University of British Columbia) for contributing the foreword to this book.

Many professional colleagues and friends have contributed directly or indirectly to this book: Ashour Ashraf (University of Bradford), Xun Yu (New York Institute of Technology), Xufeng Dong (Dalian University of Technology), Yanlei Wang (Dalian University of Technology), Weina Meng (Stevens Institute of Technology), Antonella D'Alessandro (University of Perugia), Wei Zhang (National University of Singapore), Danna Wang (Dalian University of Technology), Feng Yu (Dalian University of Technology), Shuoxuan Ding (Dalian University of Technology), Dongyu Wang (Dalian University of Technology), Liqing Zhang (East China Jiaotong University), Zhen Li (Harbin Engineering University), Siqi Ding (Harbin Institute of Technology, Shenzhen), Shuzhu Zeng (Dalian University of Technology), Jialiang Wang (Dalian University of Technology), Linwei Li (Dalian University of Technology), Liangsheng Qiu (Dalian University of Technology), Tong Sun (Dalian University of Technology), Chenyu Zhang (Dalian University of Technology), Xia Cui (Dalian University of Technology), Shan Jiang (Dalian University of Technology), Linyang Han (Dalian University of Technology), Qiaofeng Zheng (University of Cambridge), Yanfeng Ruan (Dalian University of Technology), Sichuan Shao (Dalian University of Technology), and Nueraili Maimaitituersun (Dalian University of Technology). The authors thank all of them most sincerely. We also thank Taylor & Francis/CRC Press for their enthusiastic and hard work to make the publication possible. We also thank the officer Ms. Mengying Ji of Hangzhou Linping Grand Theatre (the first example of UHPC being used in Chinese architecture, and the largest engineering project using UHPC in China) for her warm-hearted help and providing the exquisite theatre picture in Chapter 12 of this book. This book is financially supported by the National Science Foundation of China (grant nos. 52178188, 51978127, 51908103, 51578110, 51428801, 51178148, and 50808055), China Postdoctoral Science Foundation (grant nos. 2019M651116, 2022M720648, and 2022M710973), the Ministry of Science and Technology of China (grant nos. 2011BAK02B01, 2018YFC0705601, and 2017YFC0703410), and the Fundamental Research Funds for the Central Universities in China (grant no. DUT21RC(3)039).

Chapter 1

Design principle and fabrication of stainless steel wires-engineered multifunctional ultra-high performance concrete

1.1 INTRODUCTION

The core of materials science is understanding the central theme, i.e. the relationship of compositions–fabrication/processing–structures–properties/performance, thus improving the performance or designing new materials by tailoring the relationships of compositions, fabrication/processing, and structures (as shown in Figure 1.1) [1–3]. Unlike conventional structural concrete and functional materials, the design of multifunctional concrete should consider the intrinsic mechanics, durability, and processing properties as well as functional/intelligent properties, but the preparation of multifunctional concrete should be realized by using the simple and low-cost large-scale preparation method of conventional concrete. Stainless steel wires (SSWs)-engineered multifunctional ultra-high performance concrete (UHPC), as multifunctional concrete in the true sense, is a new structural-functional integrated/functional-smart integrated concrete derived from conventional UHPC.

Hence, this chapter introduces the design principle of SSWs-engineered multifunctional UHPC based on the design concept of conventional UHPC

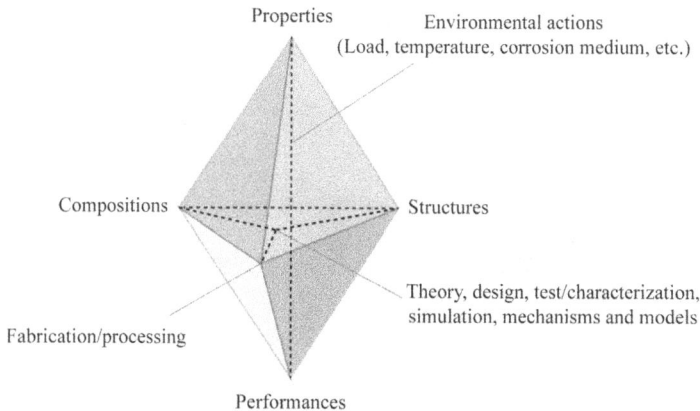

Figure 1.1 The central theme of material science and technology

DOI: 10.1201/9781003276357-1

1

and then gives the fabrication method of SSWs-engineered multifunctional UHPC combined with this design principle and the preparation of conventional UHPC.

1.2 DESIGN PRINCIPLE OF SSWS-ENGINEERED MULTIFUNCTIONAL UHPC

Multifunctional concrete is defined as a kind of structural-functional/functional-smart integrated concrete, which not only has good mechanics and durability but also has electrical, thermal, self-sensing, self-damping, electromagnetic, and other functions or intelligent characteristics [4]. The raw materials in conventional concrete (mainly including cement, water, fine aggregate, coarse aggregate, mineral admixture, chemical admixture, etc., as shown in Figure 1.2) are held together mainly relying on covalent bonds, ionic bonds, and van der Waals forces. Conventional concrete with moderate mechanical properties and durability (as shown in Table 1.1) serves only as a structural material, and it usually has no functional or intelligent properties. Incorporating functional fillers to improve the metal bond or proportion and increase electron and hole or heat conduction can endow conventional concrete with a certain function or smart performance. However, a high content of functional fillers is needed to obtain the desired function or smart performance, often causing negative effects on the mechanics and durability of conventional concrete. Hence, conventional concrete with functional fillers is basically not the real meaning of multifunctional concrete. The main reason for this is the existence of coarse aggregate. On the one hand, coarse aggregate accounts for a large proportion, about 40% volume, of concrete; on the other hand, coarse aggregate with its large particle size usually cuts off the connection between functional fillers.

UHPC is a kind of high-performance fiber-reinforced cement-based composite with a uniform compact structure and ultra-high strength/toughness as well as excellent durability, which is prepared by steam curing technology

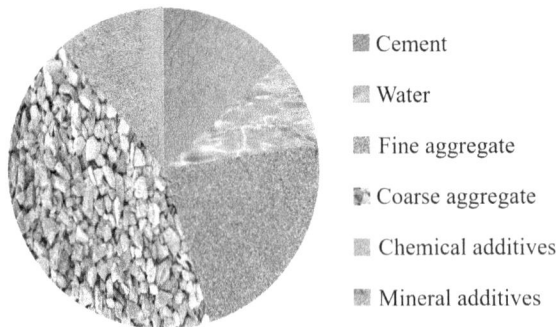

- Cement
- Water
- Fine aggregate
- Coarse aggregate
- Chemical additives
- Mineral additives

Figure 1.2 Raw materials in conventional concrete

Table 1.1 Comparison of main properties between UHPC and conventional concrete

Properties	Conventional concrete	UHPC	Comparison of UHPC and conventional concrete
Compressive strength (MPa)	20–50	120–230	~5 times
Flexural strength (MPa)	2–5	30–60	~10 times
Elastic modulus (GPa)	30–40	40–60	~1.2 times
Fracture toughness (kJ/m²)	0.12	20–40	~200 times
Creep coefficient	1.3–2.1	0.29–0.31	~1/5
Chloride ions diffusion coefficient (10^{-12} m²/s)	1.1	0.02	~1/50
Freeze–thaw peeling (g/cm²)	>1000	7	~1/140
Water absorption (kg/m³)	2.7	0.2	~1/13
Abrasion coefficient	4	1.3	~1/3
Specific compressive strength	1.7	6.5	~4 times

and the particle packing density principle. The design principle of UHPC is to minimize defects (pores and microcracks) in concrete and achieve high uniformity and compactness as well as strong matrix–aggregate interface bonding [5–7] by improving the fineness and activity of raw material (grinding quartz sand and using mineral admixtures), adopting thermal curing to promote the hydration reaction, removing coarse aggregate, using superplasticizer to ensure a low water–binder ratio, and modifying the composition and structure of hydration products (the core feature is to increase the amount of calcium silicate hydrate gel, reduce the amount of calcium hydroxide, increase the proportion of silicon oxygen bond, and reduce the proportion of calcium oxygen bond), as shown in Figure 1.3. In addition, it is equally important that steel fibers are used in the preparation of UHPC to increase the proportion of metal bonds in materials and then improve the natural deficiencies of concrete, such as brittleness, cracking, low tensile performance, and poor impact resistance [8, 9]. Therefore, UHPC has high-performance characteristics, including excellent mechanics and durability as well as working performance (as shown in Table 1.1). For example, the compressive strength of UHPC is higher than 150 MPa, which is more than three times that of ordinary concrete. The toughness of UHPC is more than 300 times that of ordinary concrete. The dead weight of a UHPC structure is only one-third or one-half that of an ordinary concrete structure [10–12]. It is worth noting that the electrical resistivity of UHPC prepared by the commonly used steel fibers with millimeter diameter and within the effective reinforcement content range is between 290 and 56.5 kΩ·cm [13]. This can be attributed to the fact that widely distributed steel fibers in UHPC cannot overlap with each other to form a conductive pathway.

Refining the diameter of commonly used steel fibers to micro scale to prepare SSWs with a high aspect ratio is expected to form a widely distributed overlapping strengthening/toughening/electrical/thermal conductive

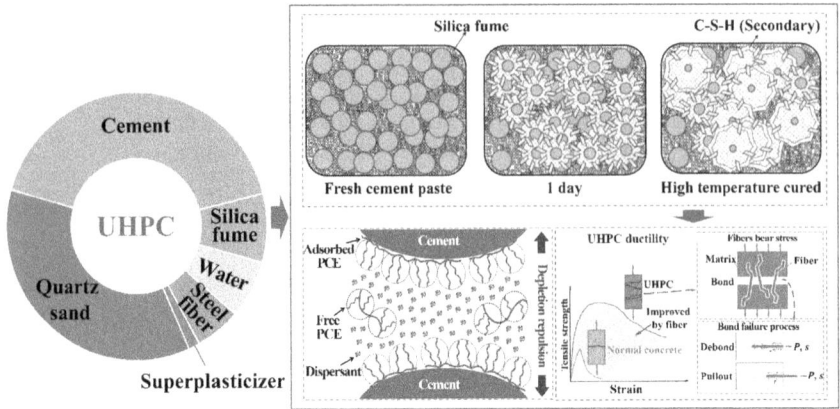

Figure 1.3 Design principle of UHPC. PCE represents polycarboxylates

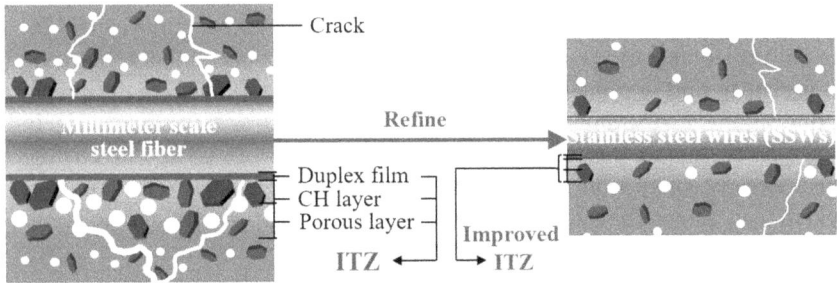

Figure 1.4 Interfacial transition zone (ITZ) between steel fibers/SSWs and concrete matrix

network in UHPC. There is a weak interface between the commonly used steel fibers and the concrete matrix [14–21]. When the diameter of a steel fiber is reduced to micro scale, the water sac that accumulates below the steel fiber becomes small, as shown in Figure 1.4. Meanwhile, the fine powder (e.g. silica fume, fly ash, ground slag) in UHPC can easily adhere to the surface of SSWs and continue to hydrate, thus enhancing the interface strength between SSWs and the concrete matrix [22–24]. In addition, if the mechanical and functional properties of concrete cannot maintain long-term stability, this kind of concrete is not multifunctional concrete in the true meaning. Steel fibers should be corrosion resistant in order to maintain the long-term stability of mechanical and functional properties. The selection of SSWs with stainless steel matrix and high corrosion resistance can solve this problem, and the dense microstructure of UHPC can protect SSWs from corrosion. The high flexibility of SSWs can solve the adverse effect of commonly used steel fibers on vehicle tires when UHPC is used as a pavement material [25]. To sum up, the design principle of the multifunctional UHPC engineered by SSWs is shown in Figure 1.5.

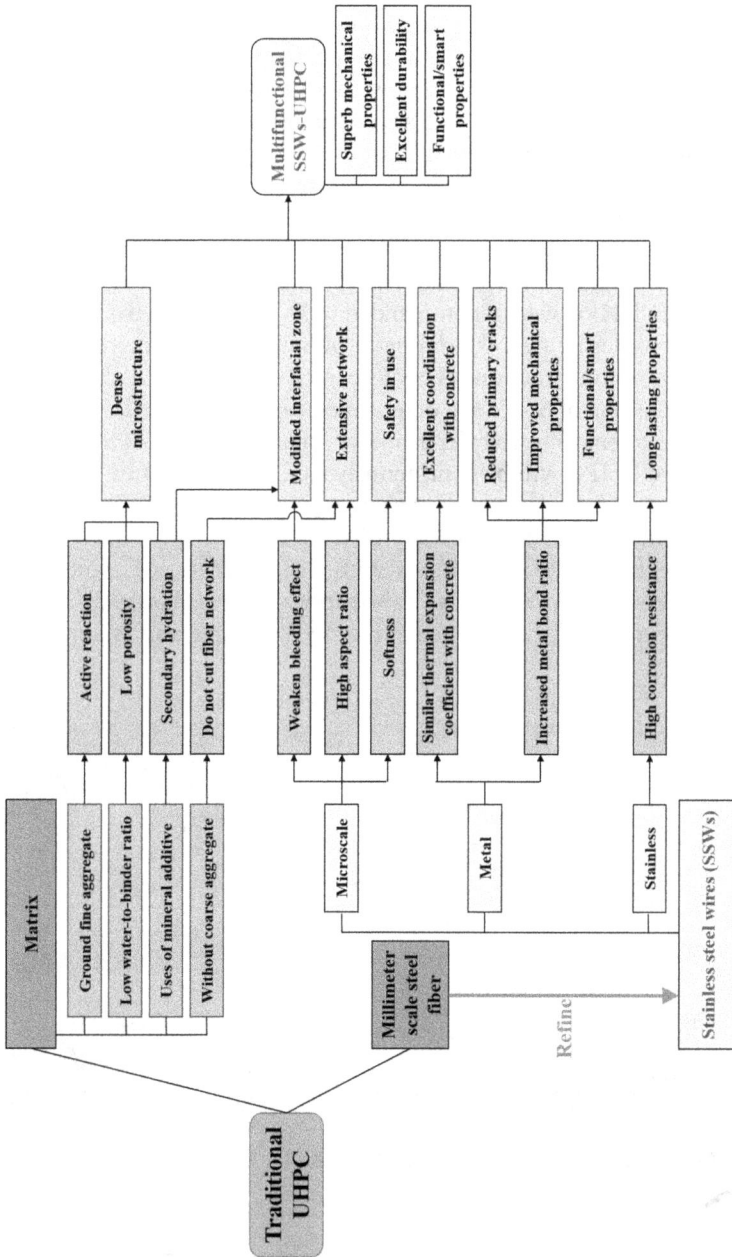

Figure 1.5 Design principle of SSWs-engineered multifunctional UHPC

1.3 FABRICATION OF SSWS-ENGINEERED MULTIFUNCTIONAL UHPC

Similarly to conventional UHPC, the raw materials of SSWs-engineered multifunctional UHPC mainly include cement, silica fume, fly ash, silica sand, and water reducing agent. Silica sand with a diameter of 0.12–0.85 mm is used to solve the blocking effect of aggregate on the SSWs' pathway. SSWs with diameter of 8 μm and 20 μm and length of 6 mm and 10 mm are incorporated to enhance UHPC, as shown in Figure 1.6 and Table 1.2. SSWs are made of 316L stainless steel by strong drawing. The surface of SSWs contains a certain amount of molybdenum, which provides a beneficial effect on corrosion resistance. It is worthwhile to note that the common volume fractions of steel fibers in UHPC were recommended to be in the range from 2.0 to 3.0 vol% [26]. Considering that the diameter of SSWs is only one-tenth that of steel fibers, the volume fractions of SSWs can be suitably reduced. Therefore, the volume fractions of SSWs are determined to be less than 1.5 vol%. The influence of different diameter, length, and volume fraction of SSWs on the performance of UHPC will be continuously studied in the future.

In order to give full play to the strengthening effect of SSWs on UHPC, and for the fine powder to be evenly distributed on the surface of SSWs, it is recommended to mix the SSWs with fine powders, e.g. silica fume, first. The following representative fabricating process has been adopted, as shown in Figure 1.7. First, silica fume, SSWs, water and water reducer

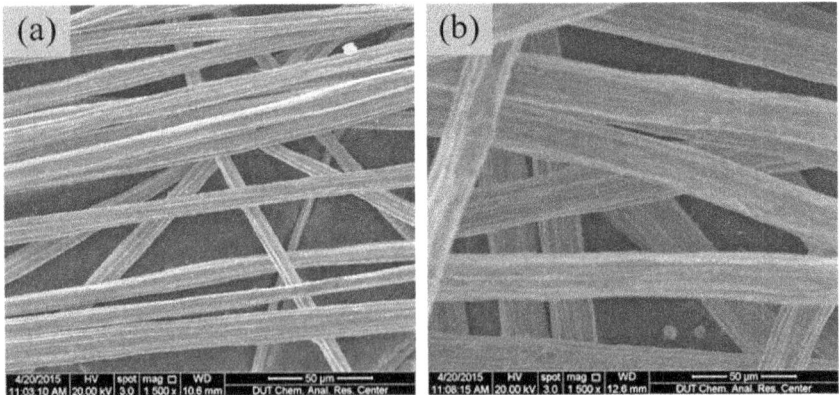

Figure 1.6 Morphology of SSWs: (a) 8 μm in diameter; (b) 20 μm in diameter

Table 1.2 Properties of SSWs

Type	Length (mm)	Diameter (μm)	Ductility	Tensile strength (MPa)
316L	6/10	8/20	>1%	≈780

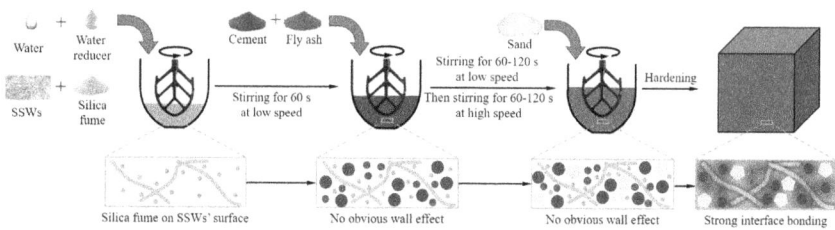

Figure 1.7 Fabrication process of SSWs-engineered multifunctional UHPC

are put into mixing pans together, stirring for 60 seconds at the low speed of 140 ± 5 r/min. Second, cement and silica fume are put into mixing pans together, stirring for 60–120 seconds at low speed and 60–120 seconds at the high speed of 285 ± 10 r/min separately. Third, quartz sand was put into the slurry, stirring for 60 seconds at low speed and 180–240 seconds at high speed.

1.4 SUMMARY

In this chapter, the design principle and fabricating process of SSWs-engineered multifunctional UHPC are introduced. The main conclusions are as follows:

1) The design principle of SSWs-engineered multifunctional UHPC is to take full advantage of SSWs and UHPC, thus obtaining a kind of multifunctional concrete in the true sense with high mechanics, excellent durability, good workability, stable function, and smart performance.
2) SSWs with a diameter of 8 μm and 20 μm, length of 6 mm and 10 mm, and content less than 1.5 vol% are incorporated into UHPC. The properties of UHPC reinforced with SSWs at other parameters of SSWs will be investigated in the future. SSWs are mixed with silica fume, cement, and fly ash first during the fabrication process to improve their dispersion and their interface with the concrete matrix.

REFERENCES

1. X. Li, K. Lu. Playing with defects in metals, *Nature Materials*. 16(7) (2017) 700–701.
2. B.G. Han, S.Q. Ding, J.L. Wang, J.P. Ou. *Nano-Engineered Cementitious Composites: Principles and Practices*. Springer, 2019.
3. B.G. Han, L.Q. Zhang, J.P. Ou. *Smart and Multifunctional Concrete Toward Sustainable Infrastructures*. Springer, 2017.

4. B.G. Han, X. Yu, J.P. Ou. *Self-Sensing Concrete in Smart Structures*. Elsevier, 2014.

5. D. Yoo, T. Oh, N. Banthia. Nanomaterials in ultra-high-performance concrete (UHPC)-A review, *Cement and Concrete Composites*. 134 (2022) 104730.

6. S. Zhang, J. Wang, G. Lin, T. Yu, D. Fernando. Stress-strain models for ultra-high performance concrete (UHPC) and ultra-high performance fiber-reinforced concrete (UHPFRC) under triaxial compression, *Construction and Building Materials*. 370(17) (2023) 130658.

7. Y.S. Liu, W.C. Tian, M.Z. Wang, B.M. Qi, W. Wang. Rapid strength formation of on-site carbon fiber reinforced high-performance concrete cured by ohmic heating, *Construction and Building Materials*. 244 (2020) 118344.

8. Y. Huang, J. Wang, Q. Wei, H. Shang, X. Liu. Creep behaviour of ultra-high-performance concrete (UHPC): A review, *Journal of Building Engineering*. 69(15) (2023) 106187.

9. C.C. Hung, H.S. Lee, S.N. Chan. Tension-stiffening effect in steel-reinforced UHPC composites: Constitutive model and effects of steel fibers, loading patterns, and rebar sizes, *Composites Part B: Engineering*. 134(1) (2018) 254–264.

10. H. Zhong, M. Chen, M.Z. Zhang. Effect of hybrid industrial and recycled steel fibers on static and dynamic mechanical properties of ultra-high performance concrete, *Construction and Building Materials*. 370 (2023) 130691.

11. A.M. Tahwia, M.A. Hamido, W.E. Elemam. Using mixture design method for developing and optimizing eco-friendly ultra-high performance concrete characteristics, *Case Studies in Construction Materials*. 18 (2023) 01807.

12. T.Y. Yin, K.N. Liu, D.Q. Fan, R. Yu. Derivation and verification of multi-level particle packing model for ultra-high performance concrete (UHPC): Modelling and experiments, *Cement and Concrete Composites*. 136 (2023) 104889.

13. Y. Bae, S. Pyo. Effect of steel fiber content on structural and electrical properties of ultra-high performance concrete (UHPC) sleepers, *Engineering Structures*. 222 (2020) 111131.

14. S.F. Dong, W. Zhang, D.N. Wang, X.Y. Wang, B.G. Han. Modifying self-sensing cement-based composites through multiscale composition, *Measurement Science and Technology*. 32(7) (2021) 074002.

15. S.F. Dong, X.Y. Wang, H.N. Xu, J.L. Wang, B.G. Han. Incorporating super-finer stainless wires to control thermal cracking of concrete structures caused by heat of hydration, *Construction and Building Materials*. 271 (2021) 121896.

16. S.F. Dong, Y.L. Wang, A. Ashour, B.G. Han, J.P. Ou. Uniaxial compressive fatigue behavior of ultra-high performance concrete reinforced with super-fine stainless wires, *International Journal of Fatigue*. 142 (2021) 105959.

17. S.F. Dong, X.F. Dong, A. Ashour, B.G. Han, J.P. Ou. Fracture and self-sensing characteristics of super-fine stainless wire reinforced reactive powder concrete, *Cement and Concrete Composites*. 105 (2020) 103427.

18. S.F. Dong, D.C. Zhou, Z. Li, X. Yu, B.G. Han. Super-fine stainless wires enabled multifunctional and smart reactive powder concrete, *Smart Materials and Structures*. 28(12) (2019) 125009.

19. S.F. Dong, D.C. Zhou, A. Ashour, B.G. Han, J.P. Ou. Flexural toughness and calculation model of super-fine stainless wire reinforced reactive powder concrete, *Cement and Concrete Composites*. 104 (2019) 103367.

20. S.F. Dong, B.G. Han, J.P. Ou, Z. Li, L.Y. Han, X. Yu. Electrically conductive behaviors and mechanisms of short-cut super-fine stainless wire reinforced reactive powder concrete, *Cement and Concrete Composites*. 72 (2016) 48–65.

21. B.G. Han, S.F. Dong, J.P. Ou, C.Y. Zhang, Y.L. Wang, X. Yu, S.Q. Ding. Microstructure related mechanical behaviors of short-cut super-fine stainless wire reinforced reactive powder concrete, *Materials and Design*. 96 (2016) 16–26.

22. S.F. Dong, D.Y. Wang, X.Y. Wang, A. D'Alessandro, S.Q. Ding, B.G. Han, J.P. Ou. Optimizing flexural cracking process of ultra-high performance concrete via incorporating microscale steel wires, *Cement and Concrete Composites*. 134 (2022) 104830.

23. S.F. Dong, X.Y. Wang, A.Ashour, B.G. Han, J.P. Ou. Enhancement and underlying mechanisms of stainless steel wires to fatigue properties of concrete under flexure, *Cement and Concrete Composites*. 126 (2022) 104372.

24. S.F. Dong, B.G. Han, X. Yu, J.P. Ou. Constitutive model and reinforcing mechanisms of uniaxial compressive property for reactive powder concrete with super-fine stainless wire, *Composites Part B: Engineering*. 166 (2019) 298–309.

25. S.X. Ding, S.F. Dong, X.Y. Wang, S.Q. Ding, B.G. Han, J. Ou. Self-heating ultra-high performance concrete with stainless steel wires for active deicing and snow-melting of transportation infrastructures, *Cement and Concrete Composites*. 138 (2023) 105005.

26. P. Song, S. Hwang. Mechanical properties of high-strength steel fiber-reinforced concrete, *Construction and Building Materials*. 18(9) (2004) 669–673.

Chapter 2

Flexural and compressive properties of stainless steel wires-engineered multifunctional ultra-high performance concrete

2.1 INTRODUCTION

Concrete, with its excellent compressive strength, has been widely used as a structural material for building the infrastructure that people need to survive. Due to its low flexural and tensile strength, normal concrete mainly bears compressive stress in the structural members [1]. Hence, compressive characteristics are not only an important factor to evaluate the performance of concrete but also the primary consideration to guide the design of concrete structures. Meanwhile, ultra-high performance concrete (UHPC) with high flexural and tensile strength is proposed to meet the requirements of structural members subjected to flexural/tensile load [2, 3]. This means that flexural resistance is also an important aspect for UHPC performance. These two aspects demonstrate that flexural and compressive properties are the two most basic mechanical characteristics of UHPC.

Therefore, the flexural (including flexural strength, flexural strain–stress/load–displacement curves, and flexural toughness) and compressive (including compressive strength after flexure, uniaxial compressive strength, elastic modulus, Poisson's ratio, and uniaxial compressive strain–stress curves) properties of stainless steel wires (SSWs)-engineered multifunctional UHPC cured at different regimes are introduced, and the uniaxial compressive constitutive model of composites based on damage theory is given. Meanwhile, the mechanisms by which SSWs reinforce UHPC are demonstrated through macroscopic, microscopic, and $Ca(OH)_2$ orientation analysis. The main contents of this chapter are shown in Figure 2.1.

2.2 FLEXURAL PROPERTIES OF SSWS-ENGINEERED MULTIFUNCTIONAL UHPC

2.2.1 Flexural strength

The flexural strength of SSWs-engineered multifunctional UHPC specimens with sizes of 40 mm × 40 mm × 160 mm under three-point load

DOI: 10.1201/9781003276357-2

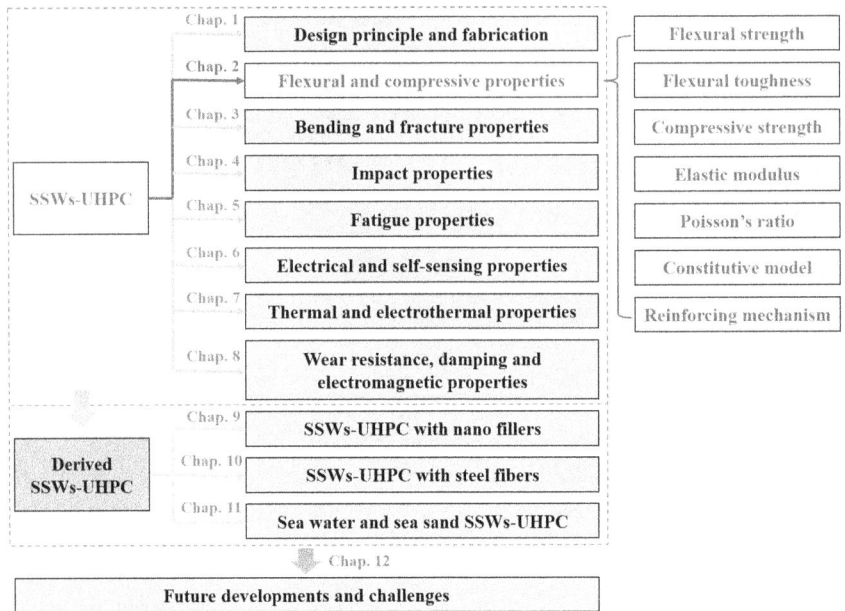

Figure 2.1 Main contents of Chapter 2

is shown in Figure 2.2. It can be seen from Figure 2.2 that the flexural strength increases with the increase of SSWs content. When the SSWs content is 1.5 vol%, the flexural strength of UHPC reinforced with SSWs at a diameter of 20 µm is increased by 100.7%, 70.5%, and 103.2% after water curing for 3 days, water curing for 28 days, and natural curing for 28 days, respectively, compared with that of UHPC without SSWs. Meanwhile, the flexural strength of UHPC reinforced with SSWs at a diameter of 8 µm is increased by 68.9%, 47.6%, and 77.1%, respectively. Under flexural load, microcracks initiate randomly at the bottom side of UHPC and then propagate and converge quickly, generating a negative effect on the flexural strength. The incorporation of SSWs can inhibit the coalescence of the microcracks, which in turn resists the propagation of the cracks and shields the crack tip from stress.

Figure 2.3 shows that the flexural strength of SSWs-engineered multifunctional UHPC changes with curing regimes. The flexural strength of UHPC without SSWs increases with water curing age but decreases with natural curing age. This phenomenon mainly depends on the hydration product characteristic of UHPC, which is dominated by C-S-H gels and has fewer calcium hydroxide crystals. As a result, microcracks are easily generated due to volume shrinkage in dry conditions [4]. When the SSWs content is low, the restriction is too small to prevent the development of shrinkage

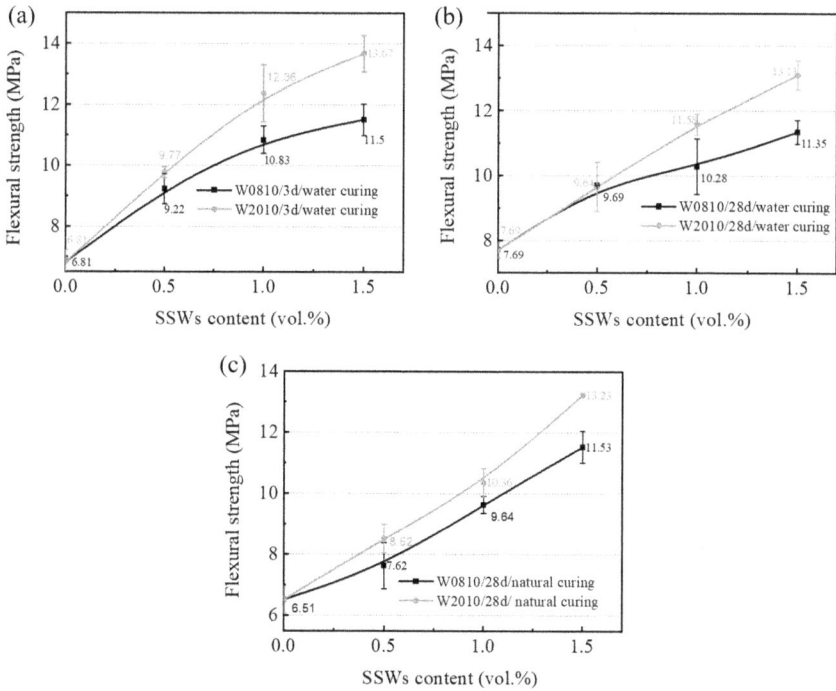

Figure 2.2 Effect of SSWs' content and diameter on the flexural strength of SSWs-engineered multifunctional UHPC cured at different regimes. (a) Water curing for 3 days; (b) Water curing for 28 days; (c) Natural curing for 28 days. W0810 represents incorporating SSWs with diameter of 8 μm and length of 10 mm into UHPC, W2010 represents incorporating SSWs with diameter of 20 μm and length of 10 mm into UHPC

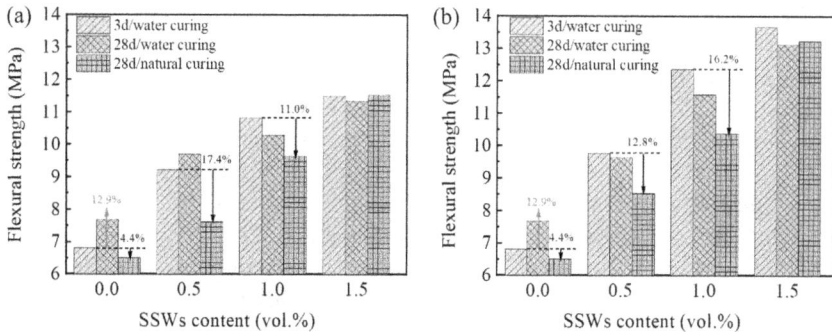

Figure 2.3 Effect of curing regimes on the flexural strength of SSWs-engineered multifunctional UHPC. (a) W0810; (b) W2010

cracks, and the incorporation of SSWs makes the porosity larger than that of composites without SSWs. These reasons are why the flexural strength after natural curing for 28 days falls significantly below that after water curing for 3 days (numbers illustrated in Figure 2.3). However, for UHPC reinforced with 1.5 vol% SSWs at a diameter of 8 and 20 µm, the SSWs are close enough to prevent crack convergence. The flexural strength of these after water curing for 28 days is slightly lower than that after water curing for 3 days, within the error of 10%. This can be attributed to the effect of high porosity due to the incorporation of SSWs.

2.2.2 Flexural load–displacement curves

The flexural toughness of composites is represented by the integral area under flexural load–displacement curves [5]. The flexural load–displacement curves and the corresponding flexural toughness of SSWs-engineered UHPC are shown in Figure 2.4. In this study, the load–displacement curves include a load adjustment process because of specimen size and experimental conditions. It can be seen from Figure 2.4 that the load–displacement curve of UHPC without SSWs falls sharply after reaching the ultimate load, which represents typical brittle characteristics. The incorporation of SSWs at a diameter of 8 µm does not change the brittleness. When the content of SSWs at a diameter of 20 µm is 1.0 or 1.5 vol%, the flexural load–displacement curves decrease slowly, i.e. the SSWs-engineered multifunctional UHPC still has a certain carrying capacity in the case of large deformation. This means that the failure characteristic of composites is transforming from brittleness to plasticity.

Figures 2.4(b), (d), and (e) show that the flexural toughness of SSWs-engineered multifunctional UHPC decreases significantly with curing age from 3 to 28 days for both water curing and natural curing, which can be attributed to the effect of water erosion and shrinkage microcracks. The results also show that the flexural toughness increases with the increase of SSWs content. Compared with UHPC without SSWs, the flexural toughness of UHPC reinforced with 1.0 vol%/1.5 vol% SSWs at a diameter of 20 µm is increased by 272.2% and 442.2%, respectively, after water curing for 3 days; 111.7% and 173.0%, respectively, after water curing for 28 days; and 142.2% and 293.2%, respectively, after natural curing for 28 days. These increments can be attributed to the pull-out effect of SSWs.

The pull-out process of SSWs is a toughening process [6, 7], which can be represented by Figure 2.5 [8]. Figure 2.5 shows the interaction between crack development and SSWs' bridging effect. Under flexural load, the cracks first appear at the bottom midspan of specimens; then, a small number of SSWs debond from the UHPC matrix, but most of them can still bridge microcracks. After that, the stress redistribution appears, and the SSWs begin to be pulled out. The pull-out of SSWs increases the load of neighboring SSWs across cracks, which results in more SSWs being pulled out from the UHPC matrix. As the flexural load increases, flexural toughness continues to increase with greater formation of cracks.

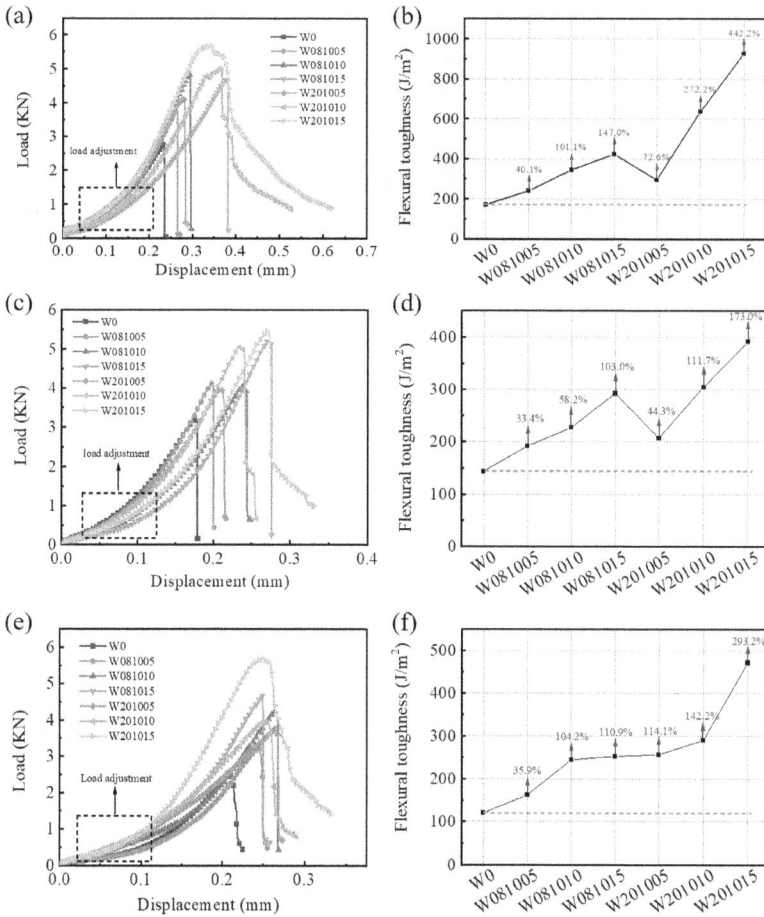

Figure 2.4 Flexural load–displacement curves and flexural toughness of SSWs-engineered multifunctional UHPC. (a) Load–displacement curves, water curing for 3 days; (b) Flexural toughness, water curing for 3 days; (c) Load–displacement, water curing for 28 days; (d) Flexural toughness, water curing for 28 days; (e) Load–displacement, natural curing for 28 days; (f) Flexural toughness, natural curing for 28 days. W081005 represents incorporating SSWs with diameter of 8 um, length of 10 mm and content of 0.5 vol% into UHPC, and other symbols have the same naming rules

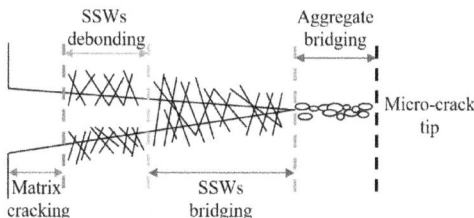

Figure 2.5 Relationship between crack development and SSWs

2.2.3 Flexural stress–strain curves

Under flexural load, the relationship between the bottom midspan flexural stress and strain is shown in Figure 2.6. Figure 2.6 shows that the initial ascending stage of the flexural stress–strain curves does not vary with the SSWs incorporation, which shows that SSWs have no effect on the tensile modulus of UHPC. After the elastic stage, the rising rate of stress–strain curves decreases, and the curves gradually close to x-axial, showing that the specimens have gone into the plastic deformation region. With the increase of SSWs content and diameter, the linear stage, i.e. proportional limit, increases, and the distance between the proportional limit and peak stress increases too. Meanwhile, the descending stage of the flexural stress–strain curves becomes slow. The experimental results also show that the peak stress of SSWs-engineered multifunctional UHPC increases with the increase of SSWs content. For UHPC reinforced with 1.5 vol% SSWs at a diameter of 8 μm and UHPC reinforced with 1.0 vol%/1.5 vol% SSWs at a diameter of 20 μm, the peak strain increases significantly. Compared with

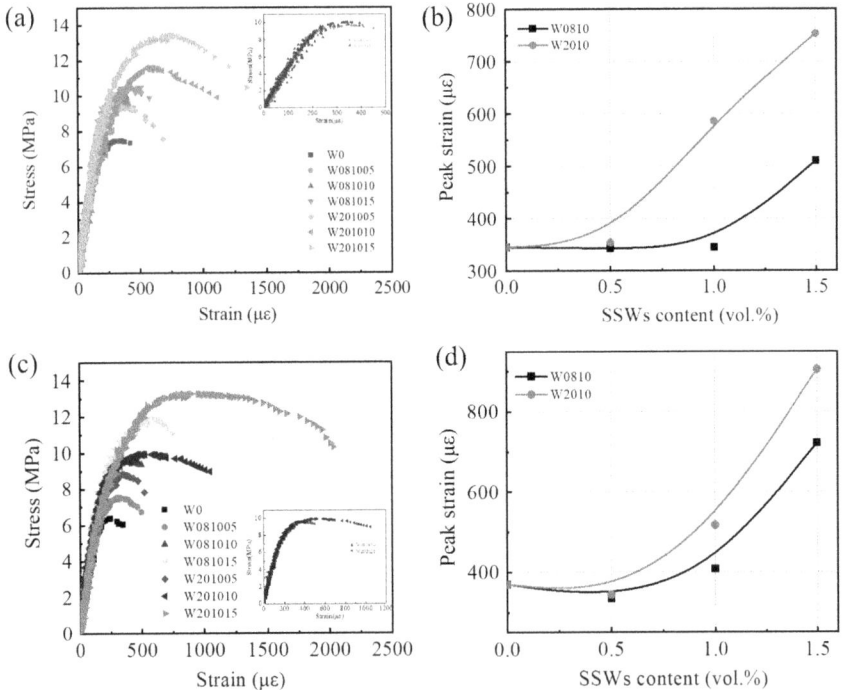

Figure 2.6 Flexural stress-strain curves and peak strain of SSWs-engineered multifunctional UHPC. (a) Flexural stress-strain curves, water curing for 28 days; (b) Flexural peak strain, water curing for 28 days; (c) Flexural stress-strain curves, natural curing for 28 days; (d) Flexural peak strain, natural curing for 28 days

UHPC without SSWs, the peak strains are increased by 47.8% and 95.1% for UHPC reinforced with 1.5 vol% SSWs at a diameter of 8 µm after water curing and natural curing for 28 days. Meanwhile, the increment ratios can reach 69.8% and 39.7% for UHPC reinforced with 1.0 vol% SSWs at a diameter of 20 µm, and 118.3% and 144.6% for UHPC reinforced with 1.5 vol% SSWs at a diameter of 20 µm, respectively, after water curing and natural curing for 28 days. The peak strains of UHPC reinforced with 1.5 vol% SSWs at a diameter of 8 and 20 µm after natural curing for 28 days are larger than those after water curing for 28 days due to the microcracks generated by shrinkage.

During the experiments, the specimens of UHPC without SSWs suddenly split in half after peak stress, but the specimens of SSWs-engineered multifunctional UHPC remained intact after failure, with one main large crack on the bottom. With the increase of SSWs content, the development path of the main large crack became longer. Additionally, there was a tendency for multiple cracking to appear. This phenomenon is consistent with the change rule of stress–strain curves and represents the failure characteristic transformation of SSWs-engineered multifunctional UHPC. By comparison, the SSWs have a more significant influence on flexural toughness than on flexural strength and flexural peak strain.

2.3 COMPRESSIVE PROPERTIES OF SSWS-ENGINEERED MULTIFUNCTIONAL UHPC AFTER FLEXURE

2.3.1 Compressive strength

The compressive strength of SSWs-engineered multifunctional UHPC after flexure is shown in Figure 2.7. Figure 2.7 shows that the compressive strength increases with the increase of SSWs content. This can be attributed to the SSWs restricting the internal material deterioration and crack propagation by absorbing the developed stress at the SSW's tip [9]. The results also show that the compressive strength of UHPC reinforced with SSWs at a diameter of 20 µm is higher than that of UHPC reinforced with SSWs at a diameter of 8 µm. When the content of SSWs at a diameter of 20 µm is 1.5 vol%, the compressive strength is increased by 25.1% and 18.9%, respectively, after water and natural curing for 28 days. Compared with Figure 2.2, the rangeability of compressive strength is far lower than that of flexural strength, which is in accordance with previous research [10].

It can be seen from Figure 2.8 that the compressive strength of SSWs-engineered multifunctional UHPC increases with curing age, while the compressive strength after natural curing for 28 days is higher than that after water curing for 28 days. This indicates that the microcracks mentioned previously have no adverse effect on compressive strength [11]. However, the water curing has a negative impact on compressive strength, especially

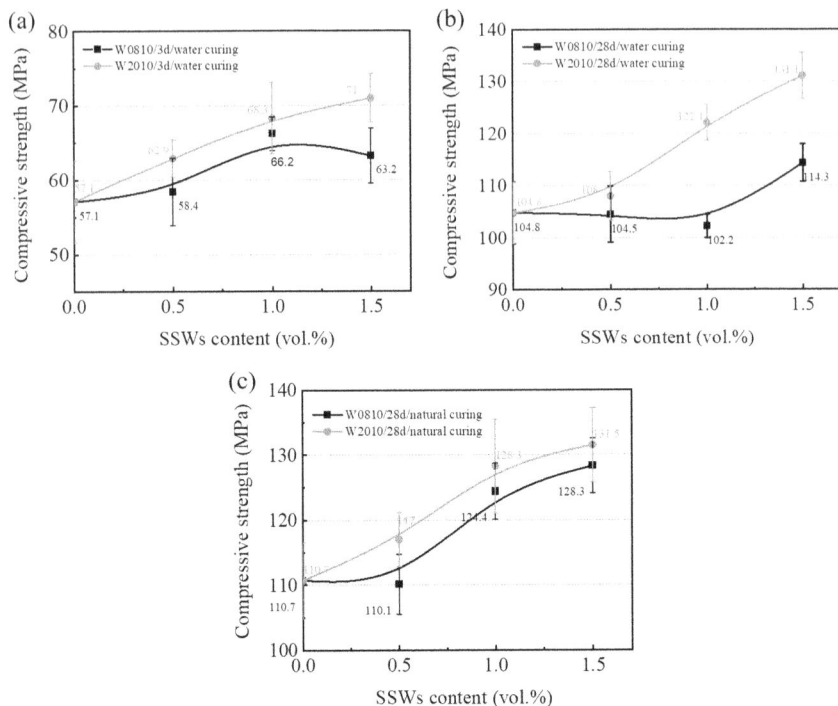

Figure 2.7 Effect of SSWs' content and diameter on the compressive strength of SSWs-engineered multifunctional UHPC cured at different regimes. (a) Water curing for 3 days; (b) Water curing for 28 days; (c) Natural curing for 28 days

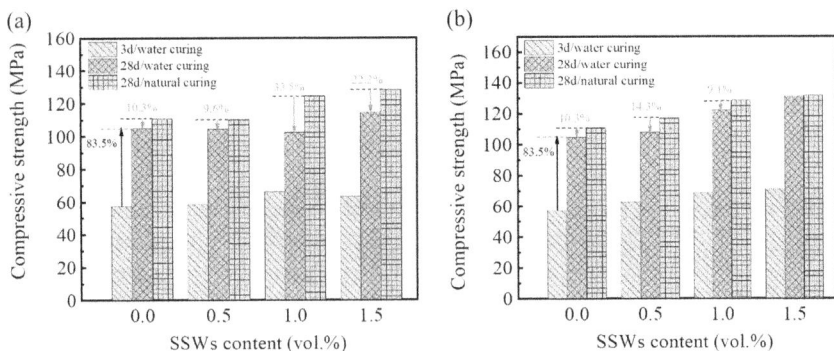

Figure 2.8 Effect of curing regimes on the compressive strength of SSWs-engineered multifunctional UHPC. (a) W0810; (b) W2010

for UHPC reinforced with 1.0 vol%/1.5 vol% SSWs at a diameter of 8 μm. Because of high aspect ratio, the molding becomes slightly difficult and the compactness is relatively lowered when the SSWs content is high. Therefore, the specimens of UHPC reinforced with 1.0 vol%/1.5 vol% SSWs at a diameter of 8 μm tend to be more influenced by water erosion. The compressive strength of UHPC without SSWs after water curing is increased by 83.5% with age from 3 to 28 days, which can be attributed to the increase of hydration products and structure compactness. By comparison, the change of compressive strength is mainly determined by the UHPC matrix.

2.3.2 Compressive load–displacement curves

The compressive load–displacement curves of SSWs-engineered multifunctional UHPC after flexure are shown in Figure 2.9. The area under the compressive load–deformation curve is called *compressive toughness* [12]. It can be seen from Figure 2.9 that the SSWs-engineered multifunctional UHPC is mostly in the elastic stage when it is subjected to relatively small compressive load. With increasing load, the ascending stage of the load–displacement curves gradually moves toward x-axial, showing that the specimens of SSWs-engineered multifunctional UHPC are entering the plastic deformation stage. The compressive load–displacement curve of UHPC without SSWs drops sharply after failure, and the specimens immediately fall to pieces. The load–displacement curves for specimens of SSWs-engineered UHPC exhibit ductile behavior with a steadier drop of load carrying capacity. This can be attributed to the lateral expansion restriction of the wires, resulting in high tolerance to axial deformation.

Figure 2.9(b), (d), and (f) show that the compressive toughness of SSWs-engineered multifunctional UHPC is substantially increased because of the addition of SSWs. The toughness of UHPC reinforced with SSWs at a diameter of 20 μm is better than that of UHPC reinforced with SSWs at a diameter of 8 μm. The toughness of SSWs-engineered multifunctional UHPC after water curing for 28 days is higher than that after natural curing for 28 days. That is to say, erosion generated by water curing has no adverse impact on the toughness of composites, although it has a negative effect on the compressive strength. However, the microcracks generated by natural curing limit the increase of compressive toughness.

During the experimental process, specimens of SSWs-engineered multifunctional UHPC make a loud noise when approaching the peak load, and the deformations become gradually larger until the specimens are completely crushed. Specimens of UHPC without SSWs break into pieces after failure, while specimens of SSWs-engineered multifunctional UHPC show insignificant fragmentation due to the confining effect of SSWs.

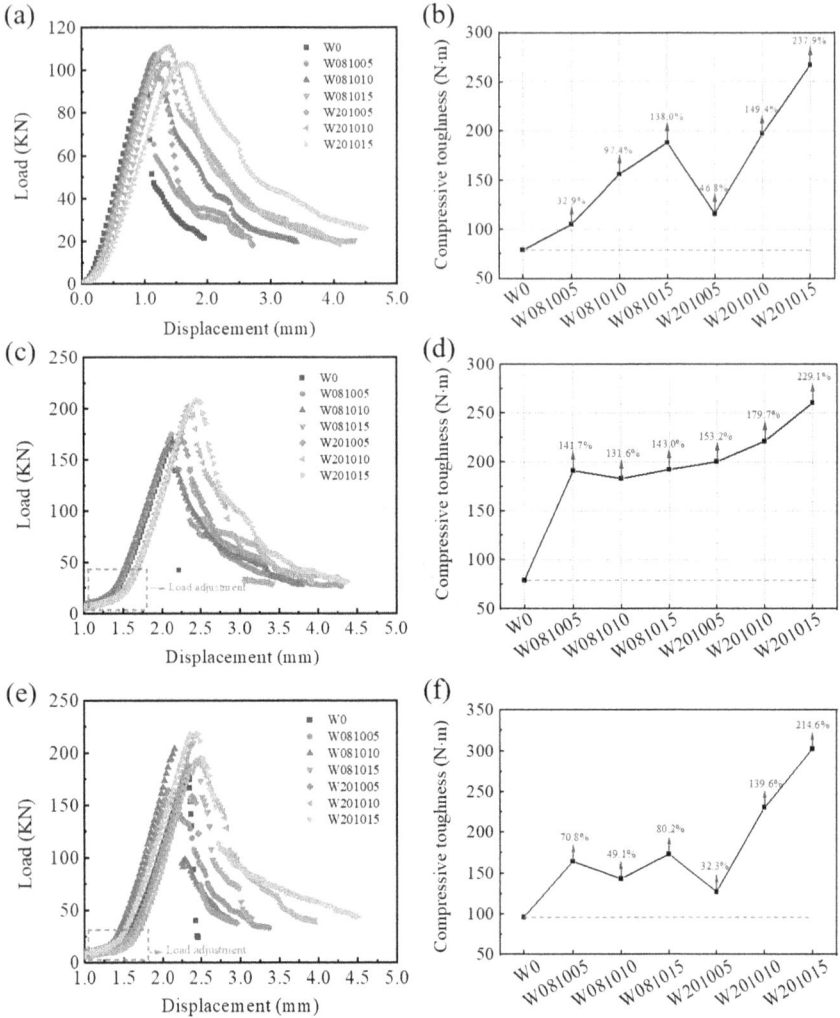

Figure 2.9 Compressive load–displacement curves and compressive toughness of SSWs-engineered multifunctional UHPC. (a) Compressive load–displacement curves, water curing for 3 days; (b) Compressive toughness, water curing for 3 days; (c) Compressive load–displacement curves, water curing for 28 days; (d) Compressive toughness, water curing for 28 days; (e) Compressive load–displacement curves, natural curing for 28 days; (f) Compressive toughness, natural curing for 28 days

2.4 UNIAXIAL COMPRESSIVE PROPERTIES OF SSWS-ENGINEERED MULTIFUNCTIONAL UHPC

2.4.1 Uniaxial compressive strength

What needs illustration is that the specimen size for testing uniaxial compressive properties was 40 mm × 40 mm × 80 mm, and SSWs with a diameter of 8 µm/20 µm and a length of 6 mm/10 mm were employed. SSWs-engineered UHPC specimens were marked as W080610, W080615, W081010, W081015, W200610, W200615, W201010, and W201015. The numbers from the first to the fourth represent the diameter and length of SSWs, while the fifth and the sixth numbers express the volume fraction of SSWs (1.0 or 1.5 vol%). In addition, two curing processes were employed. (1) The specimens were put into water (the temperature was 20 ± 1 °C) for 27 days; then they were taken out and placed in a room-temperature environment (temperature was 20 ± 1 °C and moisture was more than 50%) until the curing age reached 540 days. This curing process was called room-temperature curing. (2) The specimens were put into an accelerating curing box for 48 hours (water temperature was 90 °C, and temperature increase/decrease rate was 15 °C/h); then the specimens were placed in a room-temperature environment until curing ages reach 540 days. This curing process was called accelerated curing.

The uniaxial compressive strength of SSWs-engineered multifunctional UHPC after room-temperature curing and accelerated curing for 540 days is shown in Figure 2.10. It can be seen from Figure 2.10 that the incorporation of SSWs improves the uniaxial compressive strength of UHPC by forming a three-dimensional reinforcing network. The SSWs' length, volume fraction, and curing condition have no obvious effect on the uniaxial compressive strength of UHPC when the SSWs with a diameter of 8 µm are employed. However, the uniaxial compressive strength of UHPC

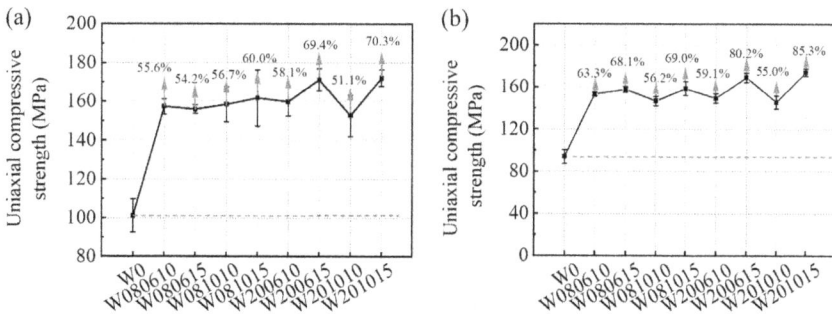

Figure 2.10 Uniaxial compressive strength of SSWs-engineered multifunctional UHPC after room-temperature and accelerated curing for 540 days. (a) Room temperature curing; (b) Accelerated curing

increases with the increase of volume fraction of SSWs with a diameter of 20 μm. Meanwhile, the effect of curing condition on the uniaxial compressive strength of UHPC reinforced with 1.5 vol% SSWs at a diameter of 20 μm and length of 6 mm/10 mm is not negligible. The increments of uniaxial compressive strength of these two composites can reach up to 69.4% and 70.3%, respectively, when room-temperature curing is adopted, and the increments are 80.2% and 85.3%, respectively, when accelerated curing is selected. The strength-enhancing efficiency increases with increasing UHPC matrix strength because the increase of matrix strength leads to a high interface bond strength between the SSWs and the UHPC matrix, which is the same as the research conclusion of Simões et al. [13].

Meanwhile, the increment of uniaxial compressive strength provided by SSWs to UHPC is considerably larger than in previous research using common steel fiber (the diameter is 0.2 mm; the length is 6 mm; the volume fraction is 4%; and the increment is only 30.4%) [14]. Contrast analysis shows that the reinforcing effect of SSWs with a diameter of 20 μm is superior to that of SSWs with a diameter of 8 μm.

The relationship between uniaxial and cubic compressive strength (with sizes of 40 mm × 40 mm × 40 mm) of SSWs-engineered multifunctional UHPC after room-temperature and accelerated curing for 540 days is demonstrated in Figure 2.11. It can be indicated from Figure 2.11 that there is a linear fitting relationship between uniaxial and cubic compressive strength of SSWs-engineered multifunctional UHPC. The linear fitting coefficient is 1.0 and 0.97 for room-temperature curing and accelerated curing, respectively, and the corresponding correlation coefficient is 0.999 and 0.995. The linear fitting coefficient is far greater than that given by GB50010-2011 <Code for design of concrete structures> (0.76–0.82) and the previous research [15], indicating that the addition of SSWs weakens the cyclohoop effect of UHPC. The SSWs can form a three-dimensional overlapping

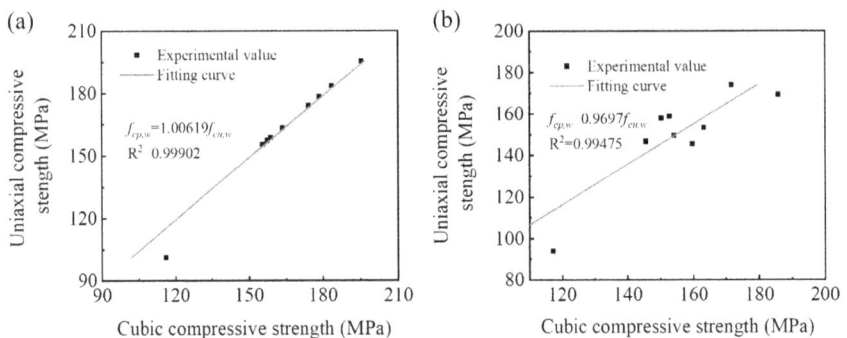

Figure 2.11 Uniaxial compressive strength versus cubic compressive strength of SSWs-engineered multifunctional UHPC after room-temperature and accelerated curing for 540 days. (a) Room temperature curing; (b) Accelerated curing

network in UHPC and play a lateral restraint role in the process of uni-axial compression. Figure 2.11 also shows that the linear fitting coefficient decreases when accelerated curing is selected, showing that the cyclo-hoop effect increases with the increase of uniaxial compressive strength.

2.4.2 Elastic modulus and Poisson's ratio

Elastic modulus is usually used to characterize the longitudinal deforma-tion capacity of composite at the elastic stage. Poisson's ratio can be used to evaluate the transverse deformation capacity of composite. Meanwhile, shear modulus reflects the resistance of composite to shear strain, and the composite rigidity increases with the increase of shear modulus. The vol-ume modulus, which can also be used for representing the deformation capacity of composite, is defined as the ratio between volume strain and average stress. The calculation of shear modulus G and volume modulus K is shown in Eqs (2.1) and (2.2) [8]:

$$G = \frac{E}{2(1+\upsilon)} \tag{2.1}$$

$$K = \frac{E}{3(1-2\upsilon)} \tag{2.2}$$

where G is shear modulus; K is volume modulus; E is elastic modulus; υ is Poisson's ratio.

The elastic modulus and Poisson's ratio of SSWs-engineered multifunc-tional UHPC after room temperature curing for 540 days are shown in Figure 2.12(a). As shown in Figure 2.12(a), the addition of SSWs with a diameter of 8 µm has less effect on the elastic modulus of UHPC. The

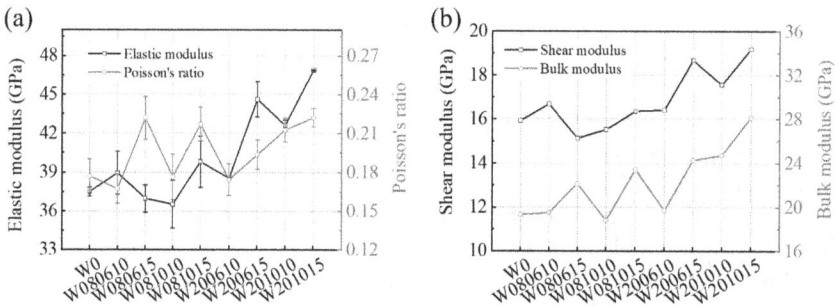

Figure 2.12 Deformation properties of SSWs-engineered multifunctional UHPC after room-temperature curing for 540 days. (a) Elastic modulus and Poisson's ratio; (b) Shear modulus and bulk modulus

elastic modulus of UHPC reinforced with 1.0 vol%/1.5 vol% SSWs at a diameter of 20 μm and length of 6 mm increases by 2.8% and 19.1%, and that of UHPC reinforced with 1.0 vol%/1.5 vol% SSWs at a diameter of 20 μm and length of 10 mm improves by 13.6% and 25.1%, respectively, compared with that of UHPC without SSWs. The elastic modulus of UHPC reinforced with 1.5 vol% SSWs at a diameter of 20 μm is 46.9 GPa, which only differs by 4.7% and 4.1%, respectively, from the results obtained by Zdeb et al. [16] (the cubic compressive strength is 236.8 MPa) and Song et al. [17], respectively. Figure 2.12(a) also shows that the Poisson's ratio increases with the increase of SSWs volume fraction because the SSWs can improve the transverse deformation capacity of UHPC. The Poisson's ratio of UHPC reinforced with 1.5 vol% SSWs at a diameter of 8 μm and length of 6 mm, reinforced with 1.5 vol% SSWs at a diameter of 8 μm and length of 10 mm, reinforced with 1.5 vol% SSWs at a diameter of 20 μm and length of 6 mm, and reinforced with 1.5 vol% SSWs at a diameter of 20 μm and length of 10 mm increased by 25.3%, 35.0%, 9.3%, and 41.2%, respectively, with respect to UHPC without SSWs. The deformation capacity of specimens of UHPC reinforced with SSWs at a diameter of 20 μm is superior to that reinforced with SSWs at a diameter of 8 μm.

The shear modulus and volume modulus of SSWs-engineered UHPC after room-temperature curing for 540 days are given in Figure 2.12(b). As indicated in Figure 2.12(b), the improvement of shear modulus and volume modulus due to the introduction of SSWs with a diameter of 20 μm is more evident than that caused by SSWs with a diameter of 8 μm with the same length and volume fraction. The shear modulus of UHPC reinforced with 1.0 vol%//1.5 vol% SSWs at a diameter of 20 μm and length of 6 mm is improved by 3.0% and 17.4%; meanwhile, that of UHPC reinforced with 1.0 vol%/1.5 vol% SSWs at a diameter of 20 μm and length of 10 mm is enhanced by 10.3% and 20.5%, respectively. Furthermore, the volume modulus of these composites grows by 14.5%, 21.2%, 25.4%, and 45.4%, respectively.

The elastic moduli and Poisson's ratios of SSWs-engineered multifunctional UHPC after accelerated curing for 540 days are shown in Figure 2.13(a). It can be observed from Figure 2.13(a) that the effect of SSWs on the elastic modulus and Poisson's ratio of UHPC is only slightly affected by the SSWs' geometry dimension and volume fraction. The elastic modulus increment of UHPC is between 40.3% and 58.2% due to the inclusion of SSWs. Figure 2.13(a) also shows that the Poisson's ratio of UHPC reinforced with 1.0 vol%/1.5 vol% SSWs at a diameter of 8 μm and length of 6 mm is increased by 56.9% and 87.8%, and that of UHPC reinforced with 1.0 vol%/1.5 vol% SSWs at a diameter of 8 μm and length of 10 mm is improved by 89.3% and 78.5%, respectively. The Poisson's ratio increments for UHPC reinforced with 1.0 vol%/1.5 vol% SSWs at a diameter of 20 μm and length of 6 mm are 114.8% and 100.8%, and those for UHPC

(a)

(b)

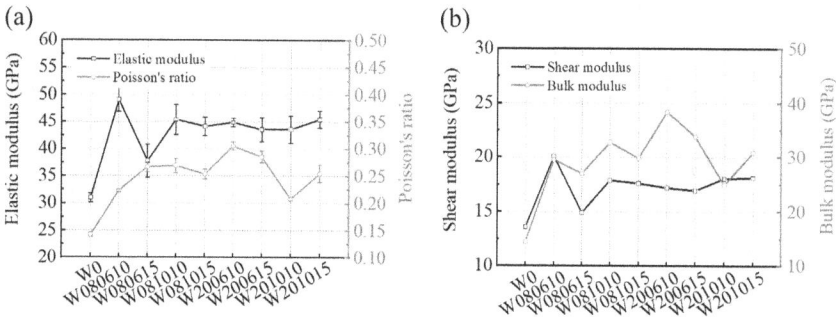

Figure 2.13 Deformation properties of SSWs-engineered multifunctional UHPC after accelerated curing for 540 days. (a) Elastic modulus and Poisson's ratio; (b) Shear modulus and bulk modulus

reinforced with 1.0 vol%/1.5 vol% SSWs at a diameter of 20 μm and length of 10 mm reach 46.5% and 79.2%, respectively. The increasing efficiency of SSWs with a diameter of 20 μm and a length of 6 mm against transverse deformation of UHPC is more obvious.

The shear modulus and volume modulus of SSWs-engineered UHPC after accelerated curing for 540 days are demonstrated in Figure 2.13(b). As demonstrated in Figure 2.13(b), the shear modulus of SSWs-engineered UHPC shows enhancement between 24.7% and 47.7% relative to UHPC without SSWs, and the increment of volume modulus can reach up to 165.6%. The improvement of volume modulus mainly comes from the increase of transverse deformation based on the three-dimensional network restraint effect of SSWs.

2.4.3 Uniaxial compressive stress–strain curves

The uniaxial compressive stress–strain curves of SSWs-engineered UHPC after room-temperature curing and accelerated curing for 540 days are shown in Figure 2.14. It is clear from Figure 2.14 that there is a direct relationship between stress and strain in UHPC without SSWs, and the stress–strain curves suddenly drop, which is not related to curing conditions. The specimens of UHPC without SSWs suddenly break into pieces after peak stress, exhibiting the typical brittleness material characteristic. The ascending stage of stress–strain curves of SSWs-engineered multifunctional UHPC consists of the elastic section and the elastic-plastic section (marked AC). The AB part in Figure 2.14 represents the stable crack propagation stage, in which the SSWs can inhibit the propagation of crack and transfer crack tip stress; the BC part represents the unstable crack propagation stage, in which the SSWs can bridge cracks and can gradually be pulled out or pulled off. The tangent slope of the BC part decreases with the increase of SSWs' diameter and volume fraction, illustrating that the SSWs with a diameter of

20 μm are more effective in bridging cracks and being pulled off compared with those with a diameter of 8 μm.

Contrast analysis between Figure 2.14(a) and (b) shows that the unstable crack propagation part of SSWs-engineered multifunctional UHPC after room-temperature curing for 540 days is longer than that after accelerated curing for 540 days, and the tangent slope of the BC part increases when accelerated curing is employed. This can be attributed to the high interface bond strength between SSWs and the UHPC matrix caused by accelerated curing, which enables the pulling off of SSWs to become easier.

The uniaxial compressive longitudinal peak strain of SSWs-engineered multifunctional UHPC after room-temperature curing and accelerated curing for 540 days is presented in Figure 2.15. Figure 2.15 shows that the uniaxial compressive longitudinal peak strain of UHPC increases

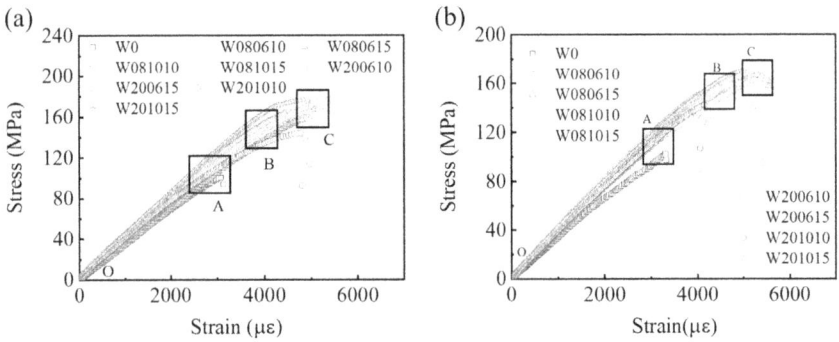

Figure 2.14 Uniaxial stress–strain curves of SSWs-engineered multifunctional UHPC after room-temperature and accelerated curing for 540 days. (a) Room temperature curing; (b) Accelerated curing

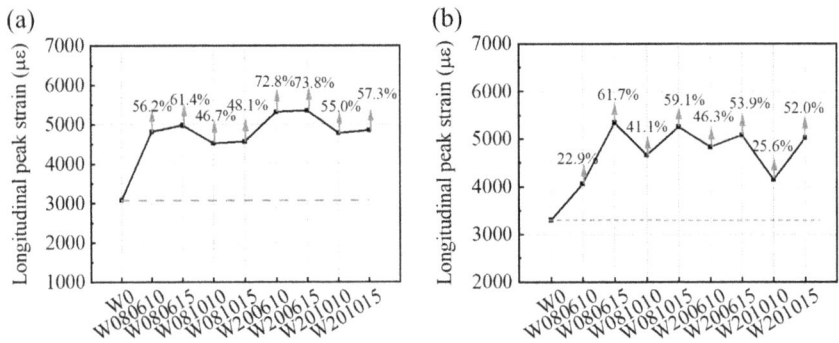

Figure 2.15 Longitudinal peak strain of SSWs-engineered multifunctional UHPC after room-temperature curing and accelerated curing for 540 days. (a) Room temperature curing; (b) Accelerated curing

with the increase of SSWs content. The minimum longitudinal peak strain value of SSWs-engineered multifunctional UHPC is 4058, which is larger than in previous research (the longitudinal peak strain is 3508 when the steel fiber volume fraction is 2.0%) [17]. The SSWs with a diameter of 20 μm and a length of 6 mm show the highest increment to longitudinal peak strain of UHPC after room-temperature curing for 540 days, as shown in Figure 2.15(a). Meanwhile, the SSWs with a diameter of 8 μm and a volume fraction of 1.5 vol% show a better modification effect on longitudinal peak strain of UHPC after accelerated curing for 540 days, as shown in Figure 2.15(b). The increase of longitudinal peak strain under room-temperature curing is mainly attributed to the role of SSWs in bridging cracks, which prevents the generation and propagation of cracks. However, the compactness structure of the UHPC matrix and the increase of crack resistance dominate the longitudinal peak strain of SSWs-engineered multifunctional UHPC under accelerated curing, resulting in the SSWs being pulled off more easily because of the increase in the interface bond strength.

2.4.4 Uniaxial compressive failure mode

The uniaxial compressive failure mode of SSWs-engineered multifunctional UHPC can be illustrated via Figure 2.16. The specimens of UHPC without SSWs present a fragmented state after failure, and this phenomenon happens rapidly. Meanwhile, the cyclo-hoop effect is relatively obvious, and the crack development is mainly hindered by aggregate, as shown in Figure 2.16(a).

The crack resistance increases due to the addition of SSWs, as shown in Figure 2.16(b). The local stress concentration phenomenon happens with the increase of uniaxial compression, and then the UHPC matrix starts to crack. The generation and propagation of cracks are postponed by the three-dimensional SSWs network. Hence, there is a multiple cracking

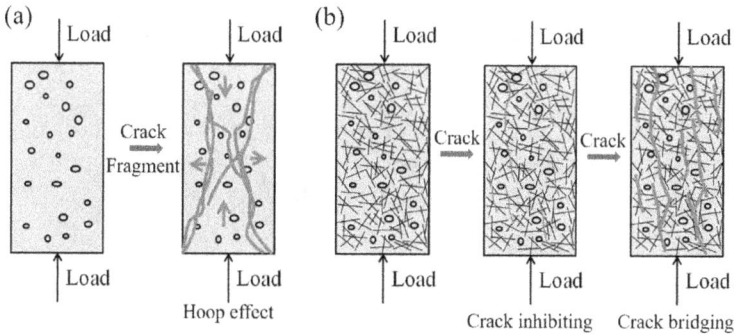

Figure 2.16 Schematic diagram of uniaxial compressive failure mode

tendency based on stress distribution, leading to the increase of load capacity. The specimens of SSWs-engineered multifunctional UHPC still maintain their integrity when failure happens. There are only two or three main cracks extending along the loading direction and a few other visible cracks on the surface. Meanwhile, the cyclo-hoop effect is weakened because of the lateral restraint by the SSWs network.

2.5 UNIAXIAL COMPRESSIVE CONSTITUTIVE MODEL FOR SSWS-ENGINEERED MULTIFUNCTIONAL UHPC

The difference between concretes caused by raw materials is mainly determined by the initial crack quantity and the crack development characteristic under load. The local stress concentration phenomenon happens in concrete with the increase of uniaxial compression; then, the concrete starts to crack, and the damage appears. The damage variable D is introduced to describe the generation and propagation of cracks in concrete [18]; $D = 0$ when there is no damage, and $D = 1$ when the concrete is completely damaged. Meanwhile, it can be assumed that damage to concrete is caused by the failure of local microelements. If the number of failure microelements is c under a certain compressive load, the damage variable D can be defined as the ratio between the failure number c and the total number N of microelements:

$$D = c / N \tag{2.3}$$

The following damage constitutive model is obtained on the basis of continuum damage mechanics [19]:

$$\sigma = E\varepsilon(1 - D) \tag{2.4}$$

As a random variable, the uniaxial stress of concrete is affected by many factors, such as pore structure, initial defect, and hydration product. The factors themselves are independent random variables with some statistical laws. Hence, the uniaxial stress of concrete can be described by a statistical distribution.

The existing research shows that the stress–strain curves, mechanical properties, and damage variables of concrete follow a Weibull distribution [20]. Assuming that the uniaxial stress of concrete obeys a Weibull distribution, the probability density function is as follows:

$$P(\varepsilon) = \frac{m}{F}\left(\frac{\varepsilon}{F}\right)^{m-1} e^{-\left(\frac{\varepsilon}{F}\right)^m} \tag{2.5}$$

where ε is concrete strain, and m and F are parameters characterizing the physical and mechanical properties of concrete.

Meanwhile, the strain caused by load is generally used to measure the damage variable D [21]. Then, the number of failure microelements generated in any interval $[\varepsilon, \varepsilon + d\varepsilon]$ is defined as $NP(x)dx$. When the uniaxial compressive strain is ε, the number of failure microelements is

$$c(\varepsilon) = \int_0^{\varepsilon} NP(x)dx = N\left[1 - e^{-\left(\frac{\varepsilon}{F}\right)^m}\right]$$

(2.6)

Therefore, the damage variable D is as follows:

$$D = \frac{c}{N} = 1 - e^{-\left(\frac{\varepsilon}{F}\right)^m}$$

(2.7)

The constitutive model of SSWs-engineered multifunctional UHPC can be expressed by Eq. (2.8):

$$\sigma = E\varepsilon(1 - D) = E\varepsilon e^{-\left(\frac{\varepsilon}{F}\right)^m}$$

(2.8)

The slope at the uniaxial peak stress point is 0: then,

$$\left.\frac{d\sigma}{d\varepsilon}\right|_{\varepsilon = \varepsilon_0} = E\left[1 - m\left(\frac{\varepsilon}{F}\right)^m\right]e^{-\left(\frac{\varepsilon}{F}\right)^m} = 0$$

(2.9)

Combining Eqs (2.7) and (2.8), the parameters m and F can be calculated:

$$m = \frac{1}{ln(\sigma_c / E\varepsilon_c)}$$

(2.10)

$$F = \varepsilon_c \left(\frac{1}{m}\right)^{-1/m}$$

(2.11)

where σ_c and ε_c are uniaxial peak stress and strain, respectively.

Equation (2.10) shows that the brittleness of concrete increases with the increase of parameter m. Therefore, the physical nature of parameter m is the concentration degree of microelement strength. Meanwhile, the parameter F represents the macroscopic statistical strength.

The constitutive model parameters of SSWs-engineered multifunctional UHPC after room-temperature curing for 540 days fitted by Eq. (2.8) are listed in Table 2.1. As shown in Table 2.1, good agreement between the experimental and simulated results is obtained by the proposed model. The addition of SSWs improves the values of parameter F, while the value of parameter m is closely related to the aspect ratio of SSWs.

Table 2.1 Constitutive model parameters of SSWs-engineered multifunctional UHPC after room-temperature curing for 540 days

Specimens	E (GPa)	ε_c (με)	F	m	R^2
W0	33.90	3084	0.00333	36.858	0.9991
W080610	40.11	4817	0.00647	5.472	0.9996
W080615	34.15	4978	0.00573	17.712	0.9987
W081010	36.52	4571	0.00500	30.092	0.9823
W081015	36.83	4569	0.00584	52.144	0.9775
W200610	37.35	5329	0.00650	8.528	0.9933
W200615	37.09	5361	0.00715	7.239	0.9982
W201010	36.37	4708	0.00553	8.527	0.9898
W201015	42.06	4851	0.00590	9.454	0.9992

The SSWs factor W is defined as Eq. (2.12):

$$W = V_f \frac{l_f}{d_f} \tag{2.12}$$

where V_f is the volume faction of SSWs; l_f is the length of SSWs; d_f is the diameter of SSWs.

In order to describe the effect of SSWs on the uniaxial compressive behavior of UHPC, the SSWs factor is introduced to express parameters m and F. This means that the SSWs inhibit the development of damage in UHPC by influencing the microelement strength concentration degree and the macroscopic statistical strength.

Then, the parameters m and F are expressed by the linear regression and quadratic polynomial regression of SSWs factor W, respectively.

When room-temperature curing is employed, the following Eqs (2.13) and (2.14) can be obtained:

$$F = -1.19218 \times 10^{-4} W + 0.00693 \tag{2.13}$$

$$m = 8.77851 - 1.21283W + 0.19203W^2 \tag{2.14}$$

The uniaxial compressive constitutive model of SSWs-engineered UHPC is shown in Eq. (2.15):

$$\sigma = e^{\left[-\left(\frac{\varepsilon}{-1.19218 \times 10^{-4} W + 0.00693} \right)^{8.77851 - 1.21283W + 0.19203W^2} \right] E\varepsilon} \tag{2.15}$$

where E is the elastic modulus, which can be obtained from Figure 2.13(a).

When accelerated curing is employed, the SSWs factor W should be revised considering the effect of the curing condition, and it is redefined as W', as shown in Eq. (2.16):

$$W' = \beta V_{\mathrm{f}} \frac{l_{\mathrm{f}}}{d_{\mathrm{f}}} \tag{2.16}$$

where $\beta = 1/1.3$ and $\beta = 1$ is the curing condition influence coefficient for F and m, respectively, due to the change of interface bond strength and pore structure. The concentration degree of microelement strength is only slightly affected by accelerated curing because of the compact structure of UHPC. However, the effect of accelerated curing on the macroscopic statistical strength of SSWs-engineered multifunctional UHPC is obvious.

Then, the revised uniaxial compressive constitutive model for SSWs-engineered multifunctional UHPC under accelerated curing is expressed by Eq. (2.17):

$$\sigma = e^{\left[-\left(\frac{\varepsilon}{-1.19218\times10^{-4}\,W'+0.00693}\right)^{8.77851-1.21283W+0.1920W^2}\right]E\varepsilon} \tag{2.17}$$

where E is the elastic modulus, which can be obtained from Figure 2.13(a).

Thereby, the theoretical uniaxial compressive stress–strain relationship of SSWs-engineered multifunctional UHPC after room-temperature curing and accelerated curing for 540 days is analyzed on the basis of Eqs (2.15) and (2.17).

The comparison of theoretical and experimental uniaxial compressive stress–strain curves after room-temperature curing for 540 days is shown in Figure 2.17, where the fitting correlation coefficient is larger than 0.97. It can be observed from Figure 2.17 that the theoretical stress–strain curve of UHPC without SSWs completely coincides with the experimental curve. Meanwhile, the theoretical formula can accurately describe the elastic-plastic section of stress–strain curves of SSWs-engineered multifunctional UHPC. The introduction of the SSWs factor W effectively improves the fitting degree of theoretical and experimental uniaxial compressive stress–strain curves.

Figure 2.18 shows the relationship between damage variable D and uniaxial compressive longitudinal strain. The specimens of UHPC without SSWs rapidly go into destruction when damage happens, exhibiting typical brittle material characteristics. However, the addition of SSWs effectively delays the generation of damage and inhibits the propagation of damage. The development process of damage variable D can be divided into three stages: non-damage stage, damage speediness development stage, and damage quickness development stage. The tangent slope of the damage speediness development stage decreases and the damage quickness development stage lengthens with the increase of SSWs' content and diameter. Meanwhile, the damage quickness development stage shortens with the increase of SSWs' length. The damage variable limit value of UHPC reinforced with 1.0 vol%/1.5 vol% SSWs at a diameter of 20 μm and length of 6 mm can reach 0.24 and 0.27, respectively. The SSWs with a diameter

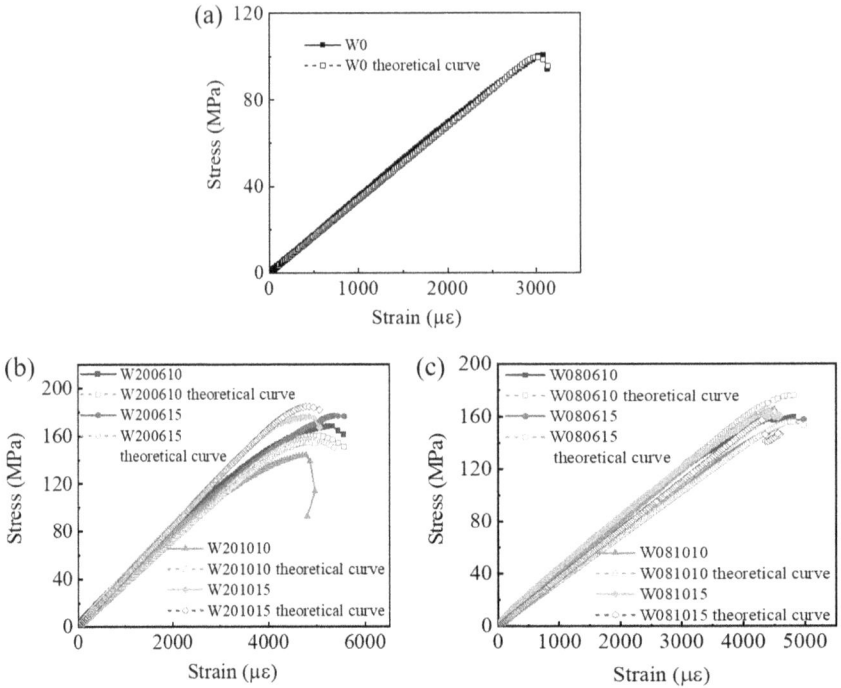

Figure 2.17 Theoretical (based on Eq. [2.15]) and experimental uniaxial compressive stress–strain curves of SSWs-engineered multifunctional UHPC after room-temperature curing for 540 days. (a) W0; (b) W2006 and W2010; (c) W0806 and W0810

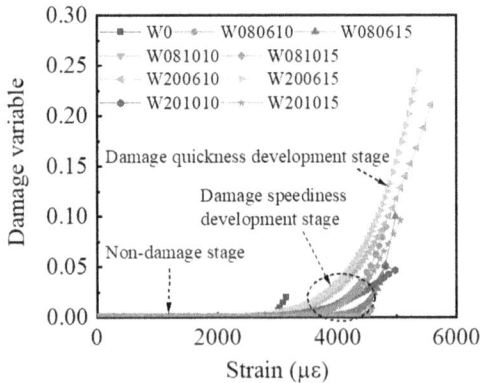

Figure 2.18 Damage variable versus uniaxial compressive longitudinal strain of SSWs-engineered multifunctional UHPC after room-temperature curing for 540 days

of 20 μm and a length of 6 mm can effectively inhibit the generation and propagation of cracks at the damage speediness development stage, and can also effectively bridge cracks and then gradually be pulled off in the damage quickness development stage.

The theoretical, revised theoretical, and experimental uniaxial compressive stress–strain curves of SSWs-engineered multifunctional UHPC after accelerated curing for 540 days are shown in Figure 2.19. The revised fitting correlation coefficient is larger than 0.98 based on Eq. (2.17). Figure 2.19

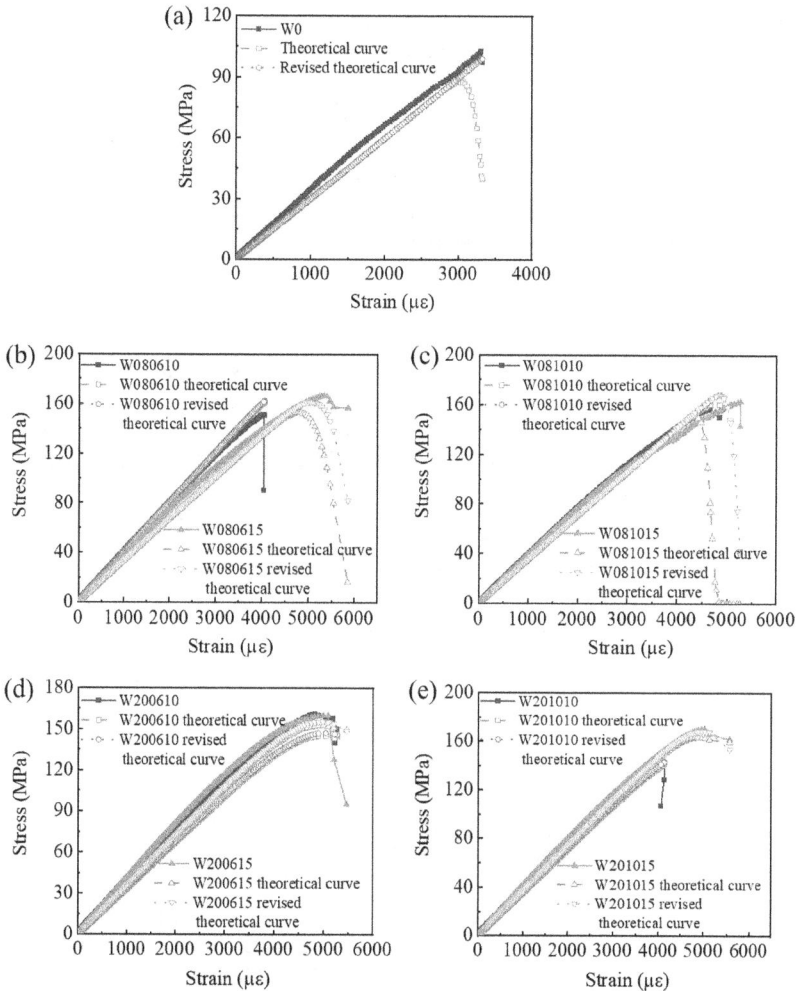

Figure 2.19 Theoretical (based on Eqs [2.15] and [2.17]) and experimental uniaxial compressive stress–strain curves of SSWs-engineered multifunctional UHPC after accelerated curing for 540 days. (a) W0; (b) W2006; (c) W2010; (d) W0806; (e) W0810

shows that the non-revised theoretical stress–strain curve of UHPC without SSWs obtained by Eq. (2.15) has an obvious descending stage, which is not consistent with the experimental result, while the revised theoretical stress–strain curve and the experimental result fit well. This means that the constitutive relationship of UHPC without SSWs is significantly affected by curing conditions. Meanwhile, the revised theoretical stress–strain curve has a higher fitting degree with the experimental result for UHPC reinforced with SSWs at a diameter of 8 μm compared with the theoretical curve. The theoretical stress–strain relationship of UHPC reinforced with 1.0 vol%/1.5 vol% SSWs at a diameter of 20 μm and length of 6 mm/10 mm is almost impervious to the curing conditions.

The relationship between the damage variable D and the uniaxial compressive longitudinal peak strain for SSWs-engineered UHPC after accelerated curing for 540 days is shown in Figure 2.20. It can be observed from Figure 2.20 that the SSWs demonstrate an evident inhibition function against the generation and propagation of damage. The damage development process still can be divided into three stages as mentioned earlier. However, the damage speediness development stage shortens and the damage quickness development stage becomes less obvious compared with that after room-temperature curing for 540 days, while the tangent slope of the damage speediness development stage decreases due to the increase of the interface bond strength and the concrete structure compactness. Meanwhile, the bridging SSWs can more easily be pulled off after damage appears, especially for SSWs with a diameter of 8 μm. The damage development characteristic related to SSWs' geometric dimension is not subject to curing conditions. The damage limit value of UHPC reinforced with 1.5 vol% SSWs at a diameter of 20 μm and length of 6 mm/10 mm after accelerated curing for 540 days is 0.2482 and 0.2580, respectively.

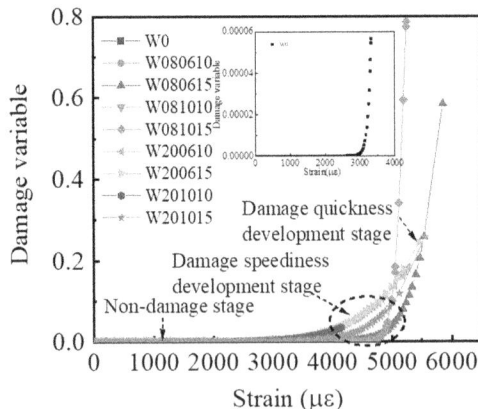

Figure 2.20 Damage variable versus uniaxial compressive longitudinal strain of SSWs-engineered multifunctional UHPC after accelerated curing for 540 days

It can be seen from this analysis that the established theoretical and revised theoretical uniaxial compressive constitutive model can accurately describe the stress–strain relationship of SSWs-engineered multifunctional UHPC under different curing conditions. The delaying effect of SSWs on damage propagation in UHPC can be proved by the damage variable development process.

2.6 STRENGTHENING MECHANISMS OF SSWS-ENGINEERED MULTIFUNCTIONAL UHPC

2.6.1 Macroscopic mechanisms

SSWs have a strong interface with the UHPC matrix, which is the same as for the common steel fiber used for other research. Meanwhile, the linear expansion coefficient of SSWs and the UHPC matrix is similar, ensuring that there is no transverse deformation between the two. Therefore, the composite theory considering the SSWs factor W (as shown in Eq. [2.12]) can be used to deduce the uniaxial compressive strength of SSWs-engineered UHPC under different curing conditions. The fitting relationship between uniaxial compressive strength and the SSWs factor W of SSWs-engineered UHPC is presented in Figure 2.21. It can be observed from Figure 2.21 that the uniaxial compressive strength of SSWs-engineered multifunctional UHPC can be expressed as the quadratic polynomial of the SSWs factor W based on the UHPC matrix strength. The uniaxial compressive strength first increases and then decreases with the increase of the SSWs factor W, and the corresponding fitting correlation coefficient is 0.988 and 0.984, respectively. By comparison, the monomial and quadratic coefficients of the fitting equation increase when accelerated curing is employed due to the increase of interface bond strength and structure compactness.

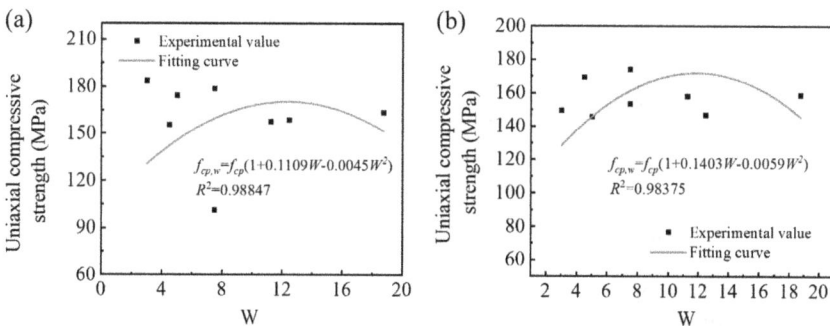

Figure 2.21 Relationship between uniaxial compressive strength and SSWs' factor of SSWs-engineered multifunctional UHPC after room-temperature curing and accelerated curing for 540 days. (a) Room temperature curing; (b) Accelerated curing

2.6.2 Microcosmic mechanisms

Scanning electron microscope (SEM) images of SSWs-engineered multifunctional UHPC after accelerated curing for 540 days are shown in Figure 2.22. It can be seen from Figure 2.22(a) and (b) that the UHPC matrix has a compact structure, and the crystal growth space is restricted, leading to high crack development resistance. The microcosmic effects caused by SSWs can be summarized as follows:

1) Interface effect. As shown in Figure 2.22(c), there is thickness and a dense hydration product layer on the SSWs' surface, illustrating that there is strong interface bond strength between the SSWs and the UHPC matrix.

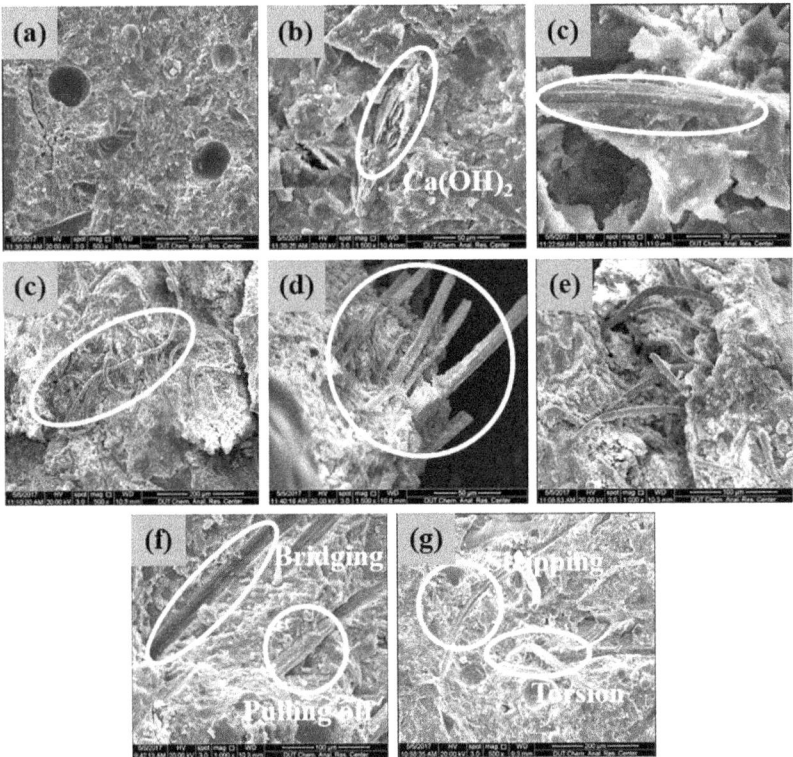

Figure 2.22 SEM images of SSWs-engineered multifunctional UHPC after accelerated curing for 540 days. (a) RPC matrix; (b) Ca(OH)$_2$; (c) Interface effect; (d) Overlapping network effect; (e) Inter-anchor effect; (f) Bridging effect; (g) Pullting-out effect; (h)Torsion, stripping and snapping effect. Notes: It is worthwhile to note that the UHPC reinforced with SSWs at a diameter of 8 μm and length of 10 mm is marked as W0810, while that reinforced with SSWs at a diameter of 20 μm and length of 10 mm is marked as W2010

2) Overlapping network effect. There is a large quantity of SSWs at a low volume fraction due to their microdiameter and high aspect ratio, and they overlap with each other to form a three-dimensional network, as shown in Figure 2.22(d). The network can effectively inhibit the generation and propagation of cracks, resulting in a circuitous crack development path. Meanwhile, the lateral deformation capacity of UHPC is enhanced, and the structural integrity is maintained when failure happens.

3) Inter-anchor effect. The cementitious materials permeate into the inter-space of bundled SSWs to form an inter-anchor interface, as shown in Figure 2.22(e). The bundled SSWs with an inter-anchor interface can form a zone of strong resistance to cracks. This phenomenon is quite different from the existing research on bundled glass fiber, in which the spaces among glass filaments remain largely vacant, and the effective bond is small [22].

4) Bridging effect. The UHPC matrix first reaches limit tensile stress and then cracks with the increase of compressive loading. Figure 2.22(f) shows that the SSWs can bridge cracks and transfer crack tip stress when the cracks encounter SSWs, leading to stress distribution and the improvement of load capacity.

5) Torsion, stripping, and pulling off effect. It can be observed from Figure 2.22(g) that the bridging SSWs will be pulled off with the propagation of cracks, and the surrounding SSWs will be twisted or stripped due to the development direction of cracks.

2.6.3 Hydration product property

In order to study the effect of SSWs on UHPC hydration product structure, the X-ray diffraction (XRD) pattern of SSWs-engineered multifunctional UHPC after room-temperature curing for 3 days was explored, as demonstrated in Figure 2.23. There is no new hydration product in SSWs-engineered multifunctional UHPC, and the diffraction peak intensity of the $Ca(OH)_2$ crystal phase decreases with the increase of the SSWs volume fraction. This means that the $Ca(OH)_2$ content decreases [23, 24] due to the addition of SSWs.

The orientation index of $Ca(OH)_2$ is calculated on the basis of existing research [25], as listed in Table 2.2. Table 2.2 shows that the $Ca(OH)_2$ orientation index of SSWs-engineered multifunctional UHPC decreases compared with that of UHPC without SSWs. The orientation index of UHPC reinforced with 1.0 vol%/1.5 vol% SSWs at a diameter of 8 μm and length of 10 mm and that reinforced with 1.0 vol%/1.5 vol% SSWs at a diameter of 20 μm and length of 10 mm is reduced by 24.2%, 14.4%, 21.2%, and 19.5%, respectively.

The XRD pattern of SSWs-engineered multifunctional UHPC after accelerated curing for 3 days was explored, as shown in Figure 2.24. Figure 2.24 shows that the XRD pattern of SSWs-engineered UHPC features the same

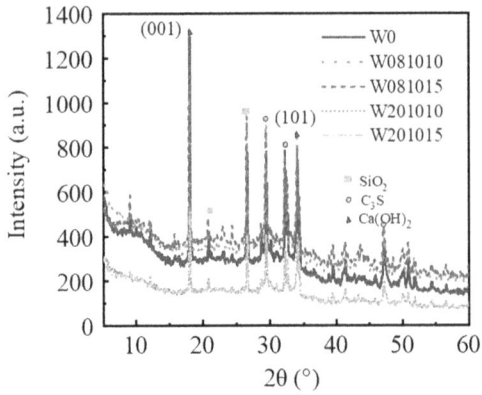

Figure 2.23 XRD pattern of SSWs-engineered multifunctional UHPC after room-temperature curing for 3 days

Table 2.2 Ca(OH)$_2$ diffraction peak and orientation index of SSWs-engineered UHPC after room-temperature curing for 3 days

Specimens	001	101	Orientation index
W0	1297	742	2.36
W081010	1120	844	1.79
W081015	1205	807	2.02
W201010	545	396	1.86
W201015	514	366	1.90

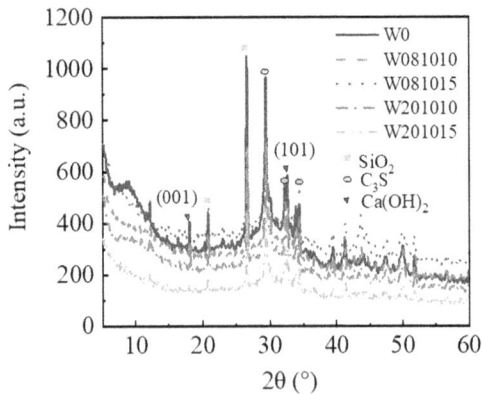

Figure 2.24 XRD pattern of SSWs-engineered multifunctional UHPC after accelerated curing for 3 days

Table 2.3 Ca(OH)$_2$ diffraction peak and orientation index of SSWs-engineered UHPC after accelerated curing for 3 days

Specimens	001	101	Orientation index
W0	406	464	1.18
W081010	340	392	1.17
W081015	377	466	1.09
W201010	294	326	1.11
W201015	158	201	1.05

characteristics as that after room-temperature curing for 3 days, while the diffraction peak intensity of Ca(OH)$_2$ crystal phase continues to decrease, indicating the increase of hydration degree and speed due to accelerated curing. The orientation index of Ca(OH)$_2$ after accelerated curing for 3 days is listed in Table 2.3. Table 2.3 shows that the orientation index of Ca(OH)$_2$ decreases with the increase of SSWs volume fraction and also decreases compared with that after room-temperature curing for 3 days. The orientation index of UHPC reinforced with 1.5 vol% SSWs at a diameter of 8 μm/20 μm and length of 10 mm is reduced by 7.6% and 11.0%, respectively.

The decrease of Ca(OH)$_2$ content and orientation index represents the narrowing of crystal growth space and the increase of structure compactness. The excellent microstructure of SSWs-engineered UHPC is quite different from the porous structure of concrete caused by common steel fiber [26, 27]. This is one of the reasons why the uniaxial compressive strength, deformation capacity, and damage inhibition ability of SSWs-engineered UHPC increase.

2.7 SUMMARY

In this chapter, the mechanical properties and strengthening mechanisms of SSWs-engineered multifunctional UHPC, including flexural properties, compressive properties after flexure, and uniaxial compressive properties, were introduced, and the uniaxial compressive constitutive model and strengthening mechanisms were presented. The main conclusions are as follows:

1) The flexural properties of SSSWs-engineered multifunctional UHPC have been improved significantly due to the addition of SSWs. When the content of SSWs with a diameter of 20 μm is 1.5 vol%, the increase of flexural strength reaches 103.2%, while the maximum flexural toughness increases by 442.2% compared with that of UHPC without SSWs. The increment of flexural strain can reach 144.6%. Natural

curing appears to have a more adverse effect on flexural strength than on flexural toughness and flexural strain. The compressive strength and toughness after flexure of SSWs-engineered multifunctional UHPC can increase by 18.9% and 214.6%, respectively, after natural curing for 28 days. The curing conditions have a different influence on compressive strength and toughness.

2) The uniaxial compressive strength of UHPC is increased by 80.2% and 85.3%, respectively, after accelerated curing for 540 days due to the addition of 1.5 vol% SSWs at a diameter of 20 μm and length of 6 mm or 10 mm. The strengthening effect of SSWs with a diameter of 20 μm is superior to that of SSWs with a diameter of 8 μm. The linear fitting coefficient between uniaxial and cubic compressive strength of SSWs-engineered multifunctional UHPC is higher than 0.97, which represents the weakness of the cyclo-hoop effect. The increment of elasticity modulus, Poisson's ratio, shear modulus, and volume modulus of UHPC can reach 58.2%, 114.8%, 47.7%, and 165.6%, respectively, due to the introduction of SSWs.

3) The ascending stage of uniaxial compressive stress–strain curves of SSWs-engineered multifunctional UHPC features a more obvious elastic-plastic part when room-temperature curing is employed. The tangent slope of the elastic-plastic part decreases with the increase of SSWs' volume fraction and diameter, and the longitudinal peak strain can be enhanced by 73.8%. By introducing the SSWs factor and curing condition factor, the uniaxial compressive constitutive model based on continuum damage theory can well describe the stress–strain relationship of SSWs-engineered multifunctional UHPC.

4) The uniaxial compressive strength of SSWs-engineered multifunctional UHPC can be expressed as the quadratic polynomial of the SSWs factor based on the UHPC matrix strength. SSWs can have an enhancing effect on UHPC through optimizing the interface, inhibiting the initiation and propagation of cracks, and improving the structure of hydration products.

REFERENCES

1. S. Dong, B. Han, X. Yu, J. Ou. Constitutive model and reinforcing mechanisms of uniaxial compressive property for reactive powder concrete with super-fine stainless wire, *Composites Part B: Engineering*. 166 (2019) 298–309.
2. B. Han, S. Dong, J. Ou, C. Zhang, Y. Wang, X. Yu, S. Ding. Microstructure related mechanical behaviors of short-cut super-fine stainless wire reinforced reactive powder concrete, *Materials and Design*. 96 (2016) 16–26.
3. B. Han, S. Dong, L. Zhang, S. Ding, S. Sun, Y. Wang. R&D of China's strategic new industries-functional materials, chapter 6: Functional civil engineering materials, China Machine Press. (2016), 195–298. (in Chinese).

4. Y.S. Yoon, D.Y. Yoo, S.W. Kim, J.J. Park. Drying shrinkage cracking characteristics of ultra-high-performance fibre reinforced concrete with expansive and shrinkage reducing agents, *Magazine of Concrete Research*. 65(4) (2013) 248–256.

5. S. Dong, D. Wang, X. Wang, A. D'Alessandro, S. Ding, B. Han, J. Ou. Optimizing flexural cracking process of ultra-high performance concrete via incorporating microscale steel wires, *Cement and Concrete Composites*. 134 (2022) 104830.

6. S.H. Park, D.J. Kim, G.S. Ryu, K.T. Koh. Tensile behavior of Ultra High Performance Hybrid Fiber Reinforced Concrete, *Cement and Concrete Composites*. 34(2) (2012) 172–184.

7. S.T. Kang, Y. Lee, Y.D. Park, J.K. Kim. Tensile fracture properties of an Ultra High Performance Fiber Reinforced Concrete (UHPFRC) with steel fiber, *Composite Structures*. 92(1) (2010) 61–71.

8. N.H. Yi, J.J. Kim, T.S. Han, Y.G. Cho, J.H. Lee. Blast-resistant characteristics of ultra-high strength concrete and reactive powder concrete, *Construction and Building Materials*. 28(1) (2012) 694–707.

9. S. Abbas, A.M. Soliman, M.L. Nehdi. Exploring mechanical and durability properties of ultra-high performance concrete incorporating various steel fiber lengths and dosages, *Construction and Building Materials*. 75 (2015) 429–441.

10. S. Aydın, B. Baradan. The effect of fiber properties on high performance alkali-activated slag/silica fume mortars, *Composites Part B: Engineering*. 45(1) (2013) 63–69.

11. S. Dong, Y. Wang, A. Ashour, B. Han, J. Ou. Uniaxial compressive fatigue behavior of ultra-high performance concrete reinforced with super-fine stainless wires, *International Journal of Fatigue*. 142 (2021) 105959.

12. O. Düğenci, T. Haktanir, F. Altun. Experimental research for the effect of high temperature on the mechanical properties of steel fiber-reinforced concrete, *Construction and Building Materials*. 75 (2015) 82–88.

13. T. Simões, C. Octávio, J. Valença, H. Costa, D. Dias-da-Costa, E. Júlio. Influence of concrete strength and steel fibre geometry on the fibre/matrix interface, *Composites Part B: Engineering*. 122 (2017) 156–164.

14. Al-Tikrite, M. N. S. Hadi. Mechanical properties of reactive powder concrete containing industrial and waste steel fibres at different ratios under compression, *Construction and Building Materials*. 154 (2017) 1024–1034.

15. Y. Ma. *Study on Constitutive Relationship of 200 MPa Reactive Powder Concrete under Uniaxial Compression*, Beijing Jiaotong University, 2006. (in Chinese).

16. T. Zdeb, I. Hager, J. Sliwinski. *Reactive Powder Concrete-Change in Compressive Strength and Modulus Od Elasticity at High Tempreature*, *Proceeding International Symposuim, Brittle Matrix Composites* 10, Warsaw, 2012.

17. J. Song, S. Liu. Properties of reactive powder concrete and its application in highway bridge, *Advances in Materials Science and Engineering*. 3 (2016) 1–7.

18. J. Mazars, G. Pyaudier-Cabot. Continuum damage theory-application to concrete, *Journal of Engineering Mechanics*. 115(2) (1989) 345–365.

19. J. Lemaitre. How to use damage mechanics, *Nuclear Engineering and Design.* 80(2) (1984) 233–245.
20. H. Zhang, B. Wang, A. Xie, Y. Qi. Experimental study on dynamic mechanical properties and constitutive model of basalt fiber reinforced concrete, *Construction and Building Materials.* 152 (2017) 154–167.
21. X. Sun, K. Zhao, Y. Li, R. Huang, Z. Ye, Y. Zhang, J. Ma. A study of strain-rate effect and fiber reinforcement effect on dynamic behavior of steel fiber reinforced concrete, *Construction and Building Materials.* 158 (2018) 657–669.
22. S. Bentur. *Mindess, Fibre Reinforced Cementitious Composites // Materials for Buildings and Structures*, Wiley-VCH Verlag GmbH & Co. KGaA, 6, 2006.
23. B. Han, S. Ding, J. Wang, J. Ou. *Nano-Engineered Cementitious Composites: Principles and Practices*, Springer, 2019.
24. B. Han, Z. Li, L. Zhang, S. Zeng, X. Yu, B. Han, J. Ou. Reactive powder concrete reinforced with Nano SiO_2-coated TiO_2, *Construction and Building Materials.* 148 (2017) 104–112.
25. Y. Qing, Z. Zenan, K. Deyu, R. Chen. Influence of Nano-SiO_2 addition on properties of hardened cement paste as compared with silica fume, *Construction and Building Materials.* 21(3) (2007) 539–545.
26. H.V. Le, D. Moon, D.J. Kim. Effects of ageing and storage conditions on the interfacial bond strength of steel fibers in mortars, *Construction and Building Materials.* 170 (2018) 129–141.
27. J.P. Hwang, M. Kim, K.Y. Ann. Porosity generation arising from steel fibre in concrete, *Construction and Building Materials.* 94(1) (2015) 433–436.

Chapter 3

Bending and fracture properties of stainless steel wires-engineered multifunctional ultra-high performance concrete

3.1 INTRODUCTION

Concrete with low tensile strength easily cracks under bending or fracture load, thus leading to structural defects or failure. Improving the ductility and crack resistance of concrete is of great importance to ensure the safety of concrete structures. Because of the excellent flexural and compressive strength demonstrated in Chapter 2, stainless steel wires (SSWs)-engineered multifunctional ultra-high performance concrete (UHPC) has potential for developing long-span structures and nuclear industrial facilities. It is more necessary for SSWs-engineered multifunctional UHPC to meet the requirements of high bending and fracture performance [1–3]. For example, opening mode cracks, i.e. mode I cracks, under fracture load are the most dangerous and most likely to cause low-stress brittle failure within structural elements [4–6]. Meanwhile, cracks in concrete structures are also typically under bending–shearing combined stress fields because of asymmetries of structural geometries and complexities of loading conditions, resulting in the occurrence of opening–sliding cracks, i.e. mixed mode I–II cracks [7–10]. Therefore, bending and fracture properties under different load forms are key indexes for evaluating the overall performances of SSWs-engineered multifunctional UHPC.

Hence, the bending properties of unnotched/notched beams and plates fabricated by SSWs-engineered multifunctional UHPC under different load forms are presented, and the mode I and mixed mode I–II fracture properties of beams are also introduced. The diameter of SSWs in this chapter is 20 µm and the length is 10 mm. The main contents of this chapter are shown in Figure 3.1.

3.2 BENDING PROPERTIES OF SSWS-ENGINEERED MULTIFUNCTIONAL UHPC

3.2.1 Four-point bending properties of SSWs-engineered multifunctional UHPC unnotched beam

Figure 3.2 shows the unnotched beam loading diagram for a four-point bending property test based on ASTM-C1609. The mid-span deflection of

DOI: 10.1201/9781003276357-3

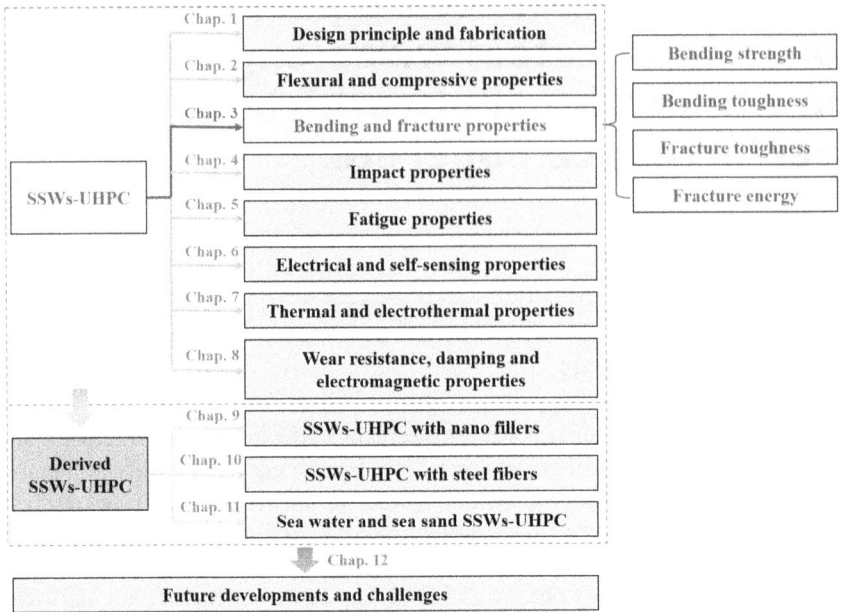

Figure 3.1 Main contents of Chapter 3

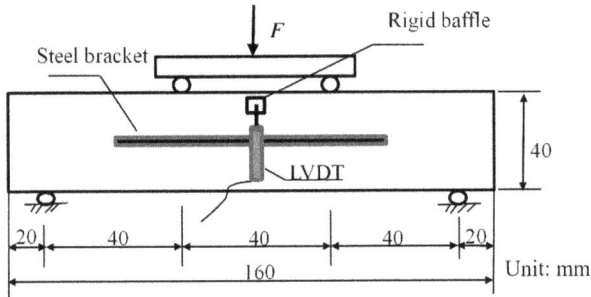

Figure 3.2 Loading diagram of bending property test by using unnotched beam under four-point bending load

unnotched beams for SSWs-engineered multifunctional UHPC was measured using a linearly varying displacement transducer (LVDT), while the displacement was the deformation value achieved by the built-in displacement sensor of an electronic universal testing machine. The bending toughness (in J/m^2) obtained by the integral area of load–deflection/displacement curves under peak load was defined as peak bending toughness, and that obtained by the total integral area of load–deflection/displacement curves was referred to limit bending toughness. There were three identical specimens for each concrete mix and element type. The load–deflection/

displacement curve closest to the average one was analyzed to obtain the crack resistance effect of SSWs. The average bending toughness value of three specimens for each concrete mix and element type was regarded as the final result if the difference between the average and the maximum and minimum values was less than 15%.

The bending strength, load–deflection curves and bending toughness of unnotched beams of SSWs-engineered multifunctional UHPC under four-point bending load are plotted in Figure 3.3. Figure 3.3(a) shows that the bending strength of UHPC is improved by 13.5% and 36.6%, respectively, due to the incorporation of 1.0 vol% (W201010) and 1.5 vol% SSWs (W201015). As shown in Figure 3.3(b), the deflection of the composites at peak load is 10.9% and 39.3% higher, respectively, than that of control UHPC, and the limit deflection (as marked in Figure 3.3(b)) of UHPC is enhanced by 18.2% and 58.9%, respectively. There is an obvious non-linear ascending stage and stationary stage before peak load on the load–deflection curve of UHPC reinforced with 1.5 vol% SSWs. During the stationary stage of the load–deflection curve, the tensile stress has been completely transferred to the SSWs, and the composite shows a

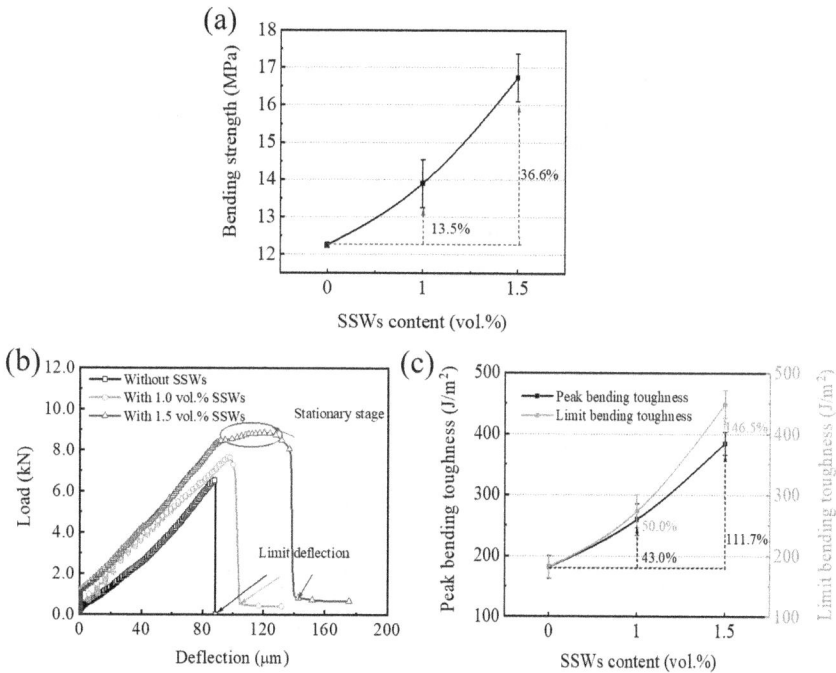

Figure 3.3 Bending strength, load–deflection curves, and bending toughness of SSWs-engineered multifunctional UHPC under four-point bending load. (a) Bending strength; (b) Load–deflection curve; (c) Bending toughness

strain hardening characteristic. The corresponding failure beams manifest a multi-cracking tendency. At the peak load, the UHPC beams without SSWs abruptly fail, losing any resistance, as indicated in Figure 3.3(b). However, the slow descending stage on the load–deflection curve is prolonged with increasing SSWs content, reflecting the gradual rupture of bridging SSWs.

It can be seen from Figure 3.3(c) that the peak bending toughness of UHPC reinforced with 1.0 vol% and 1.5 vol% SSWs is 43.0% and 111.7% higher, respectively, than that of UHPC without SSWs. Meanwhile, the limit bending toughness increments are 50.0% and 146.5% due to the addition of 1.0 vol% and 1.5 vol% SSWs, respectively. The ratios of peak bending toughness to limit bending toughness are 99.7%, 95.0%, and 85.6% for UHPC reinforced with 0%, 1.0 vol%, and 1.5 vol% SSWs, respectively, reflecting the bearing capacity enhancement of composites after cracking. This can be ascribed to the inhibiting and bridging effect of SSWs on crack development. The improvement of limit bending toughness for UHPC reinforced with 1.0 vol% SSWs is mainly attributed to the increase of peak load, while that for UHPC reinforced with 1.5 vol% SSWs is related to both the increase of peak load and the enhancement of bearing capacity after cracking.

The load–displacement curves and the corresponding bending toughness of unnotched beams for SSWs-engineered multifunctional UHPC under four-point bending load are given in Figure 3.4. Figure 3.4(a) indicates that the slope of the linear ascending stage of load–displacement curves is improved slightly due to the addition of 1.0 vol% SSWs. Meanwhile, the load–displacement curve of UHPC reinforced with 1.5 vol% SSWs possesses an obvious non-linear ascending stage. This is because the 1.0 vol% SSWs have achieved a remarkable enhancement effect on the UHPC matrix, while the inhibiting effect of SSWs on the initiation and propagation of

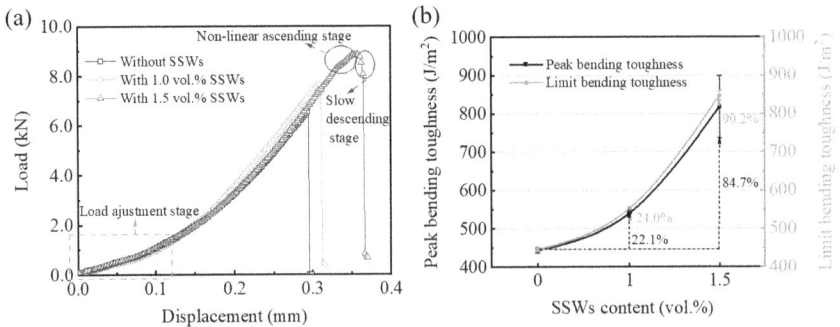

Figure 3.4 Load–displacement curves and bending toughness of SSWs-engineered multifunctional UHPC under four-point bending load. (a) Load–displacement curves; (b) Bending toughness

cracks can be fully developed at the volume fraction of 1.5 vol%. The load–displacement curve of control UHPC instantaneously drops after peak load, while that of UHPC reinforced with 1.5 vol% SSWs first exhibits a slow descending trend and then drops linearly, confirming that the bridging effect of SSWs on cracks can effectively control the propagation of macro-cracks. The peak displacements of UHPC reinforced with 1.0 vol% and 1.5 vol% SSWs are increased by 4.7% and 15.8%, respectively, compared with that of UHPC without SSWs.

As shown in Figure 3.4(b), the bending toughness of SSWs-engineered multifunctional UHPC calculated based on load–displacement curves exhibits the same variation as that calculated according to load–deflection curves, while the values of limit bending toughness are improved by 144.5%, 102.3%, and 88.6% for UHPC reinforced with 0%, 1.0 vol%, and 1.5 vol% SSWs, respectively. The improvement of deformation ability for UHPC caused by SSWs reduces the effect of equipment deformation on the values of bending toughness. The addition of 1.0 vol% and 1.5 vol% SSWs leads to a 22.1% and 84.7% increase for peak bending toughness as well as a 24. 1% and 90.2% increase for limit bending toughness, respectively. The ratios of peak bending toughness to limit bending toughness are 99.4%, 97.8%, and 96.5% for UHPC reinforced with 0 vol%, 1.0 vol%, and 1.5 vol% SSWs, respectively.

The cracking pattern of unnotched beams of SSWs-engineered multifunctional UHPC after failure is displayed in Figure 3.5. It can be observed from Figure 3.5 that the failure crack width is decreased and the propagation path of cracks is prolonged due to the incorporation of SSWs. The failure cracks become more tortuous and the failure surfaces become unsmooth with increasing SSWs content. Meanwhile, the cracks do not penetrate the entire section of beams for UHPC reinforced with SSWs, and the ruptured SSWs can be observed on the crack surface.

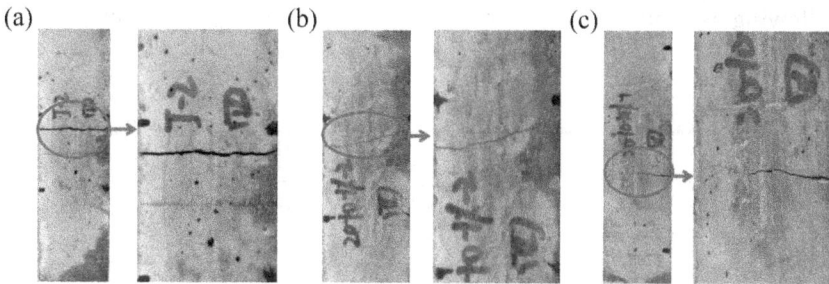

Figure 3.5 Cracking patterns of SSWs-engineered multifunctional UHPC under four-point bending load. (a) UHPC without SSWs; (b) UHPC with 1.0 vol% SSWs; (c) UHPC with 1.5 vol% SSWs

3.2.2 Three-point bending properties of SSWs-engineered multifunctional UHPC notched beam

The three-point notched beam experiment for testing bending toughness is specified based on specifications of CECS13-2009 and RILEM TC162-TDF. In order to acquire the middle notch during casting of beams, a 10-mm or 20-mm height flat steel plate having a 3-mm thickness was prepositioned at the middle of the molds. The crack length/depth ratio (a_0/h, a_0 = 10 mm or 20 mm, h = 40 mm) was 0.25 and 0.5, respectively. Before testing, two knife edges were glued to both sides of the notch in order to fix the clip gauge. It can be seen from Figure 3.6 that the clip gauge and LVDT were used to measure the cracking mouth opening displacement (CMOD) and mid-span deflection of notched beams under a three-point bending load system.

3.2.2.1 Specimens with crack length/depth ratio of 0.25

At the crack length/depth ratio of 0.25, the load–deflection curves, bending strength, and peak deflection of notched beams of SSWs-engineered multifunctional UHPC under three-point bending load are plotted in Figure 3.7. Due to the inhibiting effect of SSWs on crack development, the load–deflection curves of UHPC reinforced with SSWs exhibit a long linear ascending stage, an obvious non-linear ascending stage, and a slow descending stage, as shown in Figure 3.7(a). The peak deflection UHPC is improved due to the presence of SSWs. The stationary stage before peak load on the load–deflection curves is more obvious with increasing SSWs content, indicating the occurrence of yield state for SSWs-engineered multifunctional UHPC. Following Figure 3.7(b), it is clear that increases of 73.9% and 119.9% are obtained for the bending strength of SSWs-engineered multifunctional UHPC compared with that of UHPC without SSWs. The enhancement ratios of peak deflection caused by 1.0 vol% and 1.5 vol% SSWs are 17.3% and 95.2%, respectively.

The energy absorption and equivalent bending strength of notched beams of SSWs-engineered multifunctional UHPC are calculated considering the following assumptions. (1) Due to the non-standard size of notched beam

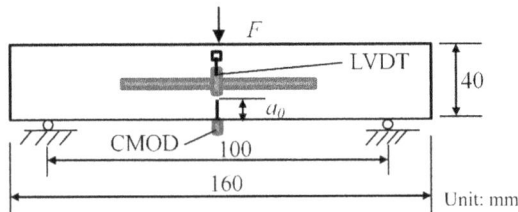

Figure 3.6 Loading diagram of bending property test by using notched beam under three-point bending load

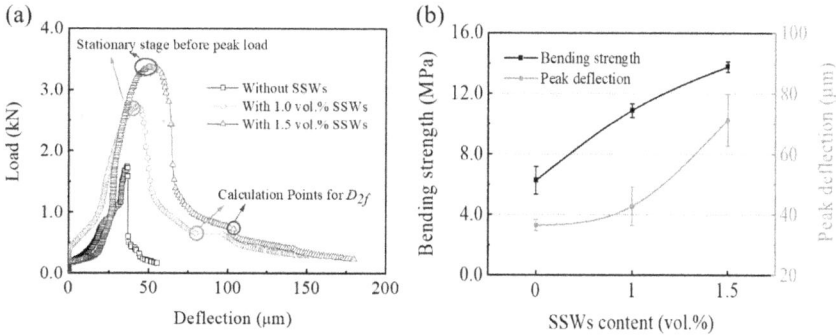

Figure 3.7 a_0/h = 0.25, load–deflection curves, bending strength, and peak deflection of notched beams for SSWs-engineered multifunctional UHPC. (a) Load–deflection curve; (b) Bending strength and peak deflection

used in this work, the integral area of the load–deflection curve for UHPC without SSWs is defined as the energy absorption of matrix cracking, D_{cr}. (2) The energy absorption calculated by load–deflection curves of SSWs-engineered multifunctional UHPC under peak load is marked as D_{1f}. (3) It can be observed from Figure 3.7(a) that it is difficult for load–deflection curves to be zero because of insufficient rigidity of the testing machine. After dropping to a certain point, there is a sudden change in the value of deflection, indicating that the beams are moving into the failure state. In order to fully reflect the toughening effect of SSWs with different content, the different failure points on load–deflection curves of UHPC reinforced with 1.0 vol% and 1.5 vol% SSWs are adopted to calculate energy absorption, D_{2f}. Meanwhile, the deflections corresponding to D_{cr} on load–deflection curves of SSWs-engineered multifunctional UHPC are denoted as δ_u. The peak deflections are denoted as δ_{u1}, and the equivalent bending strength calculated by D_{1f} is f_{eq1}. The deflection and equivalent bending strength corresponding to D_{2f} are δ_{u2} and f_{eq2}, respectively. The average value of total integral area for three beams with the same concrete mix under the load–deflection curve is denoted as $\bar{D}$. From this, the equivalent bending strength is calculated by the following Eqs (3.1) and (3.2):

$$f_{eq1} = \frac{D_{1f} - D_{cr}}{\delta_{u1} - \delta_u} \times \frac{1.5L}{Bh^2} \tag{3.1}$$

$$f_{eq2} = \frac{D_{2f} - D_{cr}}{\delta_{u2} - \delta_u} \times \frac{1.5L}{Bh^2} \tag{3.2}$$

where L is the span of the notched beam; B is the width of the notched beam; h is the effective height of the notched beam, which equals beam height minus notch height.

Table 3.1 $a_0/h = 0.25$, equivalent bending strength of SSWs-engineered multifunctional UHPC

SSWs content (vol%)	D_{cr} (N·mm)	δ_u (μm)	D_{1f} (N·mm)	δ_{u1} (μm)	f_{eq1} (MPa)	D_{2f} (N·mm)	δ_{u2} (μm)	f_{eq2} (MPa)	Standard deviation (N·mm)
0	29.9	–	–	–	–	–	–	–	2.80
1.0	29.9	29.7	64.4	43.4	8.6	114.7	89.3	5.9	11.26
1.5	29.9	36.2	74.8	50.4	13.2	157.6	104.0	7.8	10.34

The energy absorption and equivalent bending strength of UHPC composites at the crack length/depth ratio of 0.25 are listed in Table 3.1. As shown in Table 3.1, the energy absorption difference between D_{1f} and D_{cr} for UHPC reinforced with 1.5 vol% SSWs is increased by 16.1% compared with that for UHPC reinforced with 1.0 vol% SSWs, reflecting the enhancement effect of SSWs on the UHPC matrix and the inhibiting effect of SSWs on the initiation and stable propagation of cracks. Meanwhile, the increase of SSWs content leads to a 37.4% increase in energy absorption difference between D_{2f} and D_{cr}, and this difference can be mainly ascribed to the bridging effect of SSWs. The equivalent bending strength represents the ability of composites to resist deformation in the plastic stage. The equivalent bending strength corresponding to energy absorption D_{1f} and D_{2f} of UHPC reinforced with 1.5 vol% SSWs is 53.5% and 32.2% higher, respectively, than that of UHPC reinforced with 1.0% SSWs. The improvement of energy absorption and equivalent bending strength caused by SSWs can be attributed to the following phenomenon. (1) The original flaw in the UHPC matrix is reduced, and the anti-cracking ability of the matrix is enhanced, due to the micron diameter and large specific surface area of SSWs. (2) The SSWs can effectively transfer the crack tip stresses and inhibit the initiation and propagation of cracks. The propagation path of cracks is prolonged, and this endows UHPC with large plastic displacement capacity. (3) With the formation of macrocracks, tensile stress is mainly undertaken by SSWs, and the load–deflection curve displays a stationary stage before peak load and a slow descending stage after peak load. The load drops sharply with the rupture of SSWs, while the beams still remain intact when failure occurs.

The load–CMOD curves of SSWs-engineered multifunctional UHPC at the crack length/depth ratio of 0.25 are shown in Figure 3.8. It is evident from Figure 3.8 that the variation of load–CMOD curves caused by the addition of SSWs is similar to that of load–deflection curves. The load–CMOD curve of UHPC reinforced with 1.5 vol% SSWs contains three stages: a linear ascending stage, a non-linear ascending stage, and a slow descending stage. If the load–CMOD curve can also be used to characterize the bending toughness of the notched beam, the evaluation indexes are more comprehensive and reliable. However, the recommended relationship between δ and CMOD in the method of RILEM TC162-TDF and CECS 13-2009 is not applicable to SSWs-engineered multifunctional UHPC. According to fracture mechanics [4], the theoretical calculation model between bending deflection δ and CMOD of concrete beams is plotted in Figure 3.9.

$$tan\theta = \frac{\delta}{l/2} = \frac{CMOD/2}{H} \tag{3.3}$$

$$CMOD = \frac{4\delta H}{L} \tag{3.4}$$

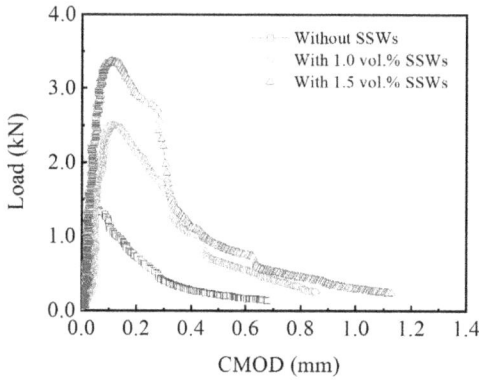

Figure 3.8 a_0/h = 0.25, load–CMOD curves of SSWs-engineered multifunctional UHPC notched specimens under three-point bending load

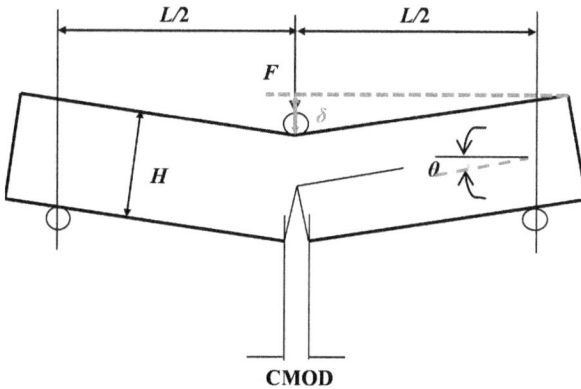

Figure 3.9 Theoretical calculation model between deflection and CMOD

It can be seen from Figure 3.9 and Eq. (3.4) that there exists a linear correlation between δ and CMOD, which can be used to obtain the characteristic points on the load–CMOD curve for calculating energy absorption and equivalent bending strength. The fitting relationship between δ and CMOD of SSWs-engineered multifunctional UHPC is given in Figure 3.10. As demonstrated in Figure 3.10, CMOD shows approximately three-stage linear behavior with δ, corresponding to the three stages of crack development: the elastic stage, the stable crack propagation stage, and the unstable crack propagation stage.

The linear slopes corresponding to these three stages are marked as K_1, K_2, and K_3, respectively, as shown in Figure 3.10. The values of K_1, K_2, and K_3 are listed in Table 3.2. Table 3.2 shows that the values of K first increase and then decrease with increasing δ. Increasing SSWs content from 1.0 vol% to 1.5 vol% leads to 80.0%, 111.0%, and 15.7% reduction for K_1, K_2 and K_3 values, indicating that a high volume fraction of SSWs can limit the opening of the crack mouth more effectively, especially in the stage of unstable crack propagation.

The characteristic values of CMOD are calculated according to the fitting results between δ and CMOD, which are listed in Table 3.3. Then, the energy absorption and equivalent bending strength of SSWs-engineered multifunctional UHPC are analyzed in accordance with load–CMOD curves, and the calculation equations are the same as Eqs (3.1) and (3.2). The calculated results are shown in Table 3.4.

As can be observed from Table 3.4, the energy absorption D_{2f} of UHPC reinforced with 1.5 vol% SSWs is 28.3% higher than that of UHPC reinforced with 1.0 vol% SSWs. The equivalent bending strength f_{eq1}' and f_{eq2}' show an increase of 43.2% and 61.2%, respectively, with the increasing SSWs content. With the increase of CMOD, the equivalent bending strength of SSWs-engineered multifunctional UHPC decreases because of the rupture of SSWs.

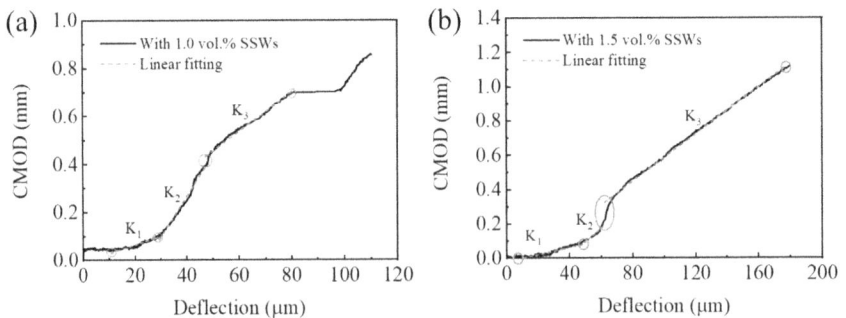

Figure 3.10 $a_0/h = 0.25$, fitting relationship between δ and CMOD of SSWs-engineered multifunctional UHPC. (a) UHPC with 1.0 vol% SSWs; (b) UHPC with 1.5 vol% SSWs

Table 3.2 a_0/h = 0.25, fitting equations between δ and CMOD of SSWs-engineered multifunctional UHPC

SSWs content (vol%)	Fitting equation	K	Deflection range	R^2
1.0	CMOD = 0.00369δ − 0.00969	K_1 = 0.00369	(10.80,30.51]	0.89845
	CMOD = 0.01680δ − 0.40692	K_2 = 0.01680	(30.51,47.31]	0.98225
	CMOD = 0.00772δ + 0.07754	K_3 = 0.00772	(47.31,108.90]	0.99408
1.5	CMOD = 0.00205δ − 0.02136	K_1 = 0.00205	(0,49.23]	0.85202
	CMOD = 0.00796δ − 0.30321	K_2 = 0.00796	(49.23,62.45]	0.91213
	CMOD = 0.00667δ − 0.07259	K_3 = 0.00667	(62.45,178.15]	0.99926

Table 3.3 a_0/h = 0.25, characteristic values of CMOD corresponding to δ

SSWs content (vol%)	δ_u (µm)	$CMOD_u$ (mm)	δ_{u1} (µm)	$CMOD_1$ (mm)	δ_{u2} (µm)	$CMOD_2$ (mm)
1.0	29.7	0.0999	43.4	0.3216	89.3	0.7671
1.5	36.2	0.0528	50.4	0.0978	104.0	0.6213

Table 3.4 a_0/h = 0.25, energy absorption and equivalent bending strength of SSWs-engineered multifunctional UHPC based on load–CMOD curves

SSWs content (vol%)	D_{cr} (kN·mm)	D_{1f} (kN·mm)	D_{2f} (kN·mm)	f_{eq1}' (MPa)	f_{eq2}' (MPa)
1.0	0.1081	0.5779	0.8928	8.8	4.9
1.5	0.0658	0.2017	1.1451	12.6	7.9

The relative errors for equivalent bending strength can be calculated using Eqs (3.5) and (3.6), and the values of relative error are given in Table 3.5. Table 3.5 shows that the relative error for UHPC reinforced with 1.0 vol% SSWs is less than 16.2%, while that for UHPC reinforced with 1.5 vol% SSWs is less than 4.5%. Therefore, the bending toughness of notched beams for SSWs-engineered multifunctional UHPC can be evaluated either by the load–deflection curve or by the load–CMOD curve. Using δ and CMOD simultaneously to evaluate the bending toughness of SSWs-engineered multifunctional UHPC improves the accuracy and reliability of test results. In addition, the testing machine is simplified and the test cost is reduced by directly using CMOD to evaluate the bending toughness of notched beams.

$$\Delta f_{eq1} = \frac{\left| f_{eq1} - f_{eq1}' \right|}{f_{eq1}} \tag{3.5}$$

Table 3.5 $a_0/h = 0.25$, relative errors of equivalent bending strength

SSWs content (vol%)	Based on δ		Based on CMOD		Relative error	
	f_{eq1} (MPa)	f_{eq2} (MPa)	f_{eq1}' (MPa)	f_{eq2}' (MPa)	Δf_{eq1}	Δf_{eq2}
1.0	8.6	5.9	8.8	4.9	2.3%	16.2%
1.5	13.2	7.8	12.6	7.9	4.5%	1.3%

$$\Delta f_{eq2} = \frac{\left| f_{eq2} - f_{eq2}' \right|}{f_{eq2}} \tag{3.6}$$

3.2.2.2 Specimens with crack length/depth ratio of 0.5

At the crack length/depth ratio of 0.5, the load–deflection curves, bending strength, peak deflection, and load–CMOD curves of notched beams for SSWs-engineered multifunctional UHPC under three-point bending load are demonstrated in Figure 3.11. Figure 3.11(a) and (c) indicate that the notched beam of UHPC without SSWs has a low bearing capacity. The load–deflection and load–CMOD curves of SSWs-engineered multifunctional

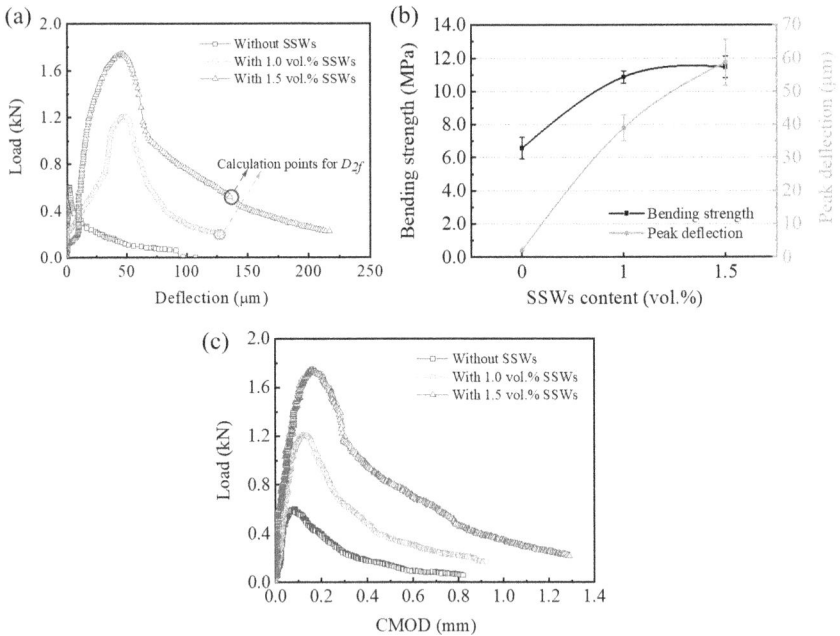

Figure 3.11 $a_0/h = 0.5$, load–deflection curves, flexural strength, peak deflection, and load–CMOD curves of SSWs-engineered multifunctional UHPC. (a) Load–deflection curve; (b) Bending strength and peak deflection; (c) Load–CMOD curve

UHPC beams possess a long linear ascending stage, an obvious non-linear ascending stage, and a slow descending stage. This toughening phenomenon caused by SSWs is not affected by the crack length/depth ratio and can be attributed to the inhibiting effect of SSWs on the initiation and propagation of cracks. It can be seen from Figure 3.11(b) that the bending strength of UHPC is improved by 65.2% and 74.4% due to the addition of 1.0 vol% and 1.5 vol% SSWs, respectively. The peak deflection of UHPC without SSWs is only 2.1 μm, increasing to 39.0 μm and 58.7 μm for UHPC reinforced with 1.0 vol% and 1.5 vol% SSWs, respectively.

The energy absorption and equivalent bending strength of SSWs-engineered UHPC at the crack length/depth ratio of 0.5 are calculated following Eqs (3.1) and (3.2), and the results are listed in Table 3.6. Table 3.6 illustrates that the deflection corresponding to matrix energy absorption D_{cr} decreases with increasing SSWs content. In the crack stable propagation stage, the energy absorption D_{1f} of UHPC reinforced with 1.5 vol% SSWs is 63.5% higher than that of UHPC reinforced with 1.0 vol% SSWs. In the crack unstable propagation stage, the increase of SSWs content leads to a 93.7% increase of energy absorption D_{2f}. The equivalent bending strength f_{eq1} and f_{eq2} are increased by 41.0% and 80.0%, respectively, because of the increasing SSWs content. SSWs can not only inhibit the initiation and propagation of cracks but also bridge cracks until ruptured, eventually resulting in enhancement of the flexural toughness of UHPC.

In accordance with Eq. (3.4), the linear fitting between δ and CMOD is carried out for notched beams with a crack length/depth ratio of 0.5. The fitting curves are plotted in Figure 3.12, and the fitting results are displayed in Table 3.7. As observed in Figure 3.12, there is a three-stage linear relationship between δ and CMOD of UHPC reinforced with 1.0 vol% and 1.5 vol% SSWs. Table 3.7 shows that the slopes (K_1, K_2, and K_3) of linear fitting are decreased by 3.0%, 29.9%, and 34.1 vol%, respectively, at different stages with the increasing SSWs content. The UHPC composite reinforced with a high content of SSWs has large deformation ability and high inhibition capacity for the opening of the crack mouth. The values of K_1, K_2, and K_3 increase with increasing deflection, indicating that the large deformation of composites speeds up the opening of the crack mouth.

Table 3.6 a_0/h=0.5, equivalent bending strength of SSWs-engineered multifunctional UHPC

SSWs content (vol%)	D_{cr} (N·mm)	δ_u (μm)	D_{1f} (N·mm)	δ_{u1} (μm)	f_{eq1} (MPa)	D_{2f} (N·mm)	δ_{u2} (μm)	f_{eq2} (MPa)	Standard deviation (N·mm)
0	15.8	–	–	–	–	–	–	–	0.70
1.0	15.8	33.6	32.9	49.0	10.7	69.8	125.1	5.5	9.26
1.5	15.8	23.7	53.8	47.3	15.1	135.2	136.3	9.9	10.60

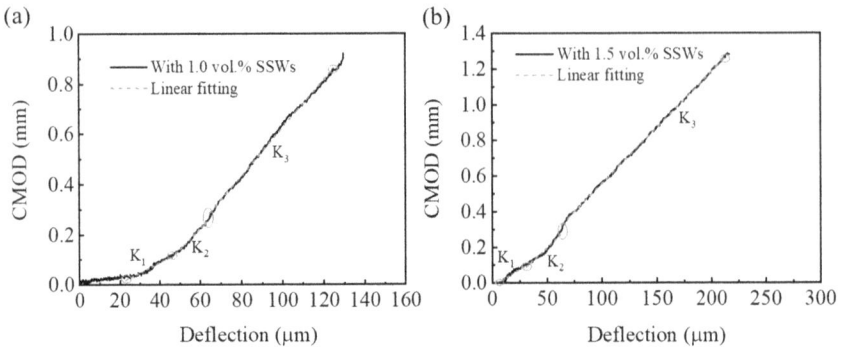

Figure 3.12 a_0/h = 0.5, fitting relationship between δ and CMOD of SSWs-engineered multifunctional UHPC. (a) UHPC with 1.0 vol% SSWs; (b) UHPC with 1.5 vol% SSWs

Table 3.7 a_0/h = 0.5, fitting equations of δ and CMOD for SSWs-engineered multifunctional UHPC

SSWs content (vol%)	Fitting equation	K	Deflection range	R^2
1.0	CMOD = 0.00406δ − 0.07464	K_1 = 0.00406	(22.78,46.06]	0.89590
	CMOD = 0.00816δ − 0.26639	K_2 = 0.00816	(46.06,64.53]	0.98663
	CMOD = 0.00957δ − 0.33150	K_3 = 0.00957	(64.53,128.32]	0.99736
1.5	CMOD = 0.00394δ − 0.02217	K_1 = 0.00394	(0,30.66]	0.83047
	CMOD = 0.00572δ − 0.08669	K_2 = 0.00572	(30.66,63.44]	0.95175
	CMOD = 0.00631δ − 0.07482	K_3 = 0.00631	(63.44,216.90]	0.99959

Based on the three-stage linear fitting relationship, the characteristic values of CMOD corresponding to different deflections are displayed in Table 3.8. The energy absorption and equivalent bending strength of SSWs-engineered multifunctional UHPC calculated by CMOD characteristic values are shown in Table 3.9. It can be found from Table 3.9 that the energy absorption D_{cr}, D_{1f}, and D_{2f} of UHPC reinforced with 1.5 vol% SSWs are 40.3%, 149.4%, and 22.5% higher, respectively, than those of UHPC reinforced with 1.0 vol% SSWs. At the stage of crack stable propagation, the toughening efficiency of SSWs is more easily affected by content. The equivalent bending strength f_{eq1}' and f_{eq2}' of SSWs-engineered multifunctional UHPC are increased by 83.8% and 34.0%, respectively, due to the increase of SSWs content.

The relative errors between equivalent bending strength obtained by load–deflection curves and that obtained by load–CMOD curves are listed in Table 3.10. Table 3.10 indicates that the relative error for UHPC

Table 3.8 a_0/h = 0.5, characteristic values of CMOD corresponding to different δ

SSWs content (vol%)	δ_u (μm)	$CMOD_u$ (mm)	δ_{ul} (μm)	$CMOD_l$ (mm)	δ_{u2} (μm)	$CMOD_2$ (mm)
1.0	33.6	0.0616	49.0	0.1334	125.1	0.8656
1.5	23.7	0.0711	47.3	0.1838	136.3	0.7854

Table 3.9 a_0/h = 0.5, energy absorption and equivalent bending strength of SSWs-engineered multifunctional UHPC based on CMOD

SSWs content (vol%)	D_{cr} (kN·mm)	D_{lf} (kN·mm)	D_{2f} (kN·mm)	f_{eq1}' (MPa)	f_{eq2}' (MPa)
1.0	0.0313	0.1118	0.4600	10.5	5.0
1.5	0.0439	0.2788	0.5635	19.3	6.7

Table 3.10 a_0/h = 0.5, relative error of equivalent bending strength

SSWs content (vol%)	Based on δ		Based on CMOD		Relative error	
	f_{eq1} (MPa)	f_{eq2} (MPa)	f_{eq1}' (Mpa)	f_{eq2}' (Mpa)	Δf_{eq1}	Δf_{eq2}
1.0	10.7	5.5	10.5	5.0	1.9%	9.1.0 vol%
1.5	15.1	9.9	19.3	6.7	27.8%	32.3%

reinforced with 1.0 vol% SSWs is smaller than 9.1%, while that for UHPC reinforced with 1.5 vol% SSWs can reach 32.3%. However, the variation of equivalent bending strength with SSWs content calculated based on load–deflection curves is similar to that calculated based on load–CMOD curves, showing that the bending toughness of SSWs-engineered multifunctional UHPC can be characterized by the limitation capacity of composites on the opening of the crack mouth.

3.2.3 Four-point bending properties of SSWs-engineered multifunctional UHPC unnotched plate

The four-point bending properties of SSWs-engineered multifunctional UHPC unnotched plate were tested according to ASTM-C1609. The mid-span deflection of plates under a four-point bending load system was also obtained using LVDT, as shown in Figure 3.13. Meanwhile, the mid-span displacement was also collected.

The load–deflection and load–displacement curves of plates for SSWs-engineered multifunctional UHPC are plotted in Figure 3.14. As shown in Figure 3.14, there are three stages in the load–deflection/displacement

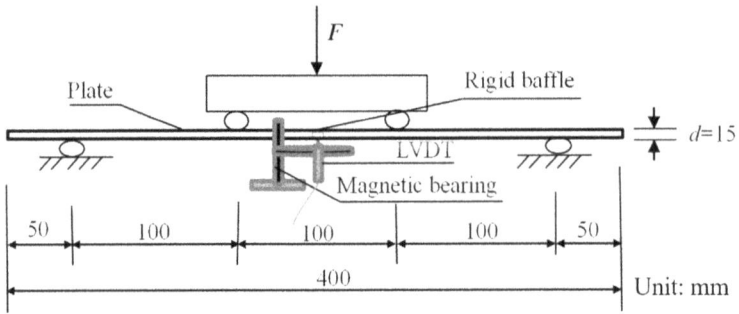

Figure 3.13 Loading diagram of plate for bending toughness test

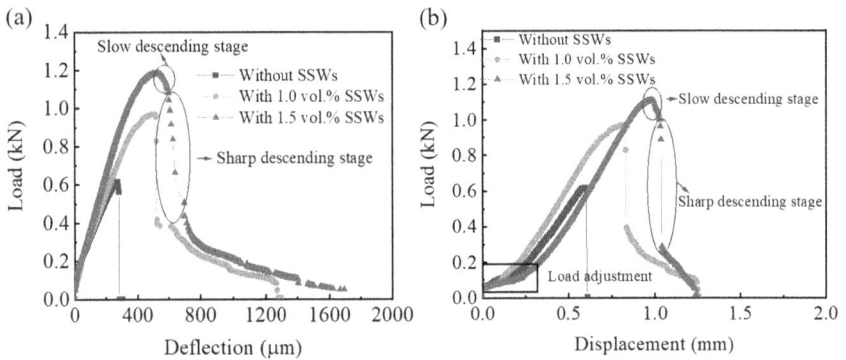

Figure 3.14 Load–deflection and load–displacement curves of plates for SSWs-engineered multifunctional UHPC. (a) Load–deflection curves; (b) Load–displacement curves

curves of plates: the elastic ascending stage, the non-linear ascending stage, and the descending stage. The increase of SSWs content leads to prolongation of the elastic ascending stage, a more obvious non-linear ascending stage, and a slower descending stage. The appearance of the non-linear ascending stage can be attributed to the inhibiting effect of SSWs on crack initiation and propagation. When the load drops by more than 80% of peak load, the load–deflection/displacement curves of UHPC reinforced with 1.5 vol% SSWs decline slowly. And then, the load–deflection/displacement curves move into an abrupt descending stage due to the rupture of SSWs. With the appearance of macrocracks, the tensile stress is transferred to the SSWs. The propagation of macrocracks is hindered by the bridging effect of the SSWs. The existence of an abrupt descending stage in load–deflection/displacement curves represents the rupture of SSWs.

The bending strength, peak deflection, and displacement of plates for SSWs-engineered multifunctional UHPC are shown in Figure 3.15. Figure 3.15(a) shows that the bending strength of UHPC reinforced with

1.0 vol% and 1.5 vol% SSWs is enhanced by 49.4% and 77.8%, respectively, compared with UHPC without SSWs. Figure 3.15(b) shows that the incorporation of 1.0 vol% and 1.5 vol% SSWs leads to a 47.1% and 94.7% increase, respectively, in peak deflection of UHPC as well as a 4.0% and 65.4% increase in peak displacement. The test results indicate that the inhibiting effect of SSWs on the initiation and propagation of cracks increases with increasing SSWs content. The peak displacement of UHPC reinforced with 0 vol%, 1.0 vol%, and 1.5 vol% SSWs is 118.5%, 103.9%, and 85.7%, respectively, higher than the peak deflection. This means that the effect of testing machine rigidity on peak deformation of SSWs-engineered multifunctional UHPC is reduced by the increasing SSWs content.

The calculation results of bending toughness for plates of SSWs-engineered multifunctional UHPC are shown in Figure 3.16. It is clear from Figure 3.16(a) that compared with UHPC without SSWs, the peak bending toughness of UHPC reinforced with 1.0 vol% and 1.5 vol% SSWs

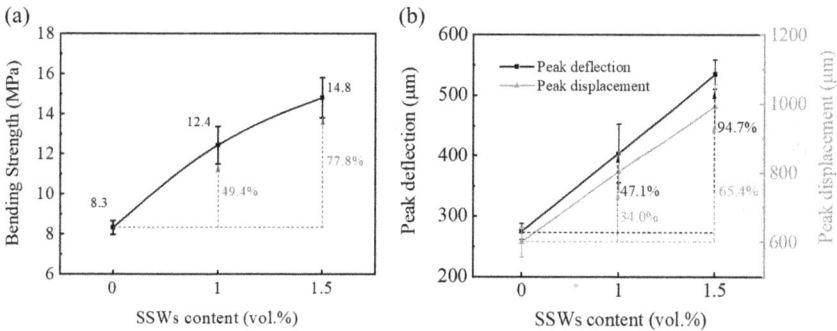

Figure 3.15 Bending strength, peak deflection, and displacement of plates for SSWs-engineered multifunctional UHPC. (a) Bending strength; (b) Peak deflection/displacement

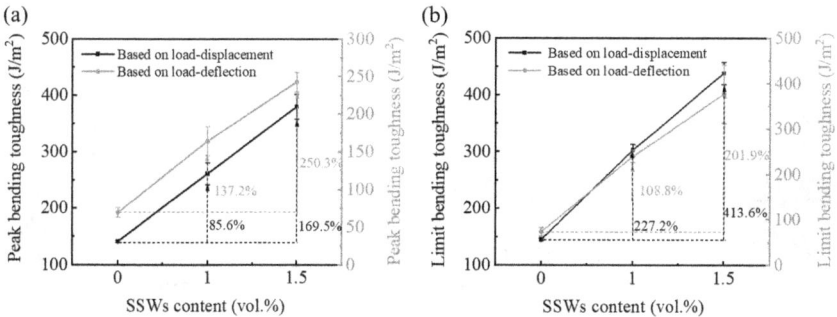

Figure 3.16 Flexural toughness of plates for SSWs-engineered multifunctional UHPC. (a) Peak bending toughness; (b) Limit bending toughness

based on load–deflection curves is enhanced by 137.2% and 250.3%, and that based on load–displacement curves is increased by 85.6% and 169.5%, respectively. The peak bending toughness of UHPC reinforced with 0 vol%, 1.0 vol%, and 1.5 vol% SSWs calculated according to load–displacement curves is 103.6%, 59.3%, and 56.6% higher, respectively, than that calculated based on load–deflection curves. The variation of peak bending toughness with SSWs content is similar to that of peak deflection and displacement. Figure 3.16(b) demonstrates that when the load–deflection curves are employed, the increments of limit bending toughness for UHPC with 1.0 vol% and 1.5 vol% SSWs are 108.8% and 201.9%, respectively, compared with UHPC without SSWs. Meanwhile, increases of 227.2% and 413.6%, respectively, are obtained for limit bending toughness of the composites when the load–displacement curves are used. The limit bending toughness of UHPC reinforced with 0 vol%, 1.0 vol%, and 1.5 vol% SSWs calculated by load–displacement curves is 98.6%, 26.4%, and 16.4% higher, respectively, than that calculated by load–deflection curves. The deformation ability enhancement of UHPC caused by SSWs prevents the test values of limit bending toughness from suffering the effect of the testing machine. The change rule of bending toughness for UHPC plates with the increase of SSWs content is the same whether the load–deflection curves or the load–displacement curves are employed.

The ratios of peak bending toughness to limit bending toughness for plates of UHPC reinforced with 0 vol%, 1.0 vol%, and 1.5 vol% SSWs are 94.4%, 68.4%, and 64.4%, respectively, on the basis of load–deflection curves, and the ratios are 97.1%, 86.3%, and 86.7%, respectively, according to load–displacement curves, showing improvement of toughness. Hence, the experimental results indicate that the toughening effect of SSWs on UHPC increases with increasing SSWs content. After the formation of macrocracks, the bridging and rupture of SSWs are the main source of toughness.

The cracking patterns of plates for SSWs-engineered UHPC are exhibited in Figure 3.17. Figure 3.17(a) shows that there are two stress concentration

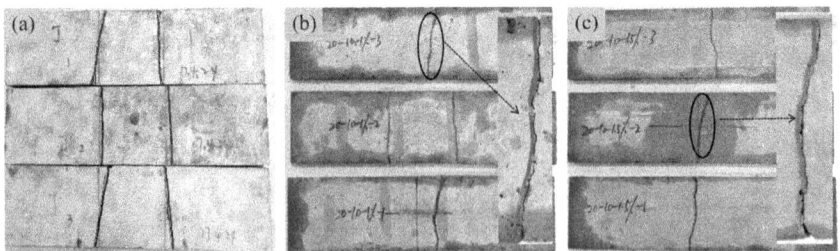

Figure 3.17 Cracking patterns of plates for SSWs-engineered multifunctional UHPC. (a) UHPC without SSWs; (b) UHPC with 1.0 vol% SSWs; (c) UHPC with 1.5 vol% SSWs

cracks on load points for failure plates of UHPC without SSWs. Figure 3.17(b) demonstrates that an oblique penetrating crack between two load points or two stress concentration cracks on the load points appears on the failure plates of UHPC reinforced with 1.0 vol% SSWs. Figure 3.17(c) shows that there is a typical tensile penetrating crack at the midspan of plates for UHPC reinforced with 1.5 vol% SSWs. The tensile stress caused by midspan bending moment plays a dominant role in the loading process, and the generation and propagation of tensile cracks are inhibited by SSWs. Hence, the tensile cracks are tortuous, and the crack path is prolonged.

3.3 FRACTURE PROPERTIES OF SSWS-ENGINEERED MULTIFUNCTIONAL UHPC

3.3.1 Three-point fracture properties of SSWs-engineered multifunctional UHPC notched beam

The schematic programs of three-point fracture tests are plotted in Figure 3.18. LVDT and a CMOD meter were used to measure the mid-span deflection and crack propagation. Meanwhile, the initial cracking load and critical effective crack length of composites were obtained by testing the strain of both sides and the upper end of the initial notch of specimens with different crack length/depth ratio.

3.3.1.1 Load–deflection and load–CMOD curves

Figure 3.19 shows the load–deflection curves of SSWs-engineered multifunctional UHPC at different crack length/depth ratios under three-point fracture load. As can be seen from Figure 3.19, the peak and limit values of deflections of UHPC are enhanced because of the incorporation of SSWs. The linear elastic stage of load–deflection curves for composites is prolonged, which can be attributed to the dense structure and refined grain of the UHPC matrix caused by SSWs. The load–deflection curves show non-linearity when the deformation of the UHPC matrix reaches the initial

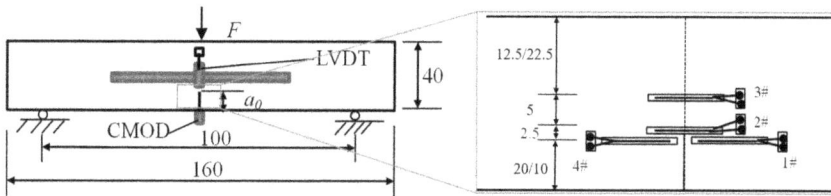

Figure 3.18 Fracture test setup

cracking strain. The non-linear ascending stage of the curves becomes more obvious with the increase of SSWs content. During this stage, the generation of new cracks and the propagation of original cracks are hindered by the UHPC matrix and SSWs together. When the cracks become saturated to form localized failure cracks, the load reaches the peak and then, decreases slowly. The bridging effect of SSWs causes a slow descending stage for the load–deflection curves of UHPC [11, 12]. The failure cracks continue to open and the SSWs are ruptured gradually with the load decrease.

The load–CMOD curves of SSWs-engineered multifunctional UHPC at different crack length/depth ratios under three-point fracture load are demonstrated in Figure 3.20. At the crack length/depth ratio of 0.25, the peak load and peak CMOD ($COMD_C$) of UHPC are improved by 84.6%/147.8% and 86.3%/74.5%, respectively, due to the addition of 1.0 vol% and 1.5 vol% SSWs. The corresponding enhancement ratios are 108.3%/190.0% and 42.3%/85.9%, respectively, at the crack length/depth ratio of 0.5. The

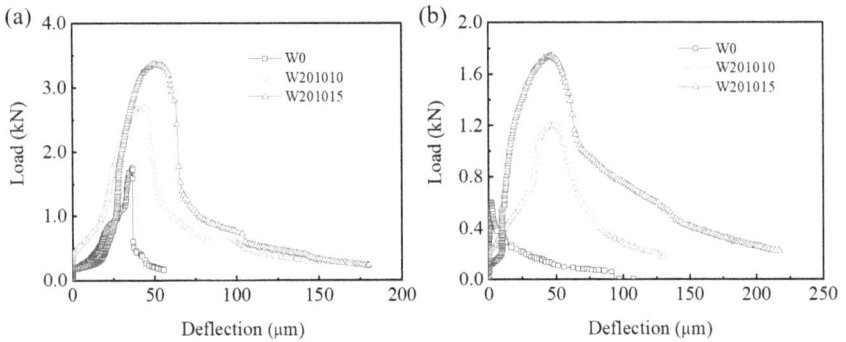

Figure 3.19 Load–deflection curves of SSWs-engineered multifunctional UHPC under three-point fracture load. (a) a_0/h=0.25; (b) a_0/h=0.5

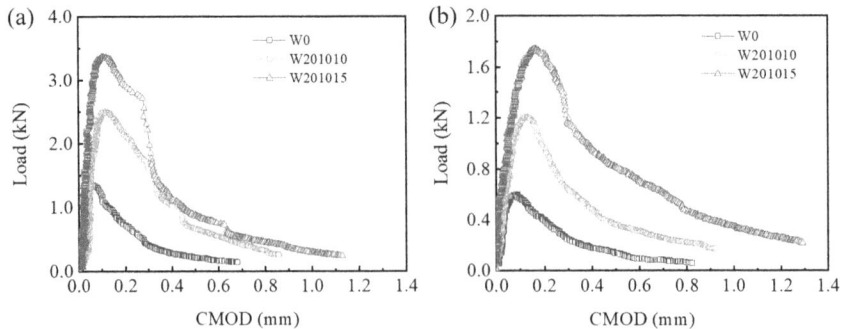

Figure 3.20 Load–CMOD curves of SSWs-engineered multifunctional UHPC under three-point fracture load. (a) a_0/h=0.25; (b) a_0/h=0.5

main features of load–CMOD curves for the composites are independent of crack length/depth ratio. The prolongation of the linear ascending stage of load–CMOD curves indicates the enhanced resistance of the UHPC matrix to cracking. With the increase of SSWs content, the slope of the non-linear ascending stage drops noticeably, the descending stage becomes gentler, and the limit value of CMOD is markedly improved. This phenomenon can be attributed to the bridging effect of SSWs in UHPC.

3.3.1.2 Measured critical effective crack length

The critical effective crack length (a_c) is defined as the crack length of concrete specimens in the failure state, which can be monitored by strain gauges arranged on the upper end of the precast crack. The variation of strains obtained by strain gauge 2# and strain gauge 3# (shown in Figure 3.18) are shown in Figure 3.21. Figure 3.21 demonstrates that SSWs-engineered multifunctional UHPC where the strain gauge is located has been cracking as the load–strain curves step into the non-linear ascending stage. The limit values of strain are enhanced with the increase of SSWs content, and the composite specimens still have a certain bearing capacity after cracking due to the bridging effect of SSWs.

The cracking load, cracking time, and critical effective crack length of SSWs-engineered UHPC are displayed in Table 3.11. The cracking loads for the locations of strain gauge 2# and strain gauge 3# increase with increasing SSWs content. At the crack length/depth ratio of 0.25, the cracking load for strain gauge 2# location of UHPC without SSWs is 0.57 kN, which increases to 1.79 kN and 2.04 kN for UHPC reinforced with 1.0 vol% and 1.5 vol% SSWs, indicating 214.0% and 257.9% enhancement, respectively. Meanwhile, the cracking load for strain gauge 3# location of composites exhibits 110.5% and 87.7% improvement, respectively. At the crack length/depth ratio of 0.5, the cracking load for the strain gauge 2# location of SSWs-engineered multifunctional UHPC exceeds that of UHPC without SSWs by 101.8% and 171.9%, respectively, and the load value for the strain gauge 3# location of composites is 105.8% and 176.3% greater than the control. The enhancement of cracking load represents the inhibiting and bridging effect of SSWs on macrocracks.

The cracking interval time between the locations of strain gauge 2# and strain gauge 3# for UHPC without SSWs is 83 s at a crack length/depth ratio of 0.25, and that between the locations of strain gauge 3# and peak load is 28.5 s, while the load difference is 0.22 kN. The unstable crack has surpassed the location of strain gauge 3#. Therefore, the measured critical crack length of control UHPC is larger than 27.5 mm. The cracking interval time between the locations of strain gauge 2# and strain gauge 3# for UHPC reinforced with 1.0 vol% SSWs is 53.5 s, and that between the locations of strain gauge 3# and peak load is 14.5 s, while the load difference is only 0.11 kN. The multiple cracking characteristic of SSWs-engineered multifunctional UHPC shortens the time of crack stable and unstable

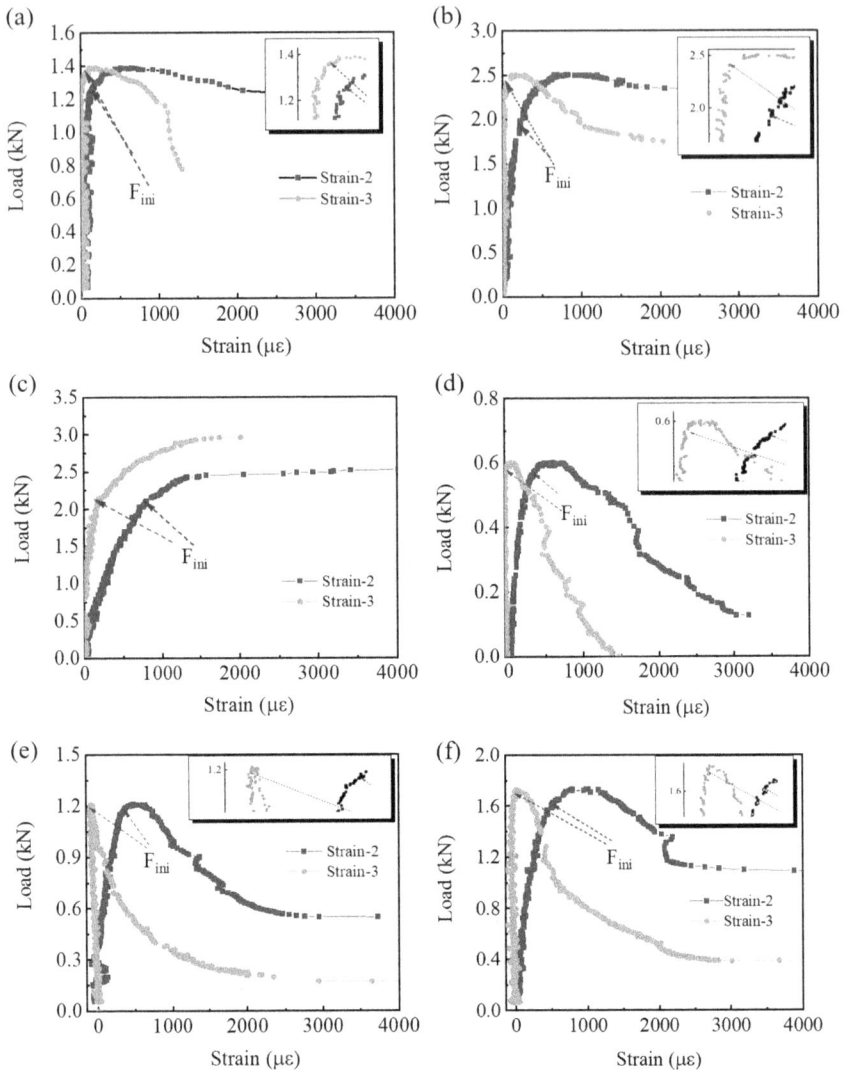

Figure 3.21 Monitoring strain of SSWs-engineered multifunctional UHPC under three-point fracture load. (a) W0, a_0/h=0.25; (b) W201010, a_0/h=0.25; (c) W201015, a_0/h=0.25; (d) W0, a_0/h=0.5; (e) W201010, a_0/h=0.5; (f) W201015, a_0/h=0.5

propagation. The measured critical crack length is also larger than 27.5 mm. The two cracking interval times for UHPC reinforced with 1.5 vol% SSWs are 6 s and 86 s, respectively. The cracking interval time between the locations of strain gauge 2# and strain gauge 3# is shortened due to the occurrence of multiple microcracks, and the prolongation of cracking interval time between the locations of strain gauge 3# and peak load can be

Table 3.11 Measured critical crack length of SSWs-engineered multifunctional UHPC

a_0/h	SSWs content (vol%)	Strain	Position (mm)	Cracking time (s)	Cracking load (kN)	Peak load (kN)	Peak load time (s)	a_c (mm)
0.25	0	Strain-2	12.5	349.5	0.57	1.36	461.0	>27.5
		Strain-3	27.5	432.5	1.14			
	1.0	Strain-2	12.5	327.5	1.79	2.51	395.5	>27.5
		Strain-3	27.5	381.0	2.40			
	1.5	Strain-2	12.5	460.5	2.04	3.37	552.5	≈27.5
		Strain-3	27.5	466.5	2.14			
0.5	0	Strain-2	22.5	347.5	0.57	0.60	352.5	<32.5
		Strain-3	37.5	351.5	0.59			
	1.0	Strain-2	22.5	363.5	1.15	1.25	379.0	≈32.5
		Strain-3	37.5	377.5	1.21			
	1.5	Strain-2	22.5	334.0	1.55	1.74	370.5	≈32.5
		Strain-3	37.5	343.0	1.63			

attributed to the bridging effect of SSWs on crack propagation. The measured critical crack length is about 27.5 mm.

At the crack length/depth ratio of 0.5, the load–strain curves of SSWs-engineered multifunctional UHPC possess a complete descending stage. Therefore, the measurement of critical effective crack length is bounded by the geometric center line of strain gauge 3#. The cracking interval time between the locations of strain gauge 2# and strain gauge 3# is 4 s, 14 s, and 9 s respectively for UHPC without SSWs, UHPC reinforced with 1.0 vol% SSWs, and UHPC reinforced with 1.5 vol% SSWs, and that between the locations of strain gauge 3# and peak load for these composites is 1 s, 1.5 s, and 27.5 s, respectively. The peak load of control UHPC is only 0.6 kN, and the critical effective crack length is smaller than 32.5 mm. The cracking loads for the locations of strain gauge 3# and peak load of UHPC reinforced with 1.0 vol% SSWs are only separated by 0.04 kN. The cracks have entered into the unstable propagation stage at the location of stain gauge 3#. Therefore, the measured critical crack length for this composite is about 32.5 mm. The differences between the cracking loads for the locations of strain gauge 3# and peak load are only 0.11 kN when 1.5 vol% SSWs are introduced, and the critical effective crack length is also about 32.5 mm. The change of cracking interval time, again, explains the crack refinement and bridging effect of SSWs on UHPC.

3.3.1.3 Initial cracking load

Stress concentration in concrete often occurs near original flaws with load increase, causing microcracks to form and develop in the frontal zones of such flaws. As these microcracks become saturated, macrocracks are formed. The initial cracking load of SSWs-engineered multifunctional UHPC is expressed by F_{ini}, which is determined by the strain gauge reading. The strains on the two sides of the precast crack are collected and marked as strain-1 and strain-4. The load–strain curves are plotted in Figure 3.22. Due to the inhibiting effect of randomly distributed SSWs on crack initiation, the composite on both sides of the crack tip might be in compression under a three-point bending load, e.g. strain-4 in Figure 3.22(c). Meanwhile,

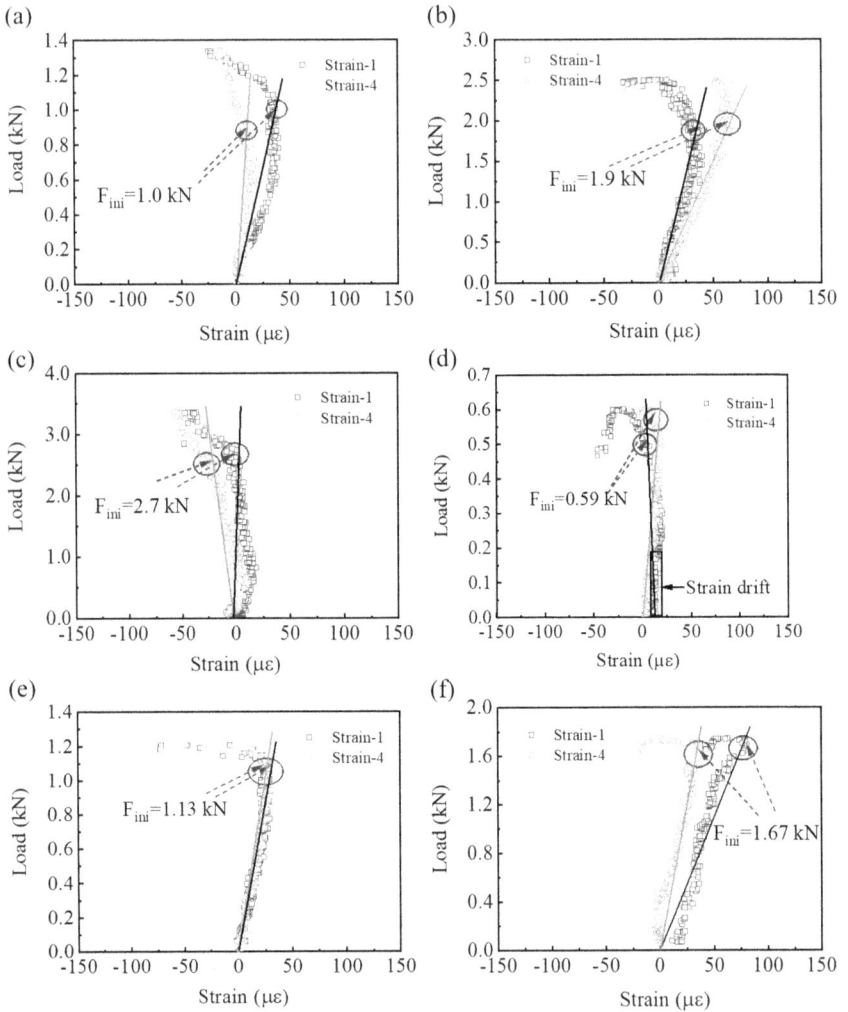

Figure 3.22 Initial cracking load of SSWs-engineered multifunctional UHPC under three-point fracture load. (a) W0, a_0/h=0.25; (b) W201010, a_0/h=0.25; (c) W201015, a_0/h=0.25; (d) W0, a_0/h=0.5; (e) W201010, a_0/h=0.5; (f) W201015, a_0/h=0.5

the strain increases linearly before the initiation of cracks and then, yields or retracts due to the release of stress and the occurrence of macrocracks. Therefore, the linear fitting is performed on strain in order to determine the maximum linear strain point. The initial cracking load is determined by the intersection point between strain and fitting line, as shown in Figure 3.22. The average value of the initial cracking load on both sides of the crack tip is taken as the final value. At the crack length/depth ratio of 0.25, the initial cracking load of UHPC without SSWs is only 1.0 kN, which increases to 1.9 kN and 2.7 kN, respectively, due to the addition of 1.0 vol% and 1.5 vol% SSWs. The initial cracking loads are improved by 91.5% and 183.1% at the crack length/depth ratio of 0.5. The increase of initial cracking load represents the reduction of original flaws and improvement of the uniformity of the UHPC matrix structure.

3.3.1.4 Fracture toughness calculated with TPFM

The two-parameter fracture model (TPFM), proposed by Jenq and Shah [4], is a modified linear elastic fracture mechanic model, which turns the macrocrack and microcrack zone into a unified critical effective crack. The fracture criterion is that the effective crack tip opening displacement reaches the critical crack tip opening displacement (CTOD$_C$). The fracture toughness K_{IC}^S is calculated at the tip of the effective crack in order to include the nonlinear crack growth prior to peak load. Meanwhile, the unloading compliance used for calculating the critical effective crack length a_c is obtained through at least one loading and unloading process. The calculation of a_c is deduced on the basis of Eq. (3.7):

$$E = \frac{6 S a V(\alpha)}{C_u t h} = \frac{6 S P_u}{CMOD_u t h} \alpha V(\alpha) \tag{3.7}$$

where:

$$V(\alpha) = 0.76 - 2.28\alpha + 3.87\alpha^2 - 2.04\alpha^3 + \frac{0.66}{(1-\alpha)^2} \tag{3.8}$$

E is elasticity modulus (the values of elasticity modulus for SSWs-engineered multifunctional UHPC can be obtained from Dong et al. [13]); S is the span of UHPC specimens; $\alpha = a_c / h$, where a_c is the critical effective crack length; C_u is the unloading compliance (when the load on the descending stage of the load–CMOD curve reaches 95% of the peak load, C_u is obtained by assuming that the unloading path returns to the starting point); P_u is 95% of peak load; $CMOD_u$ refers to the CMOD value corresponding to 95% of peak load on the descending stage of the load–CMOD curve; t is the width of the UHPC specimen; h is the height of the UHPC specimen.

The fracture toughness K_{IC}^{S} is determined by Eq. (3.9):

$$K_{IC}^{S} = 3\left(P_{max} + 0.5W\right)\frac{S\left(\pi a_{c}\right)^{0.5}}{2th^{2}}F\left(\alpha\right) \tag{3.9}$$

where P_{max} is the peak load; $W = W_{0}S/h$ (W_{0} is the dead weight of specimens, which can be ignored here); and $F(\alpha)$ is a function of α obtained from Eq. (3.10):

$$F\left(\alpha\right) = \frac{1.99 - \alpha\left(1-\alpha\right)\left(2.15 - 3.93\alpha + 2.7\alpha^{2}\right)}{\pi^{0.5}\left(1+2\alpha\right)\left(1-\alpha\right)^{1.5}} \tag{3.10}$$

The critical tip opening displacement $CTOD_{C}$ can be calculated from Eq. (3.11):

$$CTOD_{c} = \frac{6P_{max}Sa_{c}V\left(a\right)}{Eth^{2}}\left[\left(1-\beta\right)^{2} + \left(1.081 - 1.149a\right)\left(\beta - \beta^{2}\right)\right]^{0.5} \tag{3.11}$$

where $\beta = a_{0}/a_{c}$, a_{0} is the initial crack length, and a_{c} is the critical effective crack length.

The normalization parameters of TPFM are tensile strength f_{t} and brittleness index Q, calculated from Eqs (3.12) and (3.13):

$$f_{t} = \frac{1.4705\left(K_{IC}^{S}\right)^{2}}{E \times CTOD_{c}} \tag{3.12}$$

$$Q = \left(\frac{E \times CTOD_{c}}{K_{IC}^{S}}\right)^{2} \tag{3.13}$$

The fracture parameters calculated on the basis of TPFM are shown in Tables 3.12 and 3.13. As listed in Tables 3.12 and 3.13, the critical effective crack length obtained from TPFM is only slightly affected by

Table 3.12 a_{0}/h = 0.25, TPFM fracture parameters of SSWs-engineered multifunctional UHPC

SSWs content (vol%)	E (GPa)	$CMOD_{0.95}$ (mm)	$F_{un,0.95}$ (kN)	a_{c} (mm)	K_{IC}^{S} (MPa·m$^{1/2}$)	$CTOD_{C}$ (mm)	f_{t} (MPa)	Q (mm)
0	31.0	0.0849	1.29	28.0	2.488	0.0586	5.00	533.21
1.0	43.6	0.1577	2.38	29.7	5.282	0.0774	8.55	824.56
1.5	45.4	0.1565	3.20	28.5	6.263	0.1085	11.71	618.49

Table 3.13 a_0/h = 0.5, TPFM fracture parameters of SSWs-engineered multifunctional UHPC

SSWs content (vol%)	E (GPa)	CMOD$_{0.95}$ (mm)	F$_{un.0.95}$ (kN)	a_c (mm)	K$_{IC}{}^S$ (MPa·m$^{1/2}$)	CTOD$_C$ (mm)	f$_t$ (Mpa)	Q (mm)
0	31.0	0.1007	0.57	32.3	2.042	0.0428	4.62	421.90
1.0	43.6	0.1553	1.15	32.5	4.294	0.0664	9.36	454.60
1.5	45.4	0.2135	1.67	32.5	6.196	0.0913	13.60	447.95

SSWs content and is similar to the measured results (as shown in Table 3.11). Compared with UHPC without SSWs, the increases of fracture toughness $K_{IC}{}^S$ are determined as 112.3% and 151.7%, respectively, for UHPC reinforced with 1.0 vol% and 1.5 vol% SSWs at the crack length/depth ratio of 0.25. Considerable increases of 32.1 % and 85.2% for the crack tip opening displacement CTOD$_C$ are obtained. Meanwhile, the normalization tensile strength f_t and brittleness index Q are improved by 71.0%, 134.0% and 54.6%, 16.0%, respectively. At the crack length/depth ratio of 0.5, the fracture toughness $K_{IC}{}^S$ of SSWs-engineered multifunctional UHPC increases by 110.3% and 203.4% compared with that of UHPC without SSWs. The crack tip opening displacement CTOD$_C$ of composites is 55.1% and 113.3% higher than that of UHPC without SSWs. The normalization tensile strength is enhanced by 102.6% and 194.4%, respectively. The variation of normalization brittleness index at different crack length/depth ratios can be attributed to the multiple cracking and crack refinement characteristics of UHPC reinforced with high SSWs content.

As the crack length/depth ratio increases, the $K_{IC}{}^S$ of UHPC composites decreases by 18.2%, 18.7%, and 1.1%, and the CTOD$_C$ of composites is lowered by 27.0%, 14.2%, and 15.9%, respectively, for UHPC without SSWs, UHPC reinforced with 1.0 vol%, and UHPC reinforced with 1.5 vol% SSWs. When the crack length/depth ratio increases from 0.25 to 0.50, the normalization tensile strength of UHPC without SSWs is reduced by 7.6%, while 9.5% and 16.2% increases are obtained for the UHPC reinforced with 1.0 vol% and 1.5 vol% SSWs. This phenomenon illustrates that the toughening effect of high-content SSWs is more significant with the increase of crack length/depth ratio.

The improvement in $K_{IC}{}^S$ and CTOD$_C$ for UHPC caused by 1.0% SSWs is comparable to the enhancement effect of a 4% volume fraction of steel fibers (with a diameter of 0.22 mm) on UHPC, and the value of $K_{IC}{}^S$ for UHPC reinforced with 1.0% SSWs is much higher than that for UHPC reinforced with 4% steel fibers [14].

3.3.1.5 Fracture toughness calculated with DKFM

The double-K fracture model (DKFM) employs initiation fracture toughness K_{IC}^{ini} and unstable fracture toughness K_{IC}^{un} to describe the initial cracking and unstable state of concrete [6]. The initiation fracture toughness K_{IC}^{ini} is calculated by initial crack length a_0 and initial crack load P_{ini}, indicating the ability of concrete to resist an external load before the generation of cracks. The value of K_{IC}^{ini} can be evaluated using Eq. (3.14):

$$K_{IC}^{ini} = \frac{3P_{ini}S}{2th^2} \sqrt{a_0} F(\alpha_0) \tag{3.14}$$

where $\alpha_0 = a_0 / h$; a_0 is the initial crack length of UHPC specimens; h is the height of specimens; S is the span of specimens; P_{ini} is the initial cracking load; the expression of $F(\alpha_0)$ is the same as Eq. (3.10). The critical effective crack length a_c can be obtained from Eq. (3.15):

$$a_c = \frac{\left[\gamma^{1.5} + m(\beta)\gamma\right]h}{\left[\gamma^2 + m_2(\beta)\gamma^{1.5} + m_3(\beta)\gamma + m_4(\beta)\right]^{0.75}} \tag{3.15}$$

where $\gamma = \dfrac{CMOD_C tE}{6F_{un}}$, $m_1(\beta) = \beta(0.25 - 0.0505\beta^{0.5} - 0.0033\beta)$,
$m_2(\beta) = \beta^{0.5}(1.155 + 0.215\beta^{0.5} - 0.00278\beta)$, $m_3(\beta) = -1.38 + 1.75\beta$,

$m_4(\beta) = 0.506 - 1.057\beta + 0.888\beta^2$. $CMOD_C$ is the crack mouth opening displacement corresponding to peak load; E is the elasticity modulus of composites; t is the width of the UHPC specimen; h is the height of the UHPC specimen; $\beta = S/h$ is the span–depth ratio.

The unstable fracture toughness K_{IC}^{ini} is determined by critical effective crack length a_c and peak load P_{max}, representing the crack resistance of materials to external load at the critical situation. Therefore, the value of K_{IC}^{ini} can be evaluated by inserting a_c and P_{max} into Eq. (3.14) instead of a_0 and P_{ini}, respectively.

As shown in Table 3.14, the critical effective crack length a_c is only slightly affected by the addition of SSWs at the crack length/depth ratio of 0.25. The values of a_c are close to the measured results shown in Table 3.11. The increase of initiation fracture toughness K_{IC}^{ini} reaches 89.2% and 160.0%, and the unstable fracture toughness K_{IC}^{un} is enhanced by 130.8% and 152.4%, respectively, due to the addition of 1.0 vol% and 1.5 vol% SSWs. The improvement of fracture toughness reflects the inhibiting and bridging effect of SSWs on the generation and propagation of cracks in UHPC.

Table 3.14 $a_0/h = 0.25$, DKFM fracture parameters of SSWs-engineered multifunctional UHPC

SSWs content (vol%)	E (GPa)	$CMOD_c$ (mm)	F_{ini} (kN)	F_{un} (kN)	a_c (mm)	K_{IC}^{ini} (MPa·m$^{1/2}$)	K_{IC}^{un} (MPa·m$^{1/2}$)
0	31.0	0.0619	1.0	1.39	26.3	0.427	2.041
1.0	43.6	0.1153	1.9	2.51	28.2	0.807	4.696
1.5	45.4	0.1080	2.7	3.37	26.5	1.109	5.151

Table 3.15 $a_0/h = 0.5$, DKFM fracture parameters of SSWs-engineered multifunctional UHPC

SSWs content (vol%)	E (GPa)	$CMOD_c$ (mm)	F_{ini} (kN)	F_{un} (kN)	a_c (mm)	K_{IC}^{ini} (MPa·m$^{1/2}$)	K_{IC}^{un} (MPa·m$^{1/2}$)
0	31.0	0.0861	0.59	0.60	31.6	0.487	1.890
1.0	43.6	0.1225	1.13	1.25	31.6	0.940	3.792
1.5	45.4	0.1601	1.67	1.74	31.4	1.389	5.277

Table 3.15 illustrates that at the crack length/depth ratio of 0.5, the critical effective crack length a_c is also comparable to the measured results shown in Table 3.11. The initiation fracture toughness K_{IC}^{ini} of SSWs-engineered multifunctional UHPC shows an increase of 93.0% and 185.2%, and the unstable fracture toughness K_{IC}^{un} increases by 100.6% and 179.2%, respectively. The enhancement effect of SSWs on the UHPC matrix and the bridging effect of SSWs on adjacent cracks are not subject to crack length/depth ratio.

Compared with previous investigations, the K_{IC}^{ini} of SSWs-engineered multifunctional UHPC is superior to that of high-strength (C80) concrete (0.73 MPa·m$^{1/2}$) [15], and the K_{IC}^{un} of the composites exceeds that of polypropylene fiber reinforced high-strength concrete (2.18 MPa·m$^{1/2}$) [14] and normal steel fiber reinforced UHPC (with a diameter of 0.22 mm and volume fraction of 4%, 2.31 MPa·m$^{1/2}$) [14].

3.3.1.6 Fracture energy

The fracture energy is defined as the amount of energy to create one unit area of a crack and can be calculated by Eq. (3.16) [, 17]:

$$G_{\text{F-P}(\delta)} = \frac{\int P(\delta)\mathrm{d}\delta}{(h - a_0)t} \tag{3.16}$$

where $G_{\text{F-P}(\delta)}$ is the fracture energy; δ is the deflection; h is the height of the UHPC specimen; a_0 is the initial crack length in the UHPC specimen; t is the width of the UHPC specimen.

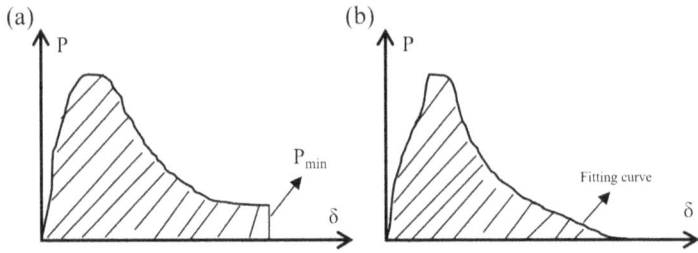

Figure 3.23 Typical load–deflection curve for notched three-point fracture beam. (a) Load–deflection curve without tail fitting; (b) Load–deflection curves with tail fitting

It should be noted that the tail of the descending stage on load–deflection curves is very gentle. It takes a long time for the load to drop to zero in an actual test. Generally, the test is stopped when the load drops to a certain value (as shown in Figure 3.23(a)), and the value of the fracture energy is significantly affected by the tail of the load–deflection curve [18]. Elices et al. [19] proposed that the experimental error can be eliminated by including the work of fracture that is not measured due to practical difficulties in capturing the tail part of the load–deflection curves. The relative complete load–deflection curves (as plotted in Figure 3.23(b)) can be obtained by fitting and modifying the tail. Then, the fracture energy can be calculated on the basis of the modified load–deflection curve using Eq. (3.17). The symbols in Eq. (3.17) have the same meaning as before.

$$G_{\text{F-N}} = \frac{\int P(\delta)^{*} \, d\delta}{(h - a_0)t} \tag{3.17}$$

In order to avoid the tail treatment influencing fracture energy, the tail of load–deflection curves is extended according to an exponential equation in this work, as demonstrated in Figure 3.24. The fracture energy values of SSWs-engineered multifunctional UHPC obtained by load–deflection curves with and without tail fitting are listed in Table 3.16.

As shown in Table 3.16, the fracture energy $G_{\text{F-P}(\delta)}$ of SSWs-engineered multifunctional UHPC calculated on the basis of Eq. (3.16) displays a notable increase compared with that of UHPC without SSWs. The inclusion of 1.0 vol% and 1.5 vol% SSWs leads to 233.8% and 440.5% increments in fracture energy at the crack length/depth ratio of 0.25, and the increments can reach 334.4% and 1017.1%, respectively, at the crack length/depth ratio of 0.5. The fracture energy of UHPC reinforced with 1.5 vol% SSWs increases with increasing crack length/depth ratio. The fracture energy $G_{\text{F-N}}$ of SSWs-engineered multifunctional UHPC calculated on the basis of Eq. (3.17) exhibits a higher value than that calculated on the basis of Eq. (3.16).

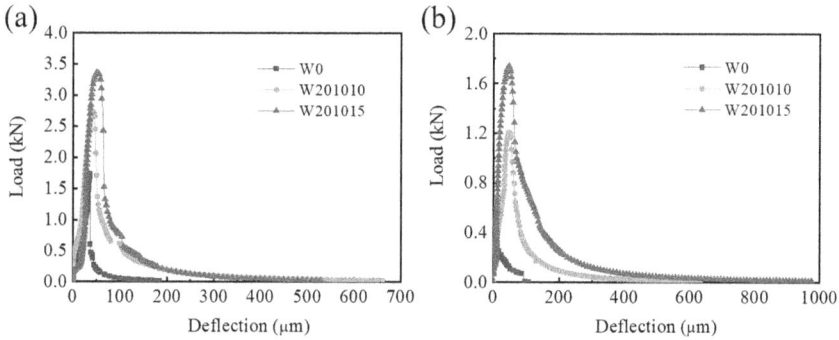

Figure 3.24 Load–deflection curves of SSWs-engineered multifunctional UHPC with tail fitting under three-point fracture load. (a) a_0/h=0.25; (b) a_0/h=0.5

Table 3.16 Fracture energy of SSWs-engineered multifunctional UHPC under three-point fracture load

a_0/h	SSWs content (vol%)	Standard deviation (J/m²)	Tail fitting equation	G_{F-N} (J/m²)	$(G_{F-N} - G_{F-P(\delta)}) /G_{F-P(\delta)}$ (%)	P_{min}/P_c (%)
0.25	0	2.80	–	32.02	–	–
	1.0	11.26	$y = 1206.39767x - 1.65811$	152.89	43.0	10.5
	1.5	10.34	$y = 4787.56857x - 1.91069$	183.45	6.0	7.0
0.5	0	0.70	–	19.70	–	–
	1.0	9.26	$y = 843.79299x - 1.72037$	118.39	38.3	14.4
	1.5	10.60	$y = 4479.37484x - 1.84429$	253.19	15.1	12.8

The error between G_{F-N} and $G_{F-P(\delta)}$ is more than 15%, and even as high as 43%, because that ratio of cut-off load P_{min} to peak load P_c on the load–deflection curve is higher than 10%. The percentage difference between G_{F-N} and $G_{F-P(\delta)}$ falls within the margin of error when the ratio of cut-off load P_{min} to peak load P_c is lower than 10%. It can be concluded that whether the load–deflection curve is tail fitted or not has no influence on the relationship between fracture energy and SSWs content. However, the cut-off load P_{min} on the load–displacement curve should be controlled within 10% of peak load in order to eliminate the test error as much as possible.

3.3.1.7 Cracking pattern

A cracking pattern schematic diagram of SSWs-engineered UHPC at the failure state is shown in Figure 3.25. It can be seen from Figure 3.25 that the cracking patterns of the composites are tortuous and blunt cracks, as the crack resistance zone is formed by SSWs and the UHPC matrix together.

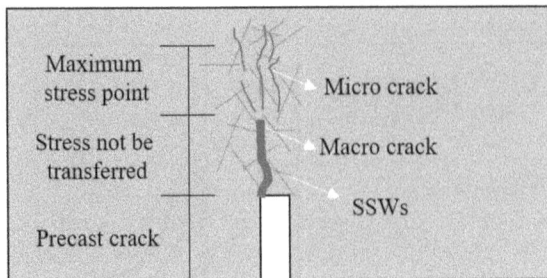

Figure 3.25 Cracking pattern schematic diagram of SSWs-engineered multifunctional UHPC under three-point fracture load

The failure cracks of composites become narrow, short, and tortuous due to the increase of SSWs content. The blunt and tortuous cracks reduce the stress concentration at the notch tip. The specimens of SSWs-engineered UHPC are not completely disconnected when failure occurs, and the bridging SSWs can be clearly observed. The failure surface of the composites is rather unsmooth due to the resistance effect of SSWs. The cracking pattern change law of SSWs-engineered UHPC is consistent with the test results of fracture toughness and fracture energy.

3.3.2 Four-point fracture properties of SSWs-engineered multifunctional UHPC notched beam

Four-point shearing beams are employed to study the mixed mode I–II fracturing characteristics of SSWs-engineered multifunctional UHPC. The shearing force Q can be calculated using Eqs (3.18) and (3.19) [20]:

$$F_1 / F_3 = (S - c)/c \tag{3.18}$$

$$Q = F_1 - F_2 = \frac{S - 2c}{S} F \tag{3.19}$$

where F is the applied load on beam specimens; F_1 represents the load at the near loading point; F_2 represents the load at the far loading point; S is the distance between the far loading point and the near loading point; c is the distance between the near loading point and the center line of the initial precast crack. The stress intensity factor $K_{I\text{-}II}$ for mixed mode I–II fracture can be calculated by Eq. (3.20) [20]:

$$K_{I\text{-}II} = \frac{Q}{t\sqrt{h}} f(a_0 / h) \tag{3.20}$$

$$f\left(\frac{a_0}{b}\right) = \left[1.442 - 5.08\left(\frac{a_0}{b} - 0.507\right)^2\right] \times \sec\frac{\pi a_0}{2b}\sqrt{\sin\frac{\pi a_0}{b}} \qquad (3.21)$$

where t is the width of the UHPC specimen; b is the height of the UHPC specimen; a_0 is the initial crack length; $f\left(\frac{a_0}{b}\right)$ is the dimensionless stress intensity factor, which is related to the cracking pattern and the initial crack length/depth ratio. The UHPC specimens are in the failure state when the concentrated load F reaches the maximum value F_m. The maximum shearing force Q_m can be obtained according to Eq. (3.19). The stress intensity factor under concentrated load F_m is the fracture toughness $K_{\text{I-II}}$ obtained from Eq. (3.22):

$$K_{\text{I-IIC}} = \frac{Q_m}{t\sqrt{b}}f\left(a_0 / b\right) \qquad (3.22)$$

3.3.2.1 Load–CMOD curves

The load–CMOD curves of SSWs-engineered multifunctional UHPC under a four-point shearing load are plotted in Figure 3.26. It can be seen from Figure 3.26 that the non-linear ascending stage of load–CMOD curves becomes more and more obvious with increasing SSWs content. At this stage, the generation and propagation of cracks are hindered by the SSWs' disordered network. When the volume fraction of SSWs is 1.5 vol%, the slope of the linear ascending stage markedly declines. Moreover, there is a stationary stage for the load–CMOD curve before peak load. Compared with UHPC without SSWs, the peak CMOD (CMOD_C) enhancement of

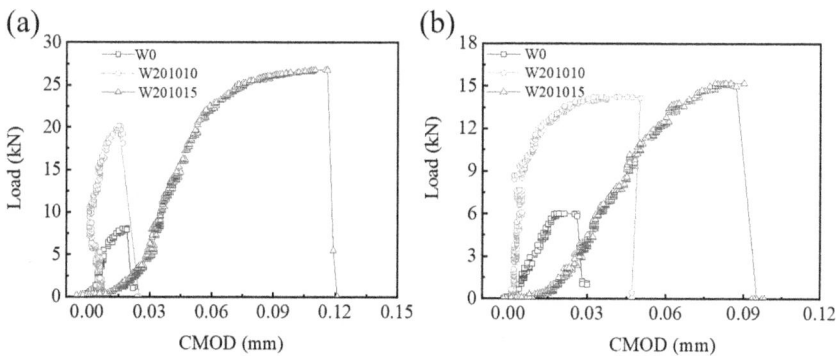

Figure 3.26 Load–CMOD curves of SSWs-engineered multifunctional UHPC under four-point shearing load. (a) $a_0/h=0.25$; (b) $a_0/h=0.5$

UHPC reinforced with 1.5 vol% SSWs reaches 625.0% at a crack length/depth ratio of 0.25. The $CMOD_C$ of UHPC reinforced with 1.0 vol% and 1.5 vol% SSWs is improved by 68.0% and 232.0%, respectively, at a crack length/depth ratio of 0.5. With the increase of crack length/depth ratio, the $CMOD_C$ of UHPC reinforced with 1.5 vol% SSWs is reduced by 28.4%. This is because the incorporation of 1.5 vol% SSWs enables the occurrence of pure shear crack.

3.3.2.2 Shearing force and mixed mode I-II fracture toughness

As demonstrated in Figure 3.27(a), the shearing force of SSWs-engineered multifunctional UHPC significantly exceeds that of UHPC without SSWs. Increments of 123.9% and 177.4% are obtained for shearing force of UHPC reinforced with 1.0 vol% and 1.5 vol% SSWs at the crack length/depth ratio of 0.25, and the shearing force increments for composites are 119.4% and 137.1%, respectively, at the crack length/depth ratio of 0.5. The shear force decreases with increasing crack length/depth ratio.

Figure 3.27(b) shows that the mixed mode I–II fracture toughness of UHPC reinforced with 1.0 vol% and 1.5 vol% SSWs is increased by 123.8% and 177.1%, respectively, compared with that of UHPC without SSWs at the crack depth ratio of 0.25, and growth rates of 118.6% and 137.3% are achieved at the crack length/depth ratio of 0.5. The mixed mode I–II fracture toughness is proportional to shear force, and they are improved by 59.5%, 54.8%, and 35.0%, respectively, for UHPC without SSWs, UHPC reinforced with 1.0 vol% SSWs, and UHPC reinforced with 1.5 vol%, as the crack length/depth ratio increases from 0.25 to 0.5. The fracture toughness values of SSWs-engineered multifunctional UHPC are larger than that for concrete and rock obtained by Mirsayar et al. [21] and Pirmohammad et al. [22].

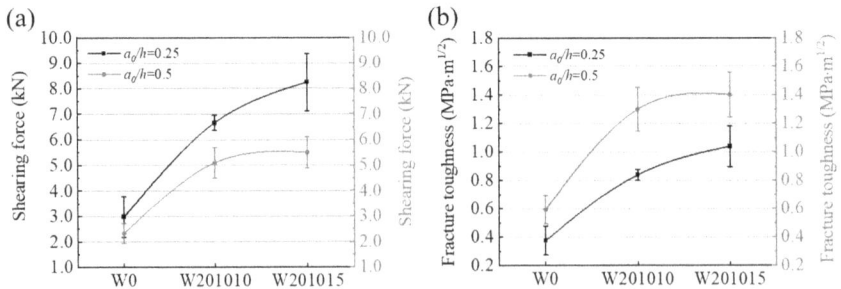

Figure 3.27 Shear force and fracture toughness of SSWs-engineered multifunctional UHPC under four-point shearing load. (a) Shearing force; (b) Fracture toughness

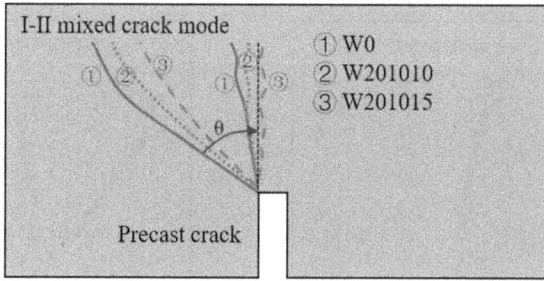

Figure 3.28 Cracking pattern of SSWs-engineered multifunctional UHPC under four-point shearing load

3.3.2.3 Cracking pattern

The failure cracks of SSWs-engineered multifunctional UHPC under a four-point shearing load are mainly I–II mixed mode type and shearing type, as shown in Figure 3.28. The I–II mixed mode type is an oblique crack running from the precast crack tip to the near loading point, and the shear type crack is a nearly vertical crack along the direction of the precast crack. Figure 3.28 also shows that the angle (θ) between the mixed mode I–II crack and the precast crack extension line decreases, and the propagation of the vertical shear crack, becomes more tortuous with increasing SSWs volume fraction. In addition, the shearing cracks at the upper end of the precast crack first slope toward the near loading point and then, turn to the parallel direction of load. The transverse shearing crack parallel to load direction occurs in the specimens of UHPC reinforced with 1.5 vol% SSWs. This can be attributed to the incorporation of a high volume fraction of SSWs, enhancing the tensile strength of UHPC and limiting the development direction of the initial crack.

3.3.2.4 Initial cracking load

The analysis of initial cracking load is performed on the SSWs-engineered multifunctional UHPC specimens with two failure cracks. The load–strain curves of the composites are drawn in Figure 3.29. It can be observed from Figure 3.29 that the strain increases with the increasing load, while the growth rate of strain is accelerated or hysteretic after the occurrence of the initial crack.

As shown in Figure 3.29, the strain on both sides of the precast crack is complex under a four-point shearing load because of the inhibiting effect of SSWs on crack initiation. However, whether the composite on both sides of the precast crack is in tension or compression, strains yield or retract due to the initiation of cracks. Therefore, the load corresponding to the yield strain point or retraction strain point is defined as the initial cracking load. The

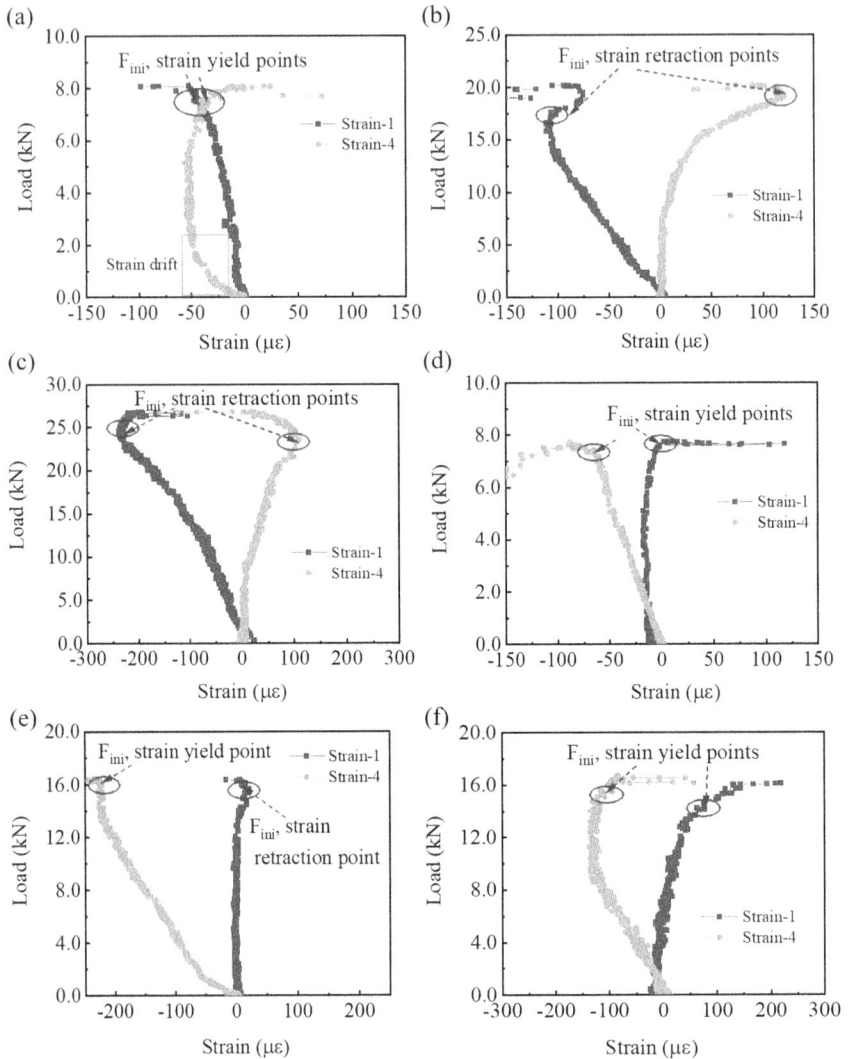

Figure 3.29 Four-point shearing initial cracking load of SSWs-engineered multifunctional UHPC. (a) W0, a_0/h=0.25; (b) W201010, a_0/h=0.25; (c) W201015, a_0/h=0.25; (d) W0, a_0/h=0.5; (e) W201010, a_0/h=0.5; (f) W201015, a_0/h=0.5

initial cracking load of SSWs-engineered UHPC is summarized in Table 3.17. As demonstrated in Table 3.17, the ratio of initial cracking load F_{ini} to peak load F_{un} is reduced by the increasing volume fraction of SSWs. This corresponds to the phenomenon that the load–CMOD curves of the composites possess a significant non-linear ascending stage. The initial cracking loads of UHPC reinforced with 1.0 vol% and 1.5 vol% SSWs are 150.0%

Table 3.17 Initial cracking load of SSWs-engineered multifunctional UHPC under four-point shearing load

$a_0/h = 0.25$				$a_0/h = 0.5$			
SSWs content (vol%)	Initial cracking load F_{ini} (kN)	Peak load F_{un} (kN)	F_{ini}/F_{un}	SSWs content (vol%)	Initial cracking load F_{ini} (kN)	Peak load F_{un} (kN)	F_{ini}/F_{un}
0	7.6	8.1	0.941	0	7.5	7.7	0.974
1.0	19.0	20.2	0.940	1.0	16.0	16.5	0.970
1.5	24.5	26.7	0.915	1.5	15.9	16.7	0.952

and 222.3% higher, respectively, than that of UHPC without SSWs at the crack length/depth ratio of 0.25. The increments of initial cracking load are calculated to be 113.3% and 112.0% at the crack length/depth ratio of 0.5. The increase of initial cracking load validates the enhancement effect of SSWs on the UHPC matrix. Meanwhile, the decrease of the ratio between initial cracking load and peak load proves the inhibiting and bridging effect of SSWs on the generation and propagation of cracks.

3.4 SUMMARY

In this chapter, the bending properties of unnotched/notched beam and plate fabricated by SSWs-engineered multifunctional UHPC under different load forms were presented, and the fracture properties of notched beams were also demonstrated. The main conclusions are as follows.

1) Incorporating 1.0 vol% and 1.5 vol% SSWs increases the four-point bending strength of unnotched UHPC beams by 13.5% and 36.6%, respectively. The corresponding increments for peak and limit bending toughness based on load–deflection curves are 43.0%/111.7% and 50.0%/146.5%. When the crack length/depth ratio is 0.25/0.50, the three-point bending strength of UHPC reinforced with 1.0 vol% and 1.5 vol% SSWs notched beams is increased by 73.9%/119.9% and 65.2%/74.4%, respectively. The four-point bending strength of UHPC plates is enhanced by 77.8% due to the incorporation of SSWs.
2) Based on the TPFM and DKFM calculation, the increments of three-point fracture toughness for a UHPC beam with crack length/depth ratio of 0.50 are determined as 203.4% and 179.2%, respectively, due to the addition of 1.5 vol% SSWs. The corresponding mode I fracture energy of SSWs-engineered multifunctional UHPC displays 440.5% and 1017.1% increments at the crack length/depth ratio of 0.25/0.50. Compared with UHPC without SSWs, the mixed mode I–II fracture toughness of UHPC reinforced with 1.5 vol% SSWs is increased by

177.1% and 137.3%, respectively, at the crack length/depth ratio of 0.25 and 0.5. The angle between the mixed mode I–II crack and the precast crack extension line is reduced, and the perpendicularity of shear cracks is improved with increased SSWs content.

REFERENCES

1. S. Dong, D. Zhou, A. Ashour, B. Han, J. Ou. Flexural toughness and calculation model of super-fine stainless wire reinforced reactive powder concrete, *Cement and Concrete Composites*. 104 (2019) 103367.
2. B. Han, L. Zhang, J. Ou. *Smart and Multifunctional Concrete Toward Sustainable Infrastructures*, Springer, 2017.
3. S. Dong, X. Dong, A. Ashour, B. Han, J. Ou. Fracture and self-sensing characteristics of super-fine stainless wire reinforced reactive powder concrete, *Cement and Concrete Composites*. 105 (2020) 103427.
4. Y. S. Jenq, S. P. Shah. A Fracture toughness criterion for concrete, *Engineering Fracture Mechanics*. 21(5) (1985) 1055–1069.
5. S. E. Swartz, N. M. Taha. Mixed mode crack propagation and fracture in concrete, *Engineering Fracture Mechanics*. 35(1–3) (1990) 137–144.
6. S. Xu, H. W. Reinhardt. A simplified method for determining double-K fracture parameters for three-point bending tests, *International Journal of Fracture*. 104(2) (2000) 181–209.
7. Y. Zhao, W. Dong, B. Xu, J. Liu. Effect of T-stress on the initial fracture toughness of concrete under I/II mixed-mode loading, *Theoretical and Applied Fracture Mechanics*. 96 (2018) 699–706.
8. L. Li, Q. Zheng, X. Wang, B. Han, J. Ou. Modifying fatigue performance of reactive powder concrete through adding pozzolanic nanofillers, *International Journal of Fatigue*. 156 (2022) 106681.
9. J. Wang, S. Dong, S. Pang, X. Yu, B. Han, J. Ou. Tailoring anti-impact properties of ultra-high performance concrete by incorporating functionalized carbon nanotubes, *Engineering*. 18 (2022) 232–245.
10. L. Li, X. Wang, H. Du, B. Han. Comparison of compressive fatigue performance of cementitious composites with different types of carbon nanotube, *International Journal of Fatigue*. 165 (2022) 107178.
11. S. Dong, D. Wang, A. Ashour, B. Han, J. Ou. Nickel plated carbon nanotubes reinforcing concrete composites: From Nano/micro structures to macro mechanical properties, *Composites Part A: Applied Science and Manufacturing*. 141 (2021) 106228.
12. J. Wang, S. Dong, C. Zhou, A. Ashour, B. Han. Investigating pore structure of nano-engineered concrete with low-field nuclear magnetic resonance, *Journal of Materials Science*. 56(1) (2021) 243–259.
13. S. Dong, B. Han, X. Yu, J. Ou. Constitutive model and reinforcing mechanisms of uniaxial compressive property for reactive powder concrete with super-fine stainless wire, *Composites Part B: Engineering*. 166 (2019) 298–309.

14. C. Su, Q. Wu, L. Weng, X. Chang. Experimental investigation of mode I fracture features of steel fiber-reinforced reactive powder concrete using semi-circular bend test, *Engineering Fracture Mechanics*. 209 (2019) 187–199.

15. W. Dong, X. Zhou, Z. Wu. On fracture process zone and crack extension resistance of concrete based on initial fracture toughness, *Construction and Building Materials*. 49 (2013) 352–363.

16. K. Yu, J. Yu, Z. Lu, Q. Chen. Determination of the softening curve and fracture toughness of high-strength concrete exposed to high temperature, *Engineering Fracture Mechanics*. 149 (2015) 156–169.

17. RILEM 50-FMC. Determination of the fracture energy of mortar and concrete by means of three-point bend tests on notched beams, *Materials and Structures*. 18(4) (1985) 287–290.

18. J. Qian. Three-point bending method for determination of fracture energy, *China Concrete and Cement Products*. 6 (1996) 20–23. (In Chinese).

19. M. Elices, G. V. Guinea, J. Planas. Measurement of the fracture energy using three-point bend tests: Part 3-influence of cutting the P–δ tail, *Materials and Structures*. 25 (1992) 137–163.

20. S. Hu, L. Hu. Experimental study of shear fracture of concrete specimens under 4-point loading, *Journal of Yangtze River Scitific Institute*. 31(9) (2014) 99–104. (In Chinese).

21. M. M. Mirsayar, A. Razmi, F. Berto. Tangential strain-based criteria for mixed-mode I/II fracture toughness of cement concrete, *Fatigue and Fracture of Engineering Materials and Structures*. 41(1) (2018) 129–137.

22. S. Pirmohammad, M. Hojjati Mengharpey. A new mixed mode I/II fracture test specimen: Numerical and experimental studies, *Theoretical and Applied Fracture Mechanics*. 97 (2018) 204–214.

Chapter 4

Impact properties of stainless steel wires-engineered multifunctional ultra-high performance concrete

4.1 INTRODUCTION

Due to the excellent flexural, compressive, bending, and fracture properties demonstrated in Chapters 2 and 3, stainless steel wires (SSWs)-engineered multifunctional ultra-high performance concrete (UHPC) can be used for building large-scale and complicated infrastructures. It is noteworthy that most large-scale and complicated infrastructure structures not only bear static load but also inevitably bear dynamic load (e.g. seismic load, impact load, and blast load) [1–3]. The impact compressive load with rapidly changing characteristics must be considered in the design of concrete structures, especially those that are in service for protection of nuclear power plants [4–7]. Therefore, investigations of the impact compressive properties of SSWs-engineered multifunctional UHPC are essential to provide valuable guidance for such special structural design.

Therefore, the impact compressive properties of SSWs-engineered multifunctional UHPC at the strain rate ranging from 94/s to 826/s obtained using a split Hopkinson pressure bar (SHPB) are introduced in this chapter. The modification mechanisms are displayed through analyzing computed tomography (CT) and scanning electron microscope (SEM) images. The impact compressive constitutive model is also given based on revised visco-elastic and continuum damage theory. The diameter of SSWs in this chapter is 20 um and the length is 10 mm. The main contents of this chapter are shown in Figure 4.1.

4.2 IMPACT COMPRESSIVE PROPERTIES OF SSWS-ENGINEERED MULTIFUNCTIONAL UHPC

4.2.1 Impact compressive strength and deformation of SSWs-engineered multifunctional UHPC

The SHPB test was used to evaluate the impact compressive behavior of SSWs-engineered multifunctional UHPC, as shown in Figure 4.2. The specimen size of SSWs-engineered multifunctional UHPC obtained by coring was φ30 mm × 15 mm. The end face of the specimen should be finely

DOI: 10.1201/9781003276357-4

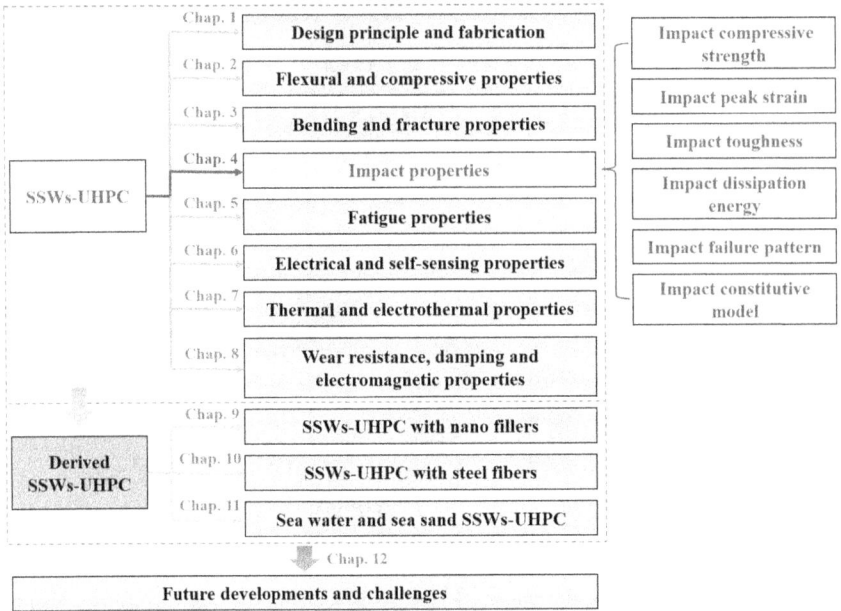

Figure 4.1 Main contents of Chapter 4

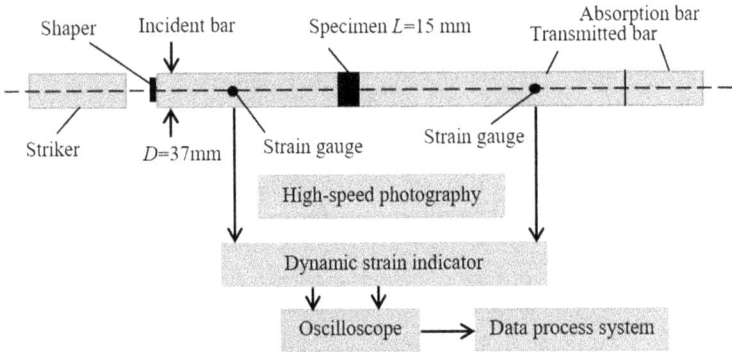

Figure 4.2 Split Hopkinson pressure bar (SHPB) device

ground with water in order to guarantee the planeness, roughness, and perpendicularity. The section diameter of the impact bar was 37.0 mm. The strain rate was controlled by launch pressure.

The average stress $\sigma_s(t)$, average stain rate $\dot{\varepsilon}_s(t)$, and average strain $\varepsilon_s(t)$ can be deduced by the three-wave method on the basis of Eqs (4.1), (4.2), and (4.3). The ratio of impact compressive strength to static compressive strength was defined as the dynamic increase factor (DIF) to illustrate the influence of SSWs on the strain rate–strengthening effect of UHPC.

The deformation performance of SSWs-engineered multifunctional UHPC was analyzed based on impact compressive peak strain and limit strain. The impact compressive toughness was calculated by the integrals of the stress–strain curve, which represents the energy absorption ability of SSWs-engineered multifunctional UHPC. Impact dissipation energy (*IDE*) calculated according to Eq. (4.4) represents the impact wave dissipation ability per unit volume material [8].

$$\sigma_s(t) = \frac{EA}{2A_s}\left[\varepsilon_i(t) + \varepsilon_r(t) + \varepsilon_t(t)\right] \tag{4.1}$$

$$\dot{\varepsilon}_s(t) = \frac{c}{l_s}\left[\varepsilon_i(t) - \varepsilon_r(t) - \varepsilon_t(t)\right] \tag{4.2}$$

$$\varepsilon_s(t) = \int_0^t \dot{\varepsilon}_s(\tau)\,d\tau \tag{4.3}$$

$$IDE = \frac{AEc}{A_s l_s}\int_0^t\left[\varepsilon_i(t) - \varepsilon_r(t) - \varepsilon_t(t)\right]dt \tag{4.4}$$

where ε_i, ε_r, and ε_t represent incident strain, reflection strain, and transmission strain, respectively; A and A_s are the section area of the incident bar and specimens, respectively; E represents the elastic modulus of the stainless incident bar; c is wave velocity in the incident bar; and l_s is the initial thickness of the specimens.

The impact compressive strength of SSWs-engineered multifunctional UHPC at the strain rate range from 94/s to 926/s is shown in Figure 4.3(a). It can be seen from Figure 4.3(a) that the impact compressive strength of SSWs-engineered multifunctional UHPC first increases and then decreases with the increase of strain rate. Compared with UHPC without SSWs (W0),

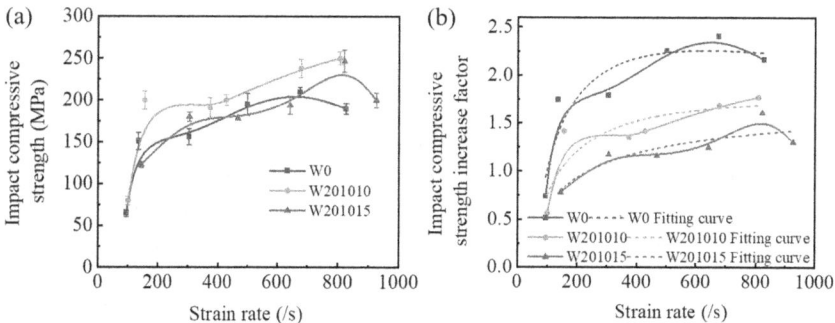

Figure 4.3 SSWs-engineered multifunctional UHPC's impact compressive strength and impact compressive strength increase factor. (a) Impact compressive strength; (b) Impact compressive strength increase factor

the impact compressive strength increment of UHPC reinforced with 1.0 vol% SSWs (W201010) can reach up to 33.7% at the strain rate of 807/s. However, the difference in impact compressive strength between UHPC reinforced with 1.5 vol% SSWs (W201015) and UHPC without SSWs is within the margin of error. The high strength of UHPC without SSWs under impact compressive load comes from the inertial confinement effect. However, the SSWs-engineered multifunctional UHPC is only slightly affected by the inertial confinement effect.

The impact compressive increase factor was defined as the ratio between impact compressive strength and static compressive strength (obtained using specimens with sizes of 40 mm × 40 mm × 40 mm). The increase factor of impact compressive strength for SSWs-engineered multifunctional UHPC at different strain rates is illustrated in Figure 4.3(b). Figure 4.3(b) demonstrates that the impact compressive strength increase factor of SSWs-engineered multifunctional UHPC decreases with increasing SSWs volume fraction. The impact compressive strength increase factor of UHPC with 1.0 vol% and 1.5 vol% SSWs is reduced by 37.3% and 55.2%, respectively, as compared with UHPC without SSWs. The increase factor of impact compressive strength for SSWs-engineered multifunctional UHPC first increases and then decreases with increasing strain rate. The increase factor of impact compressive strength for UHPC without SSWs, UHPC with 1.0 vol% SSWs, and UHPC with 1.5 vol% SSWs is 1.79, 1.35, and 1.17 at a strain rate of 304/s, 373/s, and 305/s, respectively, while the highest increase factor of impact compressive strength obtained by Ren's research [9] is 2.27 at the strain rate of 328.4/s. By comparison, the increase factor of impact compressive strength for SSWs-engineered multifunctional UHPC obtained in this chapter is smaller than in previous research on normal steel fiber (with a diameter of 0.16–0.55 mm and length of 6–30 mm) reinforced UHPC at a similar strain rate, lower than 300/s [10–13], which can be attributed to the lateral confinement effect of SSWs.

Several empirical formulas have been proposed to evaluate the strain rate–strengthening effect on the impact compressive strength of concrete [9]. However, the transition strain rate is difficult to determine, because there is no test data for SSWs-engineered multifunctional UHPC under quasi-dynamic load. Meanwhile, most of the transition strain rates in previous research have been smaller than 94/s [14, 15]. Without regard to transition strain rate, the empirical formula as expressed in Eq. (4.5) can be used to calculate the increase factor of impact compressive strength (DIF) of SSWs-engineered multifunctional UHPC at different strain rates.

$$DIF = A\left(log\dot{\varepsilon}\right)^2 + Blog\dot{\varepsilon} + C \qquad (4.5)$$

The fitting parameters between DIF and $\dot{\varepsilon}$ are summarized in Table 4.1, and the fitting curves are shown in Figure 4.3(b). As demonstrated in Table 4.1 and Figure 4.3(b), the simulation values from Eq. (4.5) agree well with the

Table 4.1 Fitting parameters between DIF and $\dot{\varepsilon}$ for SSWs-engineered multifunctional UHPC

SSWs content (vol%)	A	B	C	R^2
0	−1.9132	10.7285	−12.7851	0.7565
1.0	−0.8013	4.9574	−5.9539	0.6214
1.5	−0.3103	2.3615	−2.8559	0.6250

experimental DIF of SSWs-engineered multifunctional UHPC. The absolute values of fitting parameters decrease with increasing SSWs volume fraction, which is consistent with the change of DIF.

The strain rate–strengthening effect of SSWs-engineered multifunctional UHPC is plotted in Figure 4.4. Figure 4.4 indicates that the linear slope between strain rate and impact compressive strength of UHPC reinforced with 1.0 vol% and 1.5 vol% SSWs is reduced by 27.5% and 68.4%, respectively, compared with that of UHPC without SSWs. The strain rate–strengthening effect of UHPC is weakened significantly by the inclusion of

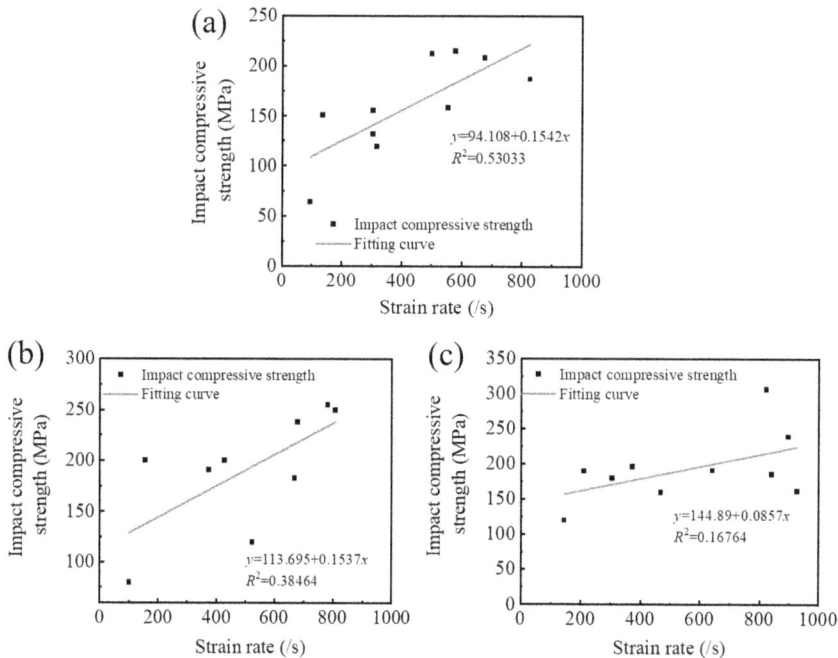

Figure 4.4 Strain rate–strengthening effect of SSWs-engineered multifunctional UHPC. (a) W0; (b) W201010; (c) W201015

SSWs. Although the fitting correlation coefficient is only 0.53, 0.38, and 0.17 for UHPC reinforced with 0 vol%, 1.0 vol%, and 1.5 vol% SSWs, respectively, the linear fitting characteristic between strain rate and compressive strength of UHPC can still obviously represent the decrease in strain rate–strengthening effect caused by SSWs.

A large number of microcracks occur in UHPC as soon as the impact compressive load is applied. These cracks do not have enough time to expand along the weak interface in UHPC; thus, they stretch in their respective regions, leading to the increase of peak stress and the occurrence of the strain rate–strengthening effect. To date, there is no consistent conclusion about the strain rate–strengthening effect of steel fiber reinforced UHPC [16–18]. However, the SSWs-engineered multifunctional UHPC is different from normal steel fiber reinforced UHPC, as shown in the following. (1) The three-dimensional network formed by SSWs in UHPC can effectively inhibit the generation and propagation of microcracks. Then, the crack tip stress can be redistributed in UHPC, leading to the increase of peak stress. This phenomenon is only slightly affected by strain rate. (2) The lateral confinement effect caused by the axial strain acceleration has been greatly offset by the restraining effect of the SSWs network on UHPC [9]. (3) The structural uniformity of UHPC has been greatly improved due to the microdiameter of SSWs, which implies that more energy will be consumed in the process of crack development. This is different from the strain rate–strengthening mechanisms of normal steel fiber reinforced UHPC [19], which achieve the purpose of consuming energy mainly by increasing the stress.

As shown in Figure 4.5, the impact compressive peak strain and limit strain are sensitive to strain rate. Figure 4.5(a) shows that the impact compressive peak strain of UHPC without SSWs increases with the strain rate, while the peak strain of UHPC reinforced with 1.0 vol% and 1.5 vol% SSWs first increases and then decreases. The impact compressive peak strain of

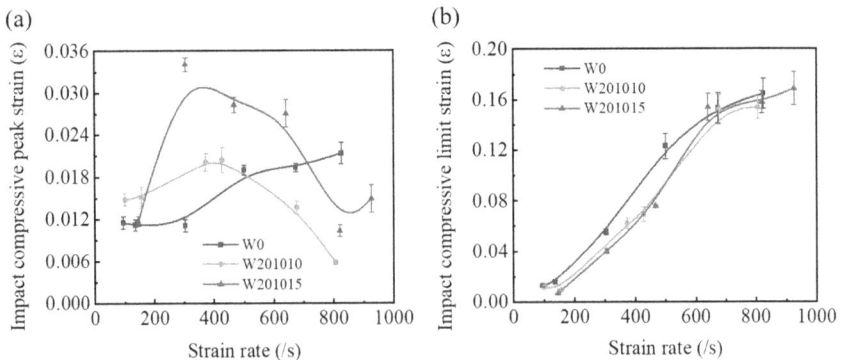

Figure 4.5 Impact peak strain and limit strain of SSWs-engineered multifunctional UHPC. (a) Peak strain; (b) Limit strain

UHPC reinforced with 1.0 vol% and 1.5 vol% SSWs is larger than that of UHPC without SSWs when the strain rate is smaller than 510/s and 710/s, respectively. The heightening efficiency of SSWs on peak strain rises as the SSWs volume fraction increases. The maximum peak strain is 34070 με for UHPC reinforced with 1.5 vol% SSWs at the strain rate of 305/s, which is significantly higher than the research results of Hou et al. [12] and Ren et al. [9].

Figure 4.5(b) demonstrates that the impact compressive limit strain of SSWs-engineered multifunctional UHPC is strain rate dependent, while the incorporation of SSWs reduces the limit strain of UHPC because of the lateral confinement effect. The error between the limit strain of UHPC reinforced with 1.0 vol% and 1.5 vol% SSWs is smaller than 35.9% at the strain rate range from 94/s to 926/s. The limit strain of SSWs-engineered multifunctional UHPC obtained in this chapter is significantly larger than previous research results on normal steel fiber reinforced UHPC at the same strain rate [10, 18, 20–23].

4.2.2 Impact compressive strain–stress curve, toughness, and dissipation energy of SSWs-engineered multifunctional UHPC

The impact compressive stress–strain curves of SSWs-engineered multifunctional UHPC at different strain rates are plotted in Figure 4.6. Figure 4.6 indicates that the stress–strain curve of SSWs-engineered multifunctional UHPC can be divided into three stages: the elastic stage, the elastic-plastic deformation stage, and the descending stage. The elastic stage ascending slope of stress–strain curves for UHPC with and without SSWs increases with the increase of strain rate due to the inertial confinement effect under high-speed impact load. The stress–strain curve of UHPC without SSWs has a slow descent stage at high strain rate. The descending stage slope of stress–strain curves for UHPC reinforced with 1.0 vol% and 1.5 vol% SSWs first decreases and then increases with the increase of strain rate. Meanwhile, there is a strain-strengthening phenomenon for UHPC reinforced with 1.0 vol% and 1.5 vol% SSWs before peak stress, as the strain rate is between 300/s and 500/s due to stress redistribution.

The total integral area is defined as the impact toughness in order to fully reflect the SSWs' toughening effect, because the limit strains of SSWs-engineered multifunctional UHPC have a large difference from each other. The impact toughness represents the energy absorbing capacity and deformability of concrete from loading to complete failure, as shown in Figure 4.7(a). Figure 4.7(a) indicates that the impact toughness of UHPC without SSWs increases with the increase of strain rate, while that of UHPC reinforced with 1.0 vol% and 1.5 vol% SSWs first increases and then decreases. The impact toughness of UHPC reinforced with 1.0 vol%

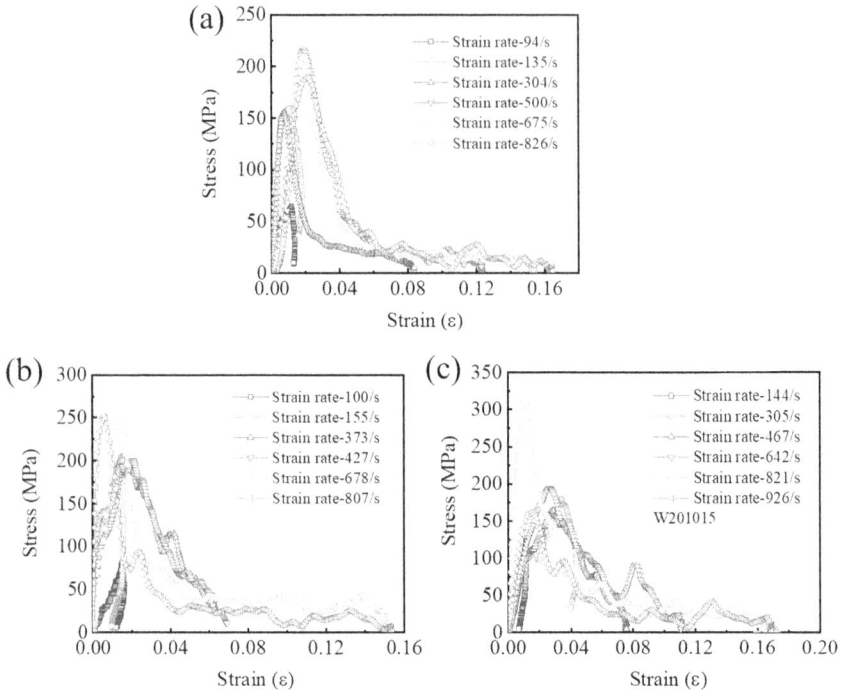

Figure 4.6 Impact compressive stress–strain curves of SSWs-engineered multifunctional UHPC' (a) W0; (b) W201010; (c) W201015

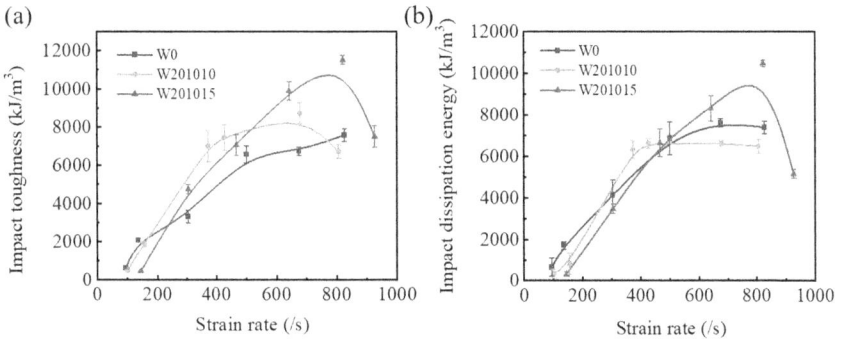

Figure 4.7 Impact toughness and impact dissipation energy of SSWs-engineered multifunctional UHPC. (a) Impact toughness; (b) Impact dissipation energy

and 1.5 vol% SSWs is enhanced by 29.8% and 43.5%, respectively, compared with that of UHPC without SSWs at the strain rate range from 180/s to 826/s. Meanwhile, the volume fraction of SSWs plays an important role in the impact toughness of UHPC when the strain rate is between 520/s and 826/s. At low-speed impact, the impact toughness of UHPC reinforced with 1.0 vol% and 1.5 vol% SSWs is lower than that of UHPC without SSWs, because the toughness contributed by specimen deformation is limited. More and more microcracks are generated in the UHPC matrix with increasing strain rate, leading to a greater inhibiting effect exerted by SSWs on the crack propagation. At high-speed impact load, the generation of microcracks becomes the main channel to consume energy; then, the toughening effect of SSWs is limited. The impact toughness of SSWs-engineered multifunctional UHPC is far greater than that in existing research. For example, the ultimate toughness of concrete with 2.0 vol% micro straight steel fibers (with a diameter of 0.2 mm and a length of 13 mm) in Hou's research [12] is 33.6 kJ/m^3, while the toughness for UHPC without and with 1.0 vol%/1.5 vol% SSWs is 2085 kJ/m^3, 1877 kJ/m^3, and 452 kJ/m^3, respectively, at a strain rate around 150/s.

The impact dissipation energy is an important toughness index which is determined by the strength and destruction degree of concrete. Figure 4.7(b) shows that the impact dissipation energy of SSWs-engineered multifunctional UHPC first increases and then features a steady state or decreases with the increase of strain rate. The impact dissipation energy of UHPC reinforced with 1.0 vol%/1.5 vol% SSWs has been improved by 53% and 58.2% at the strain rate of 373/s and 821/s, respectively, in comparison with that of UHPC without SSWs. Increasing the SSWs volume fraction provides more SSWs to resist crack generation and propagation, and more energy is consumed in the process of cracking resistance. The impact dissipation energy of UHPC without and with 1.0 vol%/1.5 vol% SSWs is 1.7 MJ/m^3, 0.82 MJ/m^3, and 0.32 MJ/m^3, respectively, while the impact dissipation energy obtained by Li et al. [18] is 1.75 MJ/m^3 at a strain rate around 160/s. The decrease of impact dissipation energy can be attributed to the reduction of degree of destruction of SSWs-engineered multifunctional UHPC at the same strain rate.

The impact compressive wave, as a compressional wave, can be translated into tensile stress in UHPC. First, the UHPC matrix will crack under tensile stress; then, layer stripping happens, and the concrete specimen gradually converts into the failure state. This destruction process can effectively dissipate impact wave energy. It can be concluded from comparison analysis that the impact dissipation energy of UHPC without SSWs is larger than the impact toughness. The reason may lie in that the specimens of UHPC without SSWs are in a triaxial compressive situation due to the inertial confinement effect, and pulverization destruction is the main failure mode. This is also one of the reasons why the strain rate–strengthening effect of UHPC without SSWs is relatively obvious. The impact dissipation energy of UHPC reinforced with 1.0 vol% and 1.5 vol% SSWs is smaller than impact toughness due to the addition of SSWs. The three-dimensional SSWs network can inhibit crack propagation in UHPC and stop the broken UHPC from falling. Therefore, the SSWs-engineered multifunctional UHPC specimens

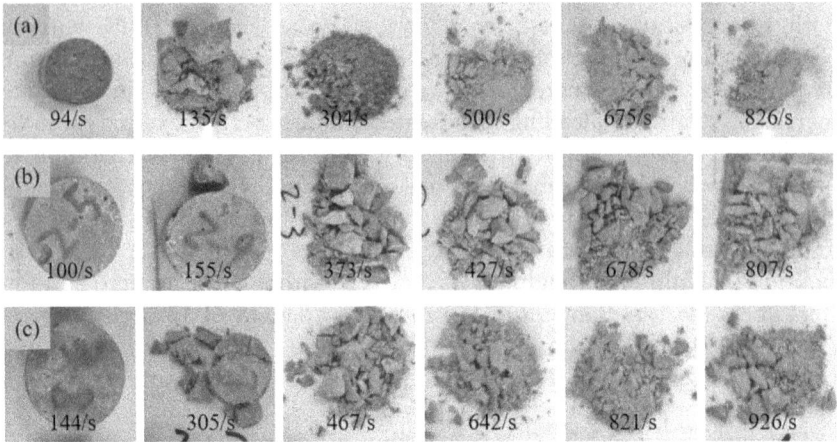

Figure 4.8 Impact compressive failure pattern of SSWs-engineered multifunctional UHPC. (a) W0; (b) W201010; (c) W201015

can only be converted into fragments when failure happens; thus, the ability to consume the stress wave is limited. This can be verified by the actual failure pattern of SSWs-engineered multifunctional UHPC, as presented in Figure 4.8.

As shown in Figure 4.8, the degree of impact compressive destruction of SSWs-engineered multifunctional UHPC increases with growing strain rate. At low-speed impact load, the specimens break into fragments after destruction. At high-speed impact load, the damage to specimens is more severe, and the specimens are crushed into smaller fragments. The specimens of UHPC without SSWs have been reduced to powder when the strain rate is 304/s, while the specimens of UHPC reinforced with 1.0 vol% and 1.5 vol% SSWs are still in the fragmented state due to the bridging function of the SSWs network when the strain rate is 807/s and 926/s, respectively. There are more SSWs to delay the crack propagation with the increase of volume fraction at the same strain rate. The degree of destruction of SSWs-engineered multifunctional UHPC is significantly reduced at the same strain rate compared with that in the existing research on normal steel fiber reinforced UHPC [10, 13, 20] because of the three-dimensional network of SSWs.

4.3 IMPACT COMPRESSIVE MODIFICATION MECHANISMS OF SSWS-ENGINEERED MULTIFUNCTIONAL UHPC

The impact modification mechanisms of SSWs-engineered multifunctional UHPC were examined by analyzing CT pictures and SEM images. There were 1002 pieces of CT pictures for each specimen before impact load. The

gray value was extracted from picture 100 to picture 900 by a MATLAB®
procedure in order to plot a gray histogram. The gray value surface of picture
500 was used to evaluate the distribution of SSWs, as given in Figure 4.9. It
can be seen from Figure 4.9 that the SSWs usually exist in a bundled state
in UHPC. The bundling of SSWs increases with increasing volume fraction of SSWs to form a three-dimensional toughening network. This SSWs
network can effectively inhibit the generation and propagation of cracks in
UHPC under impact load. Besides, Figure 4.9 also shows that the quantity
of pores in UHPC decreases due to the microdiameter of SSWs, which also
contributes to the inhibiting effect on cracks and the decrease of damage.

Gray value can represent the density of material, which is usually used
to characterize the homogeneity of composites. The gray value histogram
of CT pictures for SSWs-engineered multifunctional UHPC is extracted by
a MATLAB procedure, as shown in Figure 4.10. Figure 4.10 shows that
the gray value range of UHPC without and with 1.0 vol%/1.5 vol% SSWs
is 14–69, 26–108, and 27–110, and the corresponding pixel quantity peak
appears at the gray value of 55, 77 and 79, respectively. Therefore, the gray
value has a strong dependency on the SSWs volume fraction. The increase
of pixel quantity at high gray value characterizes the proximity of bundled
SSWs.

As presented in Figure 4.11(a), the gray value surface height of UHPC
without SSWs is 40–60. There are no obvious ups and downs characterizing the consistency of concrete density within the surface. The gray value
surface height of UHPC has been improved significantly due to the addition of SSWs, as shown in Figures 4.11(b) and (c). Meanwhile, there are
evident fluctuations on the surface of UHPC reinforced with 1.0 vol% and
1.5 vol% SSWs because of the presence of the SSWs toughening network.

The SEM images are analyzed in order to obtain the mechanisms by
which SSWs toughen UHPC. As shown in Figures 4.12(a) and (b), the
cementitious slurry can seep into the gaps between bundled SSWs to form an
inter-anchored interface, which can exert a seizing effect on broken UHPC

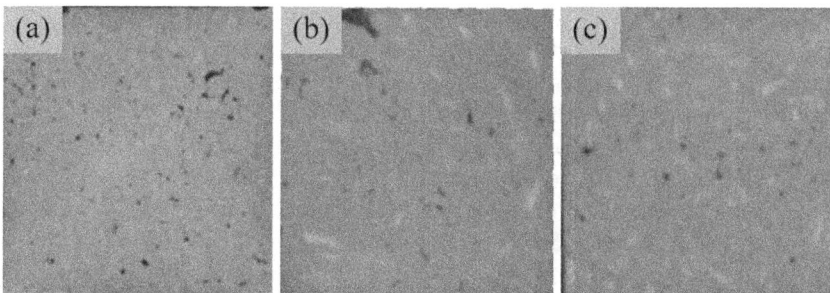

Figure 4.9 CT pictures of SSWs-engineered multifunctional UHPC. (a) W0; (b) W201010;
(c) W201015

(a)

(b)   (c)

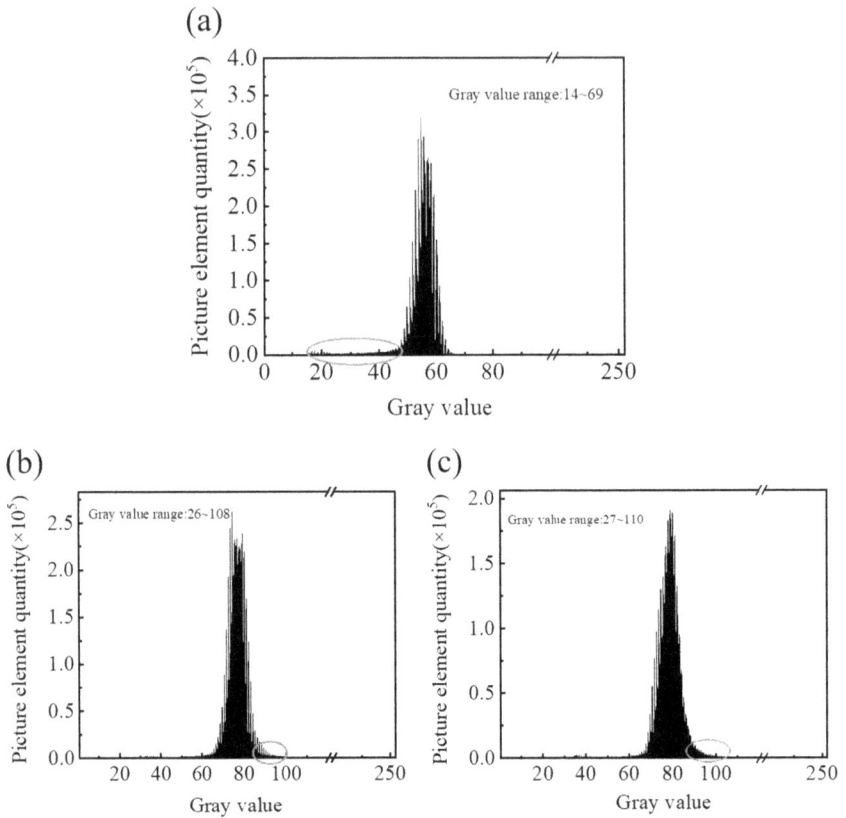

Figure 4.10 Gray value of CT pictures for SSWs-engineered multifunctional UHPC. (a) W0; (b) W201010; (c) W201015

and increases the resistance to crack propagation. In addition, the bundled SSWs can inhibit the generation and propagation of cracks until they are pulled off (as shown in Figure 4.12(c)). Meanwhile, the deflection and stripping of SSWs can also inhibit the development of cracks, as presented in Figure 4.12(d). The failure process of SSWs needs to consume a great deal of energy, so the impact toughness of SSWs-engineered multifunctional UHPC has been improved significantly at a moderate strain rate. The degree of destruction is reduced due to the confinement and pullback effect of SSWs. Meanwhile, the lateral confinement effect of the SSWs network leads to a decrease of the strain rate–strengthening effect of UHPC. Figure 4.12 also shows that the bundled SSWs are not overlapping completely, resulting in the limited toughening effect of SSWs on UHPC under high-speed impact load. This corresponds to the phenomenon that the impact toughness of UHPC reinforced with 1.0 vol% and 1.5 vol% is lower than that of UHPC without SSWs at a strain rate higher than 826/s (as shown in Figure 4.7(a)).

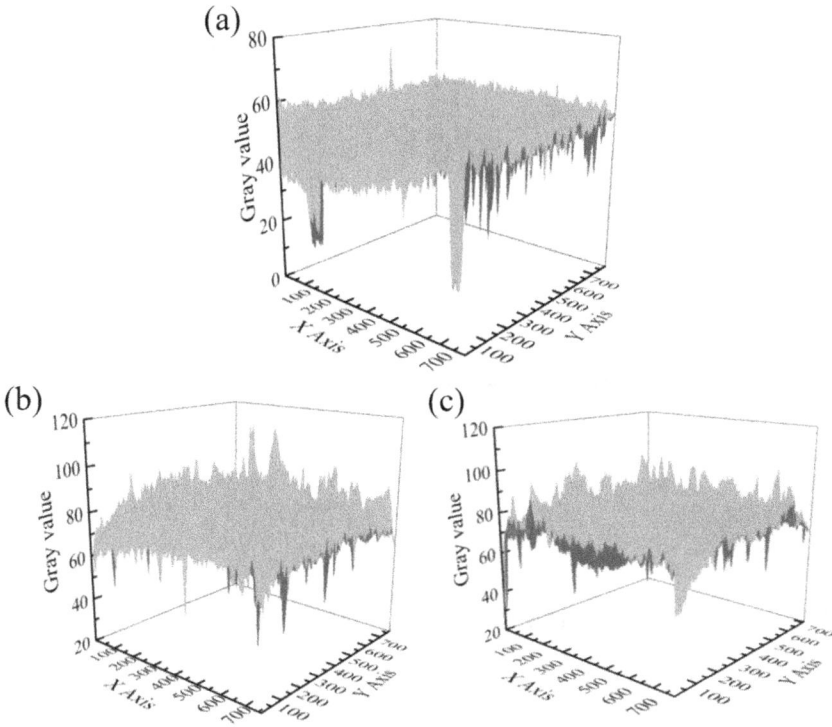

Figure 4.11 Gray value surface of SSWs-engineered multifunctional UHPC. (a) W0; (b) W201010; (c) W201015

4.4 IMPACT COMPRESSIVE CONSTITUTIVE MODEL OF SSWS-ENGINEERED MULTIFUNCTIONAL UHPC

The strain rate correlation theory should be adopted to describe the impact compressive constitutive behavior of SSWs-engineered multifunctional UHPC, because the impact compressive stress–strain curve of SSWs-engineered UHPC has visco-elastic retardation and strain rate characteristics. The strain rate correlation theory, i.e. non-linear visco-elastic theory, can be shown as Eq. (4.6):

$$f\left(\sigma,\varepsilon,\dot{\varepsilon}\right) = 0 \tag{4.6}$$

As a typical visco-elastic theory model proposed by Zhu, Wang and Tang [24], the ZWT model can be used to analyze the impact compressive constitutive relationship of SSWs-engineered multifunctional UHPC. The ZWT model is composed of a non-linear elastic spring, a low-frequency

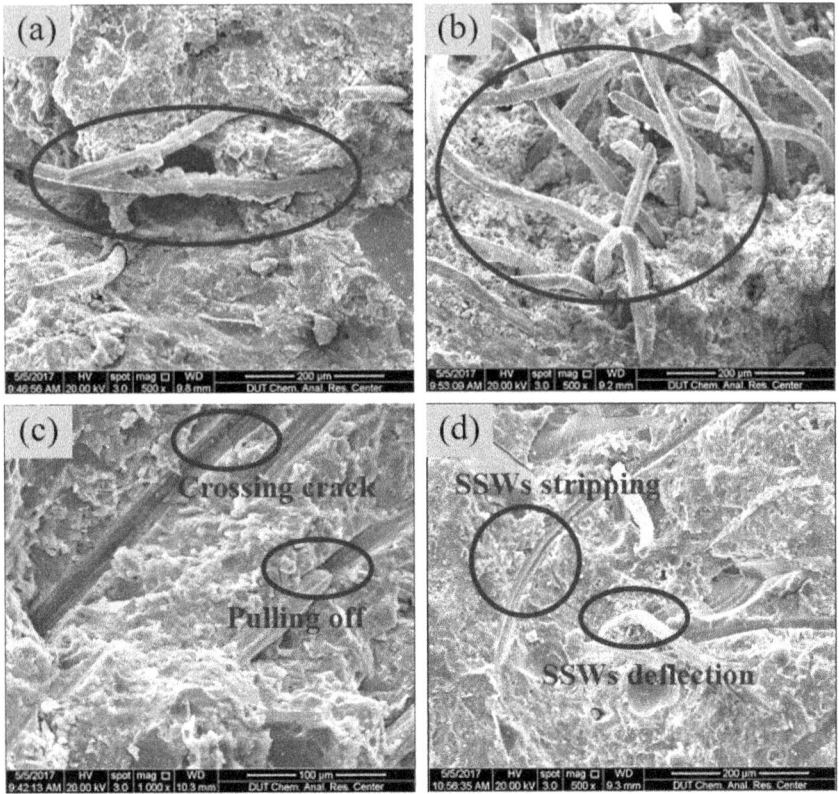

Figure 4.12 SEM images of SSWs-engineered multifunctional UHPC. (a) W201010, inter-anchored interface; (b) W201015, inter-anchored interface; (c) SSWs crossing crack and pulling off; (d) SSWs torsion and stripping

Maxwell element, and a high-frequency Maxwell element [25], as shown in Figure 4.13. It can be expressed as Eq. (4.7):

$$\sigma = f_e(\varepsilon) + E_1 \int_0^t \dot{\varepsilon} exp\left(-\frac{t-\tau}{\theta_1}\right) d\tau + E_2 \int_0^t \dot{\varepsilon} exp\left(-\frac{t-\tau}{\theta_2}\right) d\tau \qquad (4.7)$$

where $f_e(\varepsilon) = E_0\varepsilon + \lambda\varepsilon^2 + \eta\varepsilon^3$ describes the stress–strain behavior of a nonlinear elastic spring E_0; λ and η are the elastic constants; t and τ are the loading time and time variable; E_1 and θ_1 are the elastic modulus and relaxation time of the low-frequency Maxwell element, which is used to describe the material visco-elastic response at low strain rate; E_2 and θ_2 are the elastic modulus and relaxation time of the high-frequency Maxwell element, which is used to describe the material visco-elastic response at high strain rate. The range of θ_1 is from 10 s to 10^2 s, and the range of θ_2 is from 10^{-6} s to 10^{-4} s.

The visco-elastic ZWT model is simplified and revised according to the SHPB test result of SSWs-engineered multifunctional UHPC. (1) The

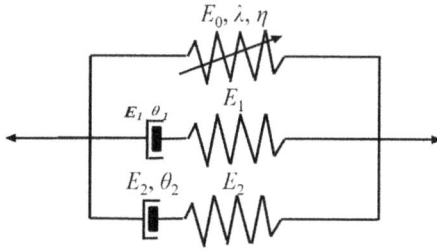

Figure 4.13 Non-linear visco-elastic ZWT model

non-linear elastic spring characterizes the stress–strain behavior of material in balance status, which is not subjected to stress. The stress–strain curve of SSWs-engineered multifunctional UHPC in the elastic phase is dominated by a linear relationship. Therefore, the non-linear elastic spring can be simplified as a linear elastic element. (2) The loading time of the SHPB test is about 10^{-6} s to 10^{-4} s; thus, the low-frequency Maxwell element does not have enough time to relax and is a negligible portion. Meanwhile, the impact compressive load can be applied at a constant strain rate on the basis of a waveform shaping technique; thus, the high-frequency Maxwell element can be simplified. Therefore, the ZWT visco-elastic model can be revised to Eq. (4.8) [9]:

$$\sigma = E'\varepsilon + E_2 \int_0^t \dot{\varepsilon} exp\left(-\frac{t-\tau}{\theta_2}\right) d\tau = E'\varepsilon + E_2\theta_2\dot{\varepsilon}\left[1 - e^{-\varepsilon/(\theta_2\dot{\varepsilon})}\right] \quad (4.8)$$

where $E' = E_0 + E_1$. (3) With the increase of impact stress/strain, damage occurs and continues to spread in concrete. Therefore, the damage factor D is introduced into the constitutive model to describe the impact process of SSWs-engineered multifunctional UHPC based on the continuum damage theory, as shown in Eq. (4.9) [26]:

$$\sigma_m = \sigma(1-D) \quad (4.9)$$

where σ_m is the stress under damage state and D is the damage factor.

The damage factor D can be expressed as the function of strain, and it follows the two-parameter Weibull statistical distribution, as indicated in Eq. (4.10) [27]:

$$D = \begin{cases} 0 & \varepsilon \leq \varepsilon_{th} \\ 1 - e^{-\left(\frac{\varepsilon-\varepsilon_{th}}{m}\right)^a} & \varepsilon > \varepsilon_{th} \end{cases} \quad (4.10)$$

where m and a are damage parameters decided by UHPC strength, SSWs volume fraction, SSW geometrical parameter, and so on. The strain threshold ε_{th} is determined by the strain rate and mechanical properties of SSWs-engineered multifunctional UHPC. ε_u is the impact peak strain.

The revised visco-elastic model can be expressed as Eq. (4.11):

$$\sigma_m = \begin{cases} E'\varepsilon + E_2\theta_2\dot{\varepsilon}\left[1 - e^{-\varepsilon/(\theta_2\dot{\varepsilon})}\right] & \varepsilon \leq \varepsilon_{th} \\ e^{-\left(\frac{\varepsilon - \varepsilon_{th}}{m}\right)^a}\left\{E'\varepsilon + E_2\theta_2\dot{\varepsilon}\left[1 - e^{-\varepsilon/(\theta_2\dot{\varepsilon})}\right]\right\} & \varepsilon > \varepsilon_{th} \end{cases} \quad (4.11)$$

Figures 4.14–4.16 demonstrate the comparison of the experimental and the corresponding impact stress–strain curves predicted by Eq. (4.11) at different strain rates. As shown in Figures 4.14–4.16, good agreements are

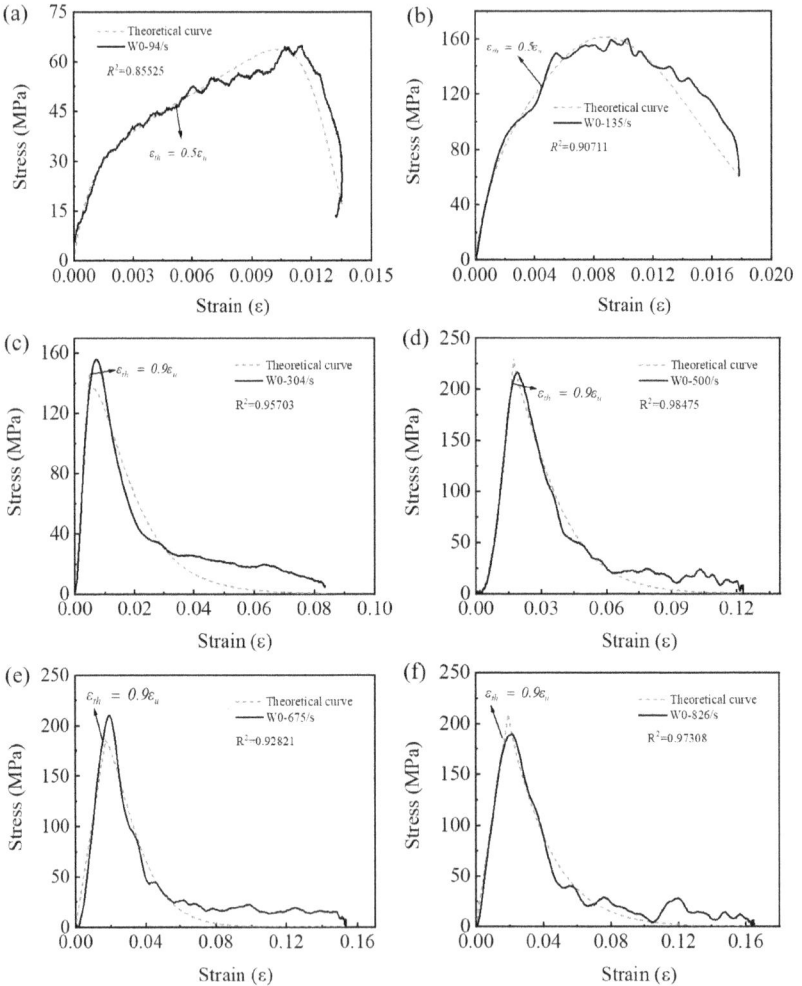

Figure 4.14 Comparison of experimental and predicted stress–strain curves of UHPC without SSWs at different strain rates. (a) W0-94/s; (b) W0-135/s; (c) W0-304/s; (d) W0-500/s; (e) W0-675/s; (f) W0-826/s

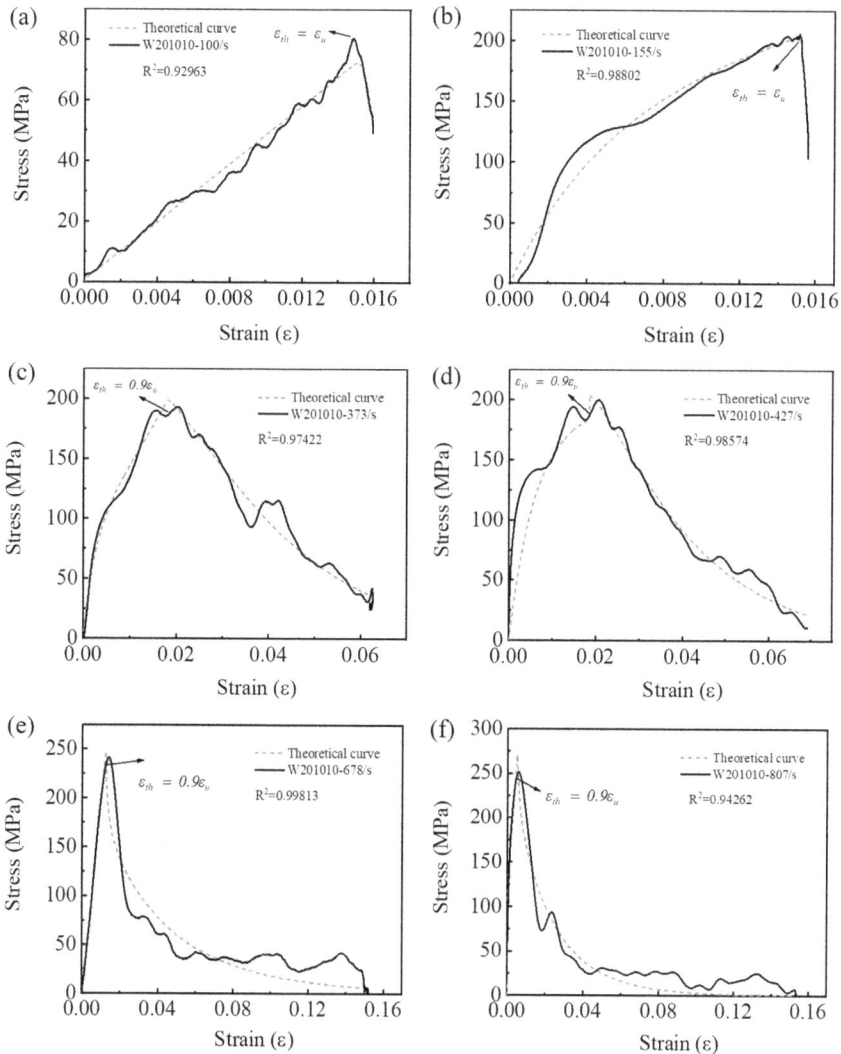

Figure 4.15 Comparison of experimental and predicted stress–strain curves of UHPC reinforced with 1.0 vol% SSWs at different strain rates. (a) W201010-100/s; (b) W201010-155/s; (c) W201010-373/s; (d) W201010-427/s; (e) W201010-678/s; (f) W201010-807/s

obtained, and the validation of the revised ZWT model in describing the impact compressive stress–strain relationship of SSWs-engineered multi-functional UHPC is verified. The fitting degree of the predicted and experimental stress–strain curves for SSWs-engineered multifunctional UHPC is closely related to the selection of strain threshold. The strain threshold is $\varepsilon_{th} = 0.5\varepsilon_u$ for UHPC without SSWs, as the strain rate is lower than 300/s,

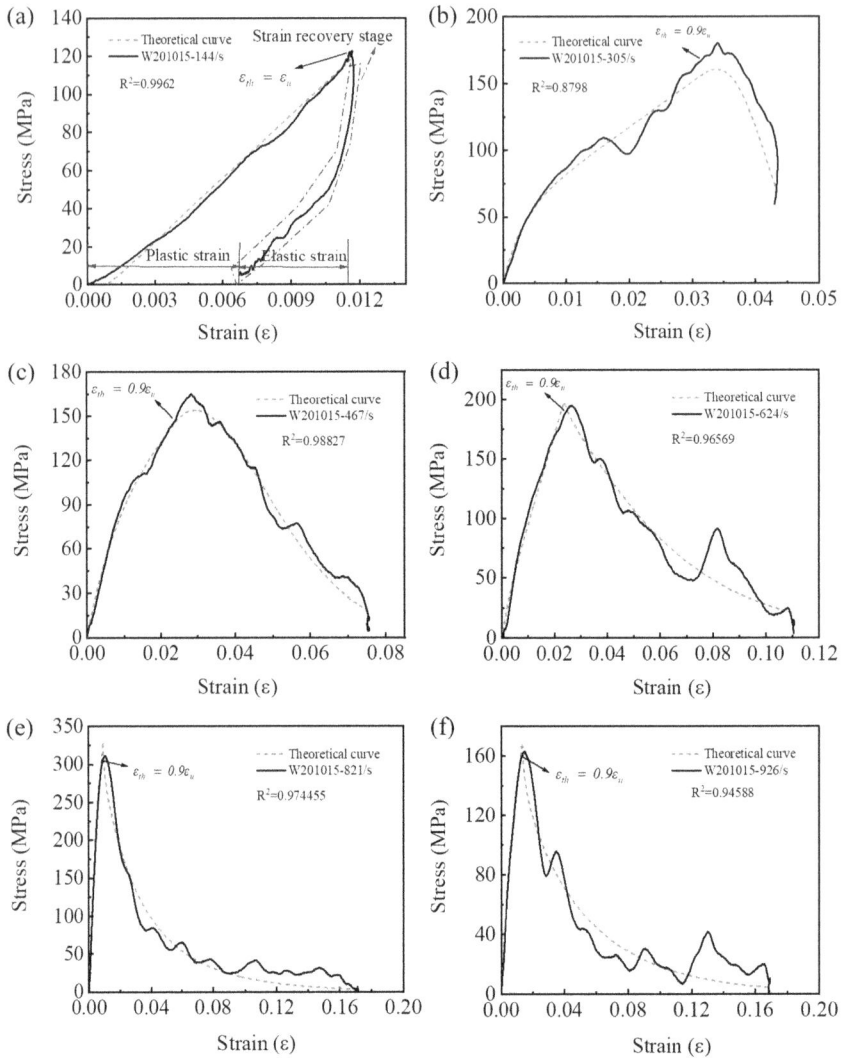

Figure 4.16 Comparison of experimental and predicted stress–strain curves of UHPC
reinforced with 1.5 vol% SSWs at different strain rates. (a) W201015-144/s;
(b) W201015-305/s; (c) W201015-467/s; (d) W201015-624/s; (e) W201015-
821/s; (f) W201015-926/s

while it is $\varepsilon_{th} = 0.9\varepsilon_u$ when the strain rate is higher than 300/s, because
that damage is delayed by the strain rate–strengthening effect. The SSWs-
engineered multifunctional UHPC specimens manifest typical visco-elastic
characteristic without damage before peak stress when the strain rate is
lower than 300/s. In particular, there is an elastic strain recovery stage

for the stress–strain curve of UHPC reinforced with 1.5 vol% SSWs at the strain rate of 144 /s, as illustrated in Figure 4.16(a). This phenomenon can be attributed to the three-dimensional lateral confinement effect of SSWs. Therefore, the strain threshold for SSWs-engineered multifunctional UHPC is $\varepsilon_{th} = \varepsilon_u$ at a strain rate lower than 300/s. Increasing strain rate makes SSWs-engineered multifunctional UHPC easier to crack. Then, the strain threshold is $\varepsilon_{th} = 0.9\varepsilon_u$ for SSWs-engineered UHPC at a strain rate higher than 300/s.

The related fitting parameters of the revised ZWT model for SSWs-engineered multifunctional UHPC are listed in Table 4.2. As can be seen from Table 4.2, the values of E' (the sum of elastic constants of non-linear elastic spring and elastic modulus of the low-frequency Maxwell element) for UHPC without and with 1.0 vol%/1.5 vol% SSWs first increase and then decrease with increasing strain rate. The values of E' for UHPC without SSWs are negative when the strain rate is larger than 300/s, illustrating that the UHPC without SSWs presents typical brittle material characteristics due to the strain rate–strengthening effect. The same variation regularity of E' and E_2 for UHPC reinforced with 1.0 vol% and 1.5 vol% SSWs is obtained with increasing strain rate. Meanwhile, the value of E' decreases

Table 4.2 Fitting parameters of revised ZWT model based on damage

SSWs content (vol%)	Strain rate (/s)	ε_{th} ($\times 10^{-3}$)	E' (GPa)	E_2 (GPa)	θ_2 (µs)	m ($\times 10^{-3}$)	α
0	94	5.7	3.6	38.1	7.9	7.3	1.5
	135	5.1	13.3	58.7	8.5	10.0	1.2
	304	5.0	37.3	−37.1	1.6	4.8	5.4
	500	17.1	32.8	−35.9	24	20.0	8.1
	675	17.0	15.9	−20.0	4.3	17.5	9.7
	826	19.2	12.5	−3.9	11.0	17.0	11.7
1.0	100	14.8	4.8	290.0	24.0	1.6	3.0
	155	15.1	3.9	29.8	33.0	0.7	1.7
	373	18.2	6.6	36.2	5.8	17.5	1.0
	427	18.4	0.5	30.5	15.0	23.3	1.1
	678	12.3	3.0	104.0	4.5	8.0	0.5
	807	5.2	29.8	550.0	0.27	5.5	0.7
1.5	144	11.7	11.1	24.6	3.0	–	–
	305	30.7	3.4	16.7	9.7	12.5	2.8
	467	25.5	2.5	10.7	18.3	27.0	1.6
	642	23.8	7.4	6.8	5.0	19.5	0.9
	821	9.3	13.9	38.0	9.0	10.0	0.7
	926	13.4	8.6	12.4	5.0	12.0	0.7

with increasing SSWs volume content because of the high deformation ability of SSWs reinforced UHPC.

The impact compressive damage factor of SSWs-engineered multifunctional UHPC is calculated based on Eq. (4.10), as shown in Figure 4.17. Figure 4.17 indicates that the addition of SSWs effectively inhibits the generation and propagation of damage in UHPC. The specimens of UHPC without SSWs are already in the pulverization state at a strain rate of 304/s. However, UHPC reinforced with 1.0 vol% and 1.5 vol% SSWs is still in a state of incomplete failure, as the strain rate is 427/s and 467/s, respectively, and the corresponding damage factor is 0.92 and 0.89. There is a damage-delaying phenomenon for UHPC without SSWs as strain rate increases due to the strain rate–strengthening effect. However, the damage of SSWs-engineered UHPC increases with the strain rate. The ascending stage slope of damage factor curves for SSWs-engineered multifunctional UHPC decreases, and the non-linear part is prolonged, with increasing SSWs volume fraction at a similar strain rate.

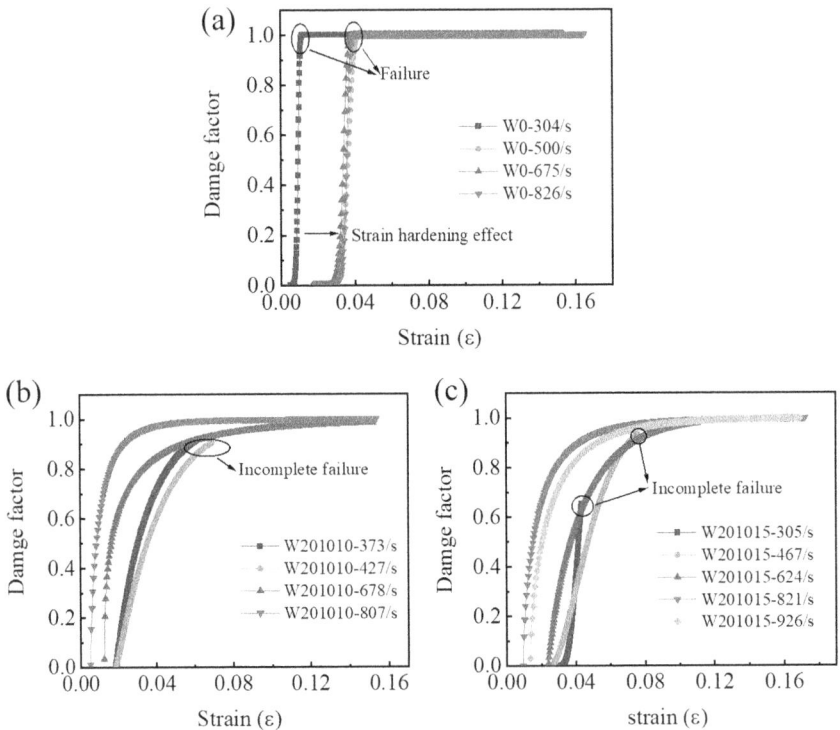

Figure 4.17 Impact compressive damage factor of SSWs-engineered multifunctional UHPC at a strain rate larger than 300/s. (a) W0; (b) W201010; (c) W201015

4.5 SUMMARY

This chapter introduced the impact compressive properties and the corresponding constitutive model of SSWs-engineered multifunctional UHPC at the strain rate of 94/s–826/s; meanwhile, the modification mechanisms of UHPC by SSWs were shown using CT and SEM analysis. The conclusions can be summarized as follows.

1) Adding SSWs has little effect on the impact compressive strength. The increase factor of impact compressive strength and the impact compressive limit strain of UHPC are reduced because of the addition of SSWs. The linear fitting coefficient between the impact compressive strength of SSWs-engineered multifunctional UHPC and the strain rate is decreased by 68.4%. The maximum impact compressive peak strain of SSWs-engineered multifunctional UHPC reaches 34070 $\mu\varepsilon$ at the strain rate of 305/s.

2) The impact compressive toughness and impact dissipation energy of SSWs-engineered multifunctional UHPC are increased by 43.5% and 58.2%, respectively, at a strain rate larger than 180/s. The impact dissipation energy of SSWs-engineered multifunctional UHPC is lower than the impact compressive toughness. Meanwhile, the degree of destruction of UHPC is decreased significantly due to the inclusion of SSWs. When the strain rate is 427/s and 467/s, the impact compressive damage factor of UHPC reinforced with 1.0 vol% and 1.5 vol% SSWs is 0.92 and 0.89, respectively.

3) The bundled SSWs with inter-anchored surface are uniformly dispersed in UHPC to inhibit the generation and propagation of cracks under impact compressive load. The constitutive model established on the basis of revised visco-elastic and continuum damage theory can well describe the impact compressive stress–strain relationship of SSWs-engineered multifunctional UHPC at different strain rates, in which the selection of strain threshold is closely related to the strain rate and the SSWs volume fraction.

REFERENCES

1. X. Cui, B. Han, Q. Zheng, X. Yu, S. Dong, L. Zhang, J. Ou. Mechanical properties and reinforcing mechanisms of cementitious composites with different types of multiwalled carbon nanotubes, *Composites Part A: Applied Science and Manufacturing*. 103 (2017) 131–147.
2. X. Wang, S. Dong, Z. Li, B. Han, J. Ou. Nanomechanical characteristics of interfacial transition zone in nano-engineered concrete, *Engineering*. 17 (2022) 99–109.

3. B. Han, Q. Zheng, S. Sun, S. Dong, L. Zhang, X. Yu, J. Ou. Enhancing mechanisms of multi-layer graphenes to cementitious composites, *Composites Part A: Applied Science and Manufacturing.* 101 (2017) 143–150.
4. X. Wang, S. Dong, A. Ashour, B. Han. Bond of nanoinclusions reinforced concrete with old concrete: Strength, reinforcing mechanisms and prediction model, *Construction and Building Materials.* 283 (2021) 122741.
5. X. Wang, S. Ding, A. Ashour, L. Qiu, Y. Wang, B. Han, J. Ou. Improving bond of fiber-reinforced polymer bars with concrete through incorporating nanomaterials, *Composites Part B: Engineering.* 239 (2022) 109960.
6. X. Wang, S. Dong, A. Ashour, W. Zhang, B. Han. Effect and mechanisms of nanomaterials on interface between aggregates and cement mortars, *Construction and Building Materials.* 240 (2020) 117942.
7. S. Dong, B. Han, X. Yu, J. Ou. Dynamic impact behaviors and constitutive model of super-fine stainless wire reinforced reactive powder concrete, *Construction and Building Materials.* 184 (2018) 602–616.
8. J. Xu, D. Zhao, F. Fan. *Dynamic Characteristics of Fiber Reinforced Concrete*, Northestern Polytechnical University Press, 2013. (In Chinese).
9. G. Ren, H. Wu, Q. Fang, J. Liu. Effects of steel fiber content and type on dynamic compressive mechanical properties of UHPCC, *Construction and Building Materials.* 164 (2018) 29–43.
10. Y. Wang, Z. Wang, X. Liang, M. An. Experimental and numerical studies on dynamic compressive behavior of reactive powder concretes, *Acta Mechanica Solida Sinica.* 21(5) (2008) 420–430.
11. Y. Tai. Uniaxial compression tests at various loading rates for reactive powder concrete, *Theoretical and Applied Fracture Mechanics.* 52(1) (2009) 14–21.
12. X. Hou, S. Cao, Q. Rong, W. Zheng, G. Li. Effects of steel fiber and strain rate on the dynamic compressive stress-strain relationship in reactive powder concrete, *Construction and Building Materials.* 170 (2018) 570–581.
13. Y. Su, J. Li, C. Wu, P. Wu, Z. Li. Influences of nano-particles on dynamic strength of ultra-high performance concrete, *Composites Part B: Engineering.* 91 (2016) 595–609.
14. Q. Li, H. Meng. About the dynamic strength enhancement of concrete-like materials in a split Hopkinson pressure Bar Test, *International Journal of Solids and Structures.* 40(2) (2002) 343–360.
15. J. W. Tedesco, C. A. Ross. Strain-rate-dependent constitutive equations for concrete, *Journal of Pressure Vessel Technology.* 120(4) (1998) 398–405.
16. Z. Wang, Z. Shi, J.Wang. On the strength and toughness properties of SFRC under static-dynamic compression, *Composites Part B: Engineering.* 42(5) (2011) 1285–1290.
17. Y. Hao, H. Hao. Dynamic compressive behaviour of spiral steel fibre reinforced concrete in split Hopkinson pressure bar tests, *Construction and Building Materials.* 48 (2013) 521–532.
18. Q. Li, X. Zhao, S. Xu, X. Gao. Influence of steel fiber on dynamic compressive behavior of hybrid fiber ultra high toughness cementitious composites at different strain rates, *Construction and Building Materials.* 125 (2016) 490–500.

19. Z. Wu, C. Shi, W. He, D. Wang. Static and dynamic compressive properties of ultra-high performance concrete (UHPC) with hybrid steel fiber reinforcements, *Cement and Concrete Composites*. 79 (2017) 148–157.
20. Z. Wang, Y. Liu, R. Shen. Stress-strain relationship of steel fiber-reinforced concrete under dynamic compression, *Construction and Building Materials*. 22(5) (2008) 811–819.
21. X. Sun, K. Zhao, Y. Li, R. Huang, Z. Ye, Y. Zhang, J. Ma. A study of strain-rate effect and fiber reinforcement effect on dynamic behavior of steel fiber-reinforced concrete, *Construction and Building Materials*. 158 (2018) 657–669.
22. H. Zhang, B. Wang, A. Xie, Y. Qi. Experimental study on dynamic mechanical properties and constitutive model of basalt fiber reinforced concrete, *Construction and Building Materials*. 152(15) (2017) 154–167.
23. Z. Wang, L. Wu, J. Wang. A study of constitutive relation and dynamic failure for SFRC in compression, *Construction and Building Materials*. 24(8) (2010) 1358–1363.
24. L. Wang, S. Shi. Research and application of ZWT nonlinear thermal visoelastic constitutive relationship. *Journal of Ningbo University: Polytechnic Edition*, 2000(B12) 141–149.
25. Q. Lu, Z. Wang, L. Wang, W. Lai, L. Yang. Analysis of linear visco-elastic spherical waves based on ZWT constitutive equation . *Explosion and Shock Waves*. 33(5) 463–470.
26. J. Mazars, G. Pijaudier-Cabot. Continuum damage theory-application to concrete. *Journal of Engineering Mechanics*. 115(2) 345–365.
27. H. Zhang, Y. Liu, H. Sun, S. Wu. Transient dynamic behavior of polypropylene fiber reinforced mortar under compressive impact loading. *Construction and Building Materials*. 111 (2016) 30–42.

Chapter 5

Fatigue properties of stainless steel wires-engineered multifunctional ultra-high performance concrete

5.1 INTRODUCTION

Concrete structures subjected to cyclic loads, such as bridge decks, tunnel linings, pavements, and offshore elements, show a sharp deterioration tendency after exposure to rapidly fluctuating and cyclic stresses due to the heterogeneous and brittle characteristics of concrete, eventually leading to the fatigue failure of structural elements even under service loads [1–3]. The fatigue failure of concrete elements severely limits the service life and increases the maintenance cost of structures [4–12]. Stainless steel wires (SSWs)-engineered multifunctional ultra-high performance concrete (UHPC) with excellent mechanical properties (shown in Chapters 2 and 3) is a preferred type of material, which can be used to improve the fatigue load resistance of structures. Evaluating the fatigue properties of SSWs-engineered multifunctional UHPC can provide design guidance for structures subjected to cyclic loads.

Therefore, the flexural and compressive fatigue properties, including fatigue life and fatigue stress–strain hysteresis curves, as well as fatigue damage of SSWs-engineered multifunctional UHPC are demonstrated in this chapter, and the underlying mechanisms are displayed by analyzing microstructures and pore structures as well as characteristics of hydration products. The main contents of this chapter are shown in Figure 5.1.

5.2 FLEXURAL FATIGUE PROPERTIES OF SSWS-ENGINEERED MULTIFUNCTIONAL UHPC

5.2.1 Flexural fatigue life and S-N curves

5.2.1.1 Flexural fatigue life

During the test of flexural fatigue properties, the specimen sizes were set as 40 mm × 40 mm × 160 mm, and the minimum flexural fatigue stress was set at 0.7 MPa to ensure that the specimens would not be unloaded during the cyclic flexural test. The maximum flexural fatigue stress equaled

DOI: 10.1201/9781003276357-5

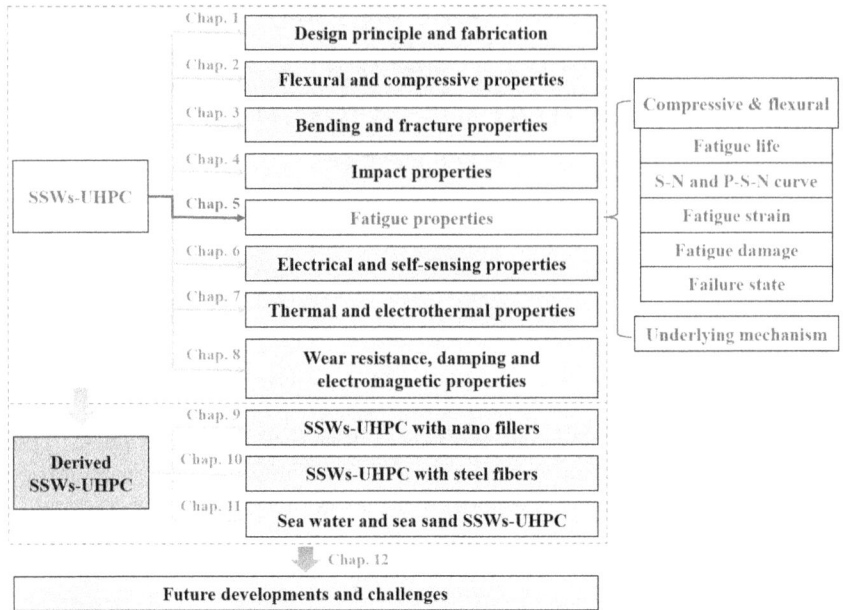

Figure 5.1 Main contents of Chapter 5

0.70, 0.80, and 0.90 times the static flexural strength after water curing for 28 days for UHPC without SSWs, UHPC reinforced with 1.0 vol% SSWs, and UHPC reinforced with 1.5 vol% SSWs, respectively (the static flexural strength of SSWs-engineered multifunctional UHPC has been presented in Figure 2.2). The loading rate of the minimum flexural fatigue stress was 0.05 mm/min. Then, the maximum fatigue stress was applied in the form of a sine wave with loading frequency of 5 Hz.

The three-point flexural fatigue life of UHPC reinforced with SSWs at a diameter of 20 μm, length of 10 mm, and content of 0.0 vol%/0.5 vol%/1.0 vol%/1.5 vol% (marked as W0, W201005, W201010, and W201015) under different maximum stress levels is shown in Table 5.1. As shown in Table 5.1, the test data corresponding to flexural fatigue life shows high scatter even under carefully controlled conditions but is within acceptable limits compared with previous studies [13–19]. It is found that the average flexural fatigue life of composites decreases with increasing maximum stress level. At the maximum stress level of 0.7, adding 0.5 vol%/1.0 vol%/1.5 vol% SSWs increases the average fatigue life of UHPC by 309.3%, 355.0%, and 636.6%, respectively. When the maximum stress level is 0.8, the average flexural fatigue life of UHPC reinforced with 0.5 vol%/1.0 vol%/1.5 vol% SSWs is 501.5%, 169.6%, and 558.3% higher, respectively, than that of UHPC without SSWs. The average flexural fatigue life of UHPC without SSWs is only 140 at the maximum stress level of 0.9, and it is improved by

Table 5.1 Flexural fatigue life of SSWs-engineered multifunctional UHPC

SSWs content (vol%)	Specimen number	Stress level	Maximum stress (MPa)	Stress amplitude (MPa)	Fatigue life (number of load cycles)	Average fatigue life
0	1	0.7	5.40	2.31	1399	61599
	2	0.7	5.40	2.31	48099	
	3	0.7	5.40	2.31	135300	
	4	0.8	6.17	2.70	264	1368
	5	0.8	6.17	2.70	673	
	6	0.8	6.17	2.70	3169	
	7	0.9	6.94	3.09	32	140
	8	0.9	6.94	3.09	65	
	9	0.9	6.94	3.09	325	
0.5	1	0.7	6.76	2.90	12426	252105
	2	0.7	6.76	2.90	225926	
	3	0.7	6.76	2.90	517965	
	4	0.8	7.72	3.38	940	8229
	5	0.8	7.72	3.38	2379	
	6	0.8	7.72	3.38	14138	
	7	0.8	7.72	3.38	15459	
	8	0.9	8.69	3.86	70	
	9	0.9	8.69	3.86	85	593
	10	0.9	8.69	3.86	629	
	11	0.9	8.69	3.86	1591	
1.0	1	0.7	8.12	3.48	28334	280299
	2	0.7	8.12	3.48	123201	
	3	0.7	8.12	3.48	689363	
	4	0.8	9.28	4.06	207	3688
	5	0.8	9.28	4.06	241	
	6	0.8	9.28	4.06	967	
	7	0.8	9.28	4.06	5485	
	8	0.8	9.28	4.06	11782	
	9	0.9	10.44	4.64	320	1375
	10	0.9	10.44	4.64	857	
	11	0.9	10.44	4.64	2949	
1.5	1	0.7	9.19	3.94	167998	453711
	2	0.7	9.19	3.94	191526	
	3	0.7	9.19	3.94	1001611	
	4	0.8	10.50	4.60	1002	9006
	5	0.8	10.50	4.60	7871	
	6	0.8	10.50	4.60	18147	
	7	0.9	11.82	5.25	986	1555
	8	0.9	11.82	5.25	1054	
	9	0.9	11.82	5.25	2625	

323.6%, 882.1%, and 1010.7% when 0.5 vol%, 1.0 vol%, and 1.5 vol% SSWs are added. It can be seen that at the maximum stress level of 0.8, the average flexural fatigue life of UHPC first increases and then decreases with increasing SSWs content, unlike the change of trend for static flexural strength and toughness.

Due to the microdiameter and large specific surface area, a low content of SSWs can improve the microstructural compactness of UHPC. Meanwhile, SSWs with low content still have a certain inhibitory effect on the development of microcracks, although the complete toughening network has not been formed, such that 0.5 vol% SSWs can effectively improve the flexural strength, toughness, and flexural fatigue life. With increasing SSWs content, the workability of concrete becomes worse, and the original macroscopic defects increase. The loading rate was slow (0.05 mm/min) for the static flexural strength/toughness test, such that the generation and propagation of microcracks caused by original defects can be suppressed in a timely way by the SSWs network. That is to say, the static flexural strength/toughness of UHPC increases with the improvement of overlapped networks and increasing content of SSWs. However, the loading speed of flexural fatigue load is high (5 Hz), resulting in accelerated generation and propagation of microcracks with increasing original defects caused by a high content of SSWs. Meanwhile, the incompletely overlapped network of SSWs is not enough to make up for the adverse effects of macroscopic defects. Therefore, the flexural fatigue life of UHPC with 1.0 vol% SSWs at the stress level of 0.8 is shorter than that of UHPC with 0.5 vol% SSWs at the same stress level. When the content of SSWs increases to 1.5 vol%, the SSWs network is sufficiently complete to effectively inhibit the initiation and propagation of microcracks under fatigue load and improve the static flexural strength, toughness, and flexural fatigue life.

In order to reveal the enhancing mechanisms of SSWs on the flexural fatigue life of UHPC, the relationships between the increment ratios of flexural fatigue life and those of flexural strength/toughness are established, as displayed in Table 5.2. The increase of flexural strength/toughness can prolong the flexural fatigue life of composites. When the linear coefficient of toughness increment ratio is greater than 1, the improvement of toughness caused by adding SSWs makes a great contribution to the prolongation of fatigue life. As shown in Table 5.2, the enhancing effect of 0.5 vol% SSWs on flexural toughness has a great influence on the increase of fatigue life of UHPC at maximum stress levels from 0.7 to 0.9, consistently with the static flexural behavior of 0.5 vol% SSWs-engineered multifunctional UHPC. For UHPC reinforced with 1.0 vol%/1.5 vol% SSWs, the increase of flexural strength plays a leading role in the improvement of fatigue life at the maximum stress levels of 0.7 and 0.8, resulting from the compacting effect of SSWs on the UHPC microstructure. When the maximum stress level is 0.9, there is already a lot of damage inside composites at the initial

Table 5.2 Relationships between increment ratios of flexural fatigue life and static flexural strength/toughness

Specimens	Stress level	Increment ratio of flexural fatigue life ΔN_f (%)	Increment ratio of static flexural strength Δf (%)	Increment ratio of static flexural toughness ΔT (%)	Relationships
W201005	0.7	3.093	0.254	1.110	$\Delta N_f = \Delta f + (2.5 - 4.3)\,\Delta T$
	0.8	5.015	0.254	1.110	
	0.9	3.236	0.254	1.110	
W201010	0.7	3.550	0.506	3.274	$\Delta N_f = \Delta f + (0.3 - 1.0)\,\Delta T$
	0.8	1.696	0.506	3.274	
W202015	0.7	6.366	0.705	5.118	
	0.8	5.583	0.705	5.118	
W201010	0.9	8.821	0.506	3.274	$\Delta N_f = \Delta f + (1.9 - 2.5)\,\Delta T$
W201015	0.9	10.107	0.705	5.118	

stage of fatigue load, such that the numbers of SSWs bridging cracks and being pulled off are essential to the fatigue life of composites.

5.2.1.2 S-N curves and P-S-N curves of flexural fatigue life

The relationships between maximum stress level (S_{max}) and number of cycles, i.e. fatigue life (N_f) (called the S-N curve), established in former studies include the following three types. Equation (5.1) is the typical Wholer equation [2], and Eq. (5.2) is the modified form of Wholer's equation considering the influence of the stress ratio [20]. Equation (5.3), a power function of Wholer's equation, has been usually adopted to describe the fatigue behavior of pavement concrete with flexural strength as the control performance [21].

$$S_{max} = A + B\log_{10}N_f \tag{5.1}$$

$$S_{max} = 1 - \beta\left(1 - R\right)\log_{10}N_f \tag{5.2}$$

$$S_{max} = C_1\left(N_f\right)^{-C_2} \tag{5.3}$$

where A, B, β, C_1, and C_2 are coefficient parameters, and R is the stress ratio between the maximum and minimum stress.

By comparing these formulas, it can be found that as S_{max} becomes small, even approaching zero, the value of N_f tends toward infinity in Eq. (5.3), and this satisfies the extreme boundary condition of the flexural fatigue performance of concrete. Meanwhile, the minimum stress was set as a fixed value in this study. Therefore, Eq. (5.3) is the most appropriate expression form to describe the relationship between the fatigue life of SSWs-engineered multifunctional UHPC and the maximum stress level. Taking the logarithm of both sides of Eq. (5.3) produces the following Eq. (5.4), highlighting a linear relationship between $\log_{10}S_{max}$ and $\log_{10}N_f$.

$$\log_{10}S_{max} = \log_{10}C_1 - C_2\log_{10}N_f \tag{5.4}$$

This linear S-N equation (5.4) for SSWs-engineered multifunctional UHPC is demonstrated in Table 5.3 and Figure 5.2. It is shown that the slope of the S-N equation decreases with increasing SSWs content, indicating the improving effect of SSWs on the fatigue performance of UHPC. However, what merits special attention is that there are three areas of concrete fatigue: low-cycle, high-cycle, and subcritical fatigue. Of these, the fatigue life corresponding to the boundary between low-cycle and high-cycle fatigue is about 10^4, and that corresponding to the boundary between high-cycle and subcritical fatigue is about 2×10^6. It can be seen from Table 5.1 that the flexural fatigue life of SSWs-engineered multifunctional UHPC at the maximum stress level of 0.7 is always larger than 10^4, and the flexural fatigue life of some specimens is

Table 5.3 Parameters of linear S-N equations of SSWs-engineered multifunctional UHPC

Group	$\log_{10}C_1$	C_2	R^2
W0	0.03998	−0.04104	0.99579
W201005	0.06776	−0.04142	0.99844
W201010	0.07228	−0.04232	0.90554
W201015	0.08258	−0.04259	0.96772

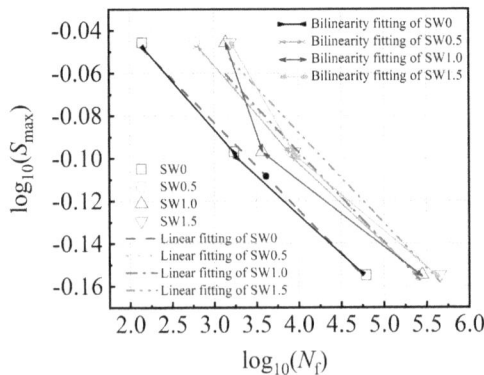

Figure 5.2 Linear and bilinear S-N curves of SSWs-engineered multifunctional UHPC. SW0, SW0.5, SW1.0, and SW1.5 represent W0, W201005, W201010, and W201015, respectively

also greater than 10^4 when the stress level is 0.8. Hence, the bilinear equation is more suitable and more accurate to describe the relationship between $\log_{10}(S_{max})$ and $\log_{10}(N_f)$, as shown in Table 5.4 and Figure 5.2. As shown in Table 5.4, when the maximum stress level ranges from 0.8 to 0.9, the linear fitting slope of the S-N equation increases with increasing SSWs dosage. This indicates that the improvement effect of SSWs on the fatigue performance of UHPC at high maximum stress level becomes more marked with increasing SSWs content. At the maximum stress level ranging from 0.7 to 0.8, the linear fitting slope of the S-N equation for UHPC reinforced with 0.5 vol% SSWs is significantly greater than that for UHPC without SSWs, indicating that 0.5 vol% SSWs has a greater enhancing effect on the fatigue performance of UHPC at the maximum stress level of 0.8 than at the maximum stress level of 0.7. This can be ascribed to 0.5 vol% SSWs having limited modification results on the UHPC matrix but possessing the function of bridging cracks during loading. The linear slope of the S-N equation for UHPC reinforced with 1.0 vol% SSWs is 7.7% lower than that for UHPC without SSWs, deriving from the outstanding controlling effect of 1.0 vol% SSWs on crack propagation. However, adding 1.5 vol% increases the linear slope of the S-N equation by 2.0%, and this is the consequence of both the enhancing effect of 1.5 vol% SSWs on matrix structure densification and crack inhibition. What needs to be pointed out is that because the linear fitting correlation coefficient for maximum stress levels of 0.7–0.9 is larger than 0.90, and there are only three maximum stress levels, Eq. (5.4) can still be used to establish the relationship between the logarithm of maximum stress level and the logarithm of fatigue life for composites.

Taking into account the fluctuation characteristics of flexural fatigue life, a two-parameter Weibull distribution with physically valid assumptions and sound experimental verification has been proposed to depict the fatigue life distribution of SSWs-engineered multifunctional UHPC [22–25], as shown in the following Eq. (5.5):

$$P\left(N_f > N_P\right) = \exp\left[-\left(\frac{N_f}{\alpha}\right)^{\theta}\right] \tag{5.5}$$

Table 5.4 Bilinear S-N fatigue equations of SSWs-engineered multifunctional UHPC

	Fatigue equations			
	Maximum stress levels of 0.8–0.9		Maximum stress levels of 0.7–0.8	
Group	$\log_{10}C_1$	C_2	$\log_{10}C_1$	C_2
W0	0.05110	−0.04697	0.02392	−0.03340
W201005	0.07861	−0.04883	0.05589	−0.03902
W201010	0.32937	−0.11951	0.01308	−0.03084
W201015	0.16853	−0.06712	0.03784	−0.03407

where P is the survival probability of concrete, N_p is the flexural fatigue life of concrete with survival probability of P; α and θ are scale and shape parameters associated with maximum stress level, respectively. Given that bilinear S-N equations possess higher accuracy compared with linear S-N equations in this study, the graphical method is no longer suitable to obtain the scale and shape parameters for Weibull distribution. Hence, the method of moments and the method of maximum likelihood are used to calculate scale and shape parameters.

The method of moments uses average fatigue life (μ) and variance of data (COV) to determine distribution parameters, as shown in the following Eqs (5.6) and (5.7) [18]:

$$\alpha = \frac{\mu}{\Gamma\left(\dfrac{1}{\theta}+1\right)} \tag{5.6}$$

$$\theta = COV^{-1.08} \tag{5.7}$$

The method of maximum likelihood has been used to acquire distribution parameters based on a probabilistic approach. The likelihood function is expressed as Eq. (5.8) [19]:

$$L(x,\alpha,\theta) = \prod_{i=1}^{n} \frac{\theta}{\alpha}\left(\frac{x_i}{\alpha}\right)^{\theta-1} \exp\left[-\left(\frac{x_i}{\alpha}\right)^{\theta}\right] \tag{5.8}$$

Equation (5.9) can be obtained by reorganizing and simplifying Eq. (5.8), as follows:

$$\frac{1}{\theta} = \frac{\displaystyle\sum_{i=1}^{n} x_i^{\theta}\ln x_i}{\displaystyle\sum_{i=1}^{n} x_i^{\theta} - \frac{1}{n}\sum_{i=1}^{n} x_i} \tag{5.9}$$

$$\alpha^{\theta} = \frac{1}{n}\sum_{i=1}^{n} x_i^{\theta} \tag{5.10}$$

Equation (5.9) is solved by an iterative method using the subsequent steps (1)–(5) [26]:

(1) Calculate the initial value of the shape parameter $\theta_0 = 1.2\mu/COV$, and the value of precision (ε) equals 0.01, where μ is the average flexural fatigue life of concrete, and COV is the standard deviation of fatigue life.

(2) The following Eq. (5.11) is used to calculate the iteration value of the shape parameter:

$$\theta_{2K+1} = \cfrac{1}{\cfrac{\displaystyle\sum_{i=1}^{n} x_i^{\theta_{2K}} \ln x_i}{\displaystyle\sum_{i=1}^{n} x_i^{\theta_{2K}}} - \cfrac{1}{n}\displaystyle\sum_{i=1}^{n} \ln x_i} \tag{5.11}$$

(3) If $|\theta_{2K+1} - \theta_{2K}| \leq \varepsilon$ is obtained, the iteration can be terminated to obtain the value of the shape parameter. Otherwise, proceed to the next step (4).

(4) Substitute $\theta_{2K+2} = (\theta_{2K+1} + \theta_{2K})/2$ into Eq. (5.11) for the next iteration step from (2) to (3).

(5) End the iteration, and get the result of the shape parameter.

After that, use Eq. (5.10) to calculate the scale parameter.

The distribution parameters of SSWs-engineered multifunctional UHPC obtained by the method of moments and the method of maximum likelihood are listed in Table 5.5. The mean values obtained by these two methods are used as the final values of the parameters.

Table 5.5 Distribution parameters of SSWs-engineered multifunctional UHPC at different stress levels

Group	Distribution parameters	0.70		0.80		0.90	
		α	θ	α	θ	α	θ
W0	Method of moments	0.899	58518	0.861	1268	0.867	131
	Maximum likelihood	1.088	64117	1.044	1394	1.052	145
	Average	0.994	61318	0.953	1331	0.959	138
W201005	Method of moments	0.993	251335	1.085	8490	0.820	533
	Maximum likelihood	1.192	270617	1.295	9071	0.998	791
	Average	1.093	260976	1.190	8781	0.909	662
W201010	Method of moments	0.769	240221	0.730	3065	0.989	1369
	Maximum likelihood	0.941	271313	0.896	3460	1.188	1461
	Average	0.855	255767	0.813	3263	1.089	1415
W201015	Method of moments	0.952	443885	1.047	9174	1.748	1746
	Maximum likelihood	1.147	476317	1.253	6668	2.012	1443
	Average	1.050	460101	1.150	7921	3.760	1595

On the basis of Eq. (5.5), the flexural fatigue failure probability P' of SSWs-engineered multifunctional UHPC is obtained from Eq. (5.12).

$$P' = 1 - P\left(N_f > N_p\right) = 1 - \exp\left(-\frac{N_f}{\alpha}\right)^{\theta} \tag{5.12}$$

According to the parameters presented in Table 5.5, the flexural fatigue life of composites at a certain stress level with specific failure probability can be obtained using the following Eq. (5.13):

$$N_f = \alpha\left|\ln\left(1 - P'\right)\right|^{1/\theta} \tag{5.13}$$

As the failure probability is between 10% and 90%, the interval of flexural fatigue life is demonstrated in Figure 5.3. The right shift distance of the interval represents the modification effect of SSWs. When the dosage of SSWs is 1.0 vol% and 1.5 vol%, the right shift distance of the fatigue life interval at the maximum stress level of 0.9 is markedly longer than that at the maximum stress level of 0.7, indicating that 1.0 vol% SSWs and 1.5 vol% SSWs possess an excellent inhibiting effect on crack development.

When the failure probability is 50%, the corresponding linear fatigue equations of the composites are displayed in Table 5.6. The increase in the slope and intercept of the fatigue equation indicates that the crack inhibition ability is enhanced by increasing SSWs content. The correlation coefficients R^2 of linear fitting are larger than 0.95 except for UHPC reinforced with 1.0 vol% SSWs, representing the accuracy of the fatigue equation. The calculated ratio of flexural fatigue limit strength to static flexural strength can be obtained by substituting the fatigue life of 2×10^6 into the fatigue equation. As shown in Table 5.6, the calculated ratio of flexural fatigue limit strength to static flexural strength for UHPC reinforced with 0.5 vol%/1.0 vol%/1.5 vol% SSWs is 0.6332, 0.6254, and 0.6420, respectively, which is 6.3%, 5.0%, and 7.8% higher than that of UHPC without SSWs. The ratio of fatigue limit strength to static flexural strength of UHPC reinforced with 1.5 vol% SSWs is higher than that of normally vibrated concrete reinforced with 1.5 vol% steel fibers (with a diameter of 1 mm and length of 30 mm)

Table 5.6 S-N fatigue equations of SSWs-engineered multifunctional UHPC at failure probability of 50%

Group	$\log_{10}C_1$	C_2	R^2	Flexural fatigue limit strength
W0	0.02970	−0.040451	0.98745	0.5954 f_b
W201005	0.06302	−0.041501	0.99915	0.6332 f_b
W201010	0.06492	−0.042651	0.87819	0.6254 f_b
W201015	0.07828	−0.042971	0.95018	0.6420 f_b

Note: f_b represents the static flexural strength of composites.

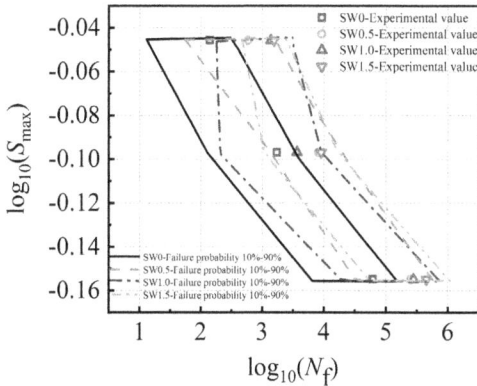

Figure 5.3 Fatigue life interval of SSWs-engineered multifunctional UHPC at failure probability of 10–90%. SW0, SW0.5, SW1.0, and SW1.5 represent W0, W201005, W201010, and W201015, respectively

(0.62) [27] but lower than that of self-compacting concrete reinforced with 1.5 vol% steel fibers (0.71) [27] and that of self-compacting concrete reinforced with 1.0 vol% hybrid fibers (0.69) [28]. However, at the maximum stress levels of 0.7 and 0.8, the flexural fatigue life of UHPC reinforced with 0.5 vol% SSWs is higher than that of concrete reinforced with 0.5 vol% smooth steel fibers (with a diameter of 0.12 mm and length of 15 mm) [16]. Meanwhile, the flexural fatigue life of UHPC reinforced with 1.5 vol% SSWs is equivalent to that of self-compacting concrete reinforced with 1.5 vol% steel fiber at the maximum stress levels of 0.7, 0.8, and 0.9 [27]. It is worth mentioning that the ratio first increases and then decreases with increasing dosage of SSWs, and this is consistent with previous studies on steel fiber or hybrid fiber reinforced concrete [27, 28].

5.2.2 Flexural fatigue strain and damage

5.2.2.1 *Flexural fatigue strain–stress hysteresis curves*

The flexural fatigue strain–stress hysteresis curves of SSWs-engineered multifunctional UHPC for the whole fatigue life are shown in Figures 5.4–5.7. The data for the curves is recorded for every 5% interval of fatigue life. It can be found that except for a specimen of UHPC without SSWs with fatigue life of 135300, and specimens of UHPC reinforced with 0.5 vol% SSWs with fatigue life of 517965, 225926, and 15459, there is a change of trend from dense to failure for the fatigue hysteresis curves of UHPC without SSWs and UHPC reinforced with 0.5 vol% SSWs at the maximum stress levels of 0.7 and 0.8 with the increase of fatigue cycles. The stress–strain hysteresis curves for specimens before failure are away from the dense areas, indicating that the strain increases instantaneously, and this is a typical manifestation of brittle material failure. This also shows that 0.5

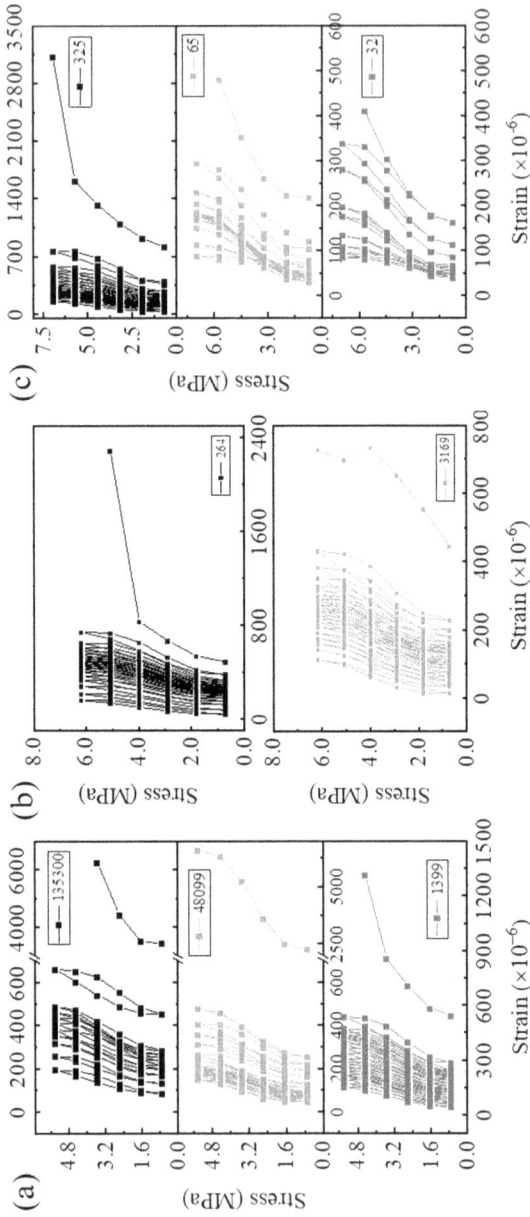

Figure 5.4 Flexural fatigue stress–strain hysteresis curves of UHPC without SSWs at maximum stress level of (a) 0.7; (b) 0.8; (c) 0.9

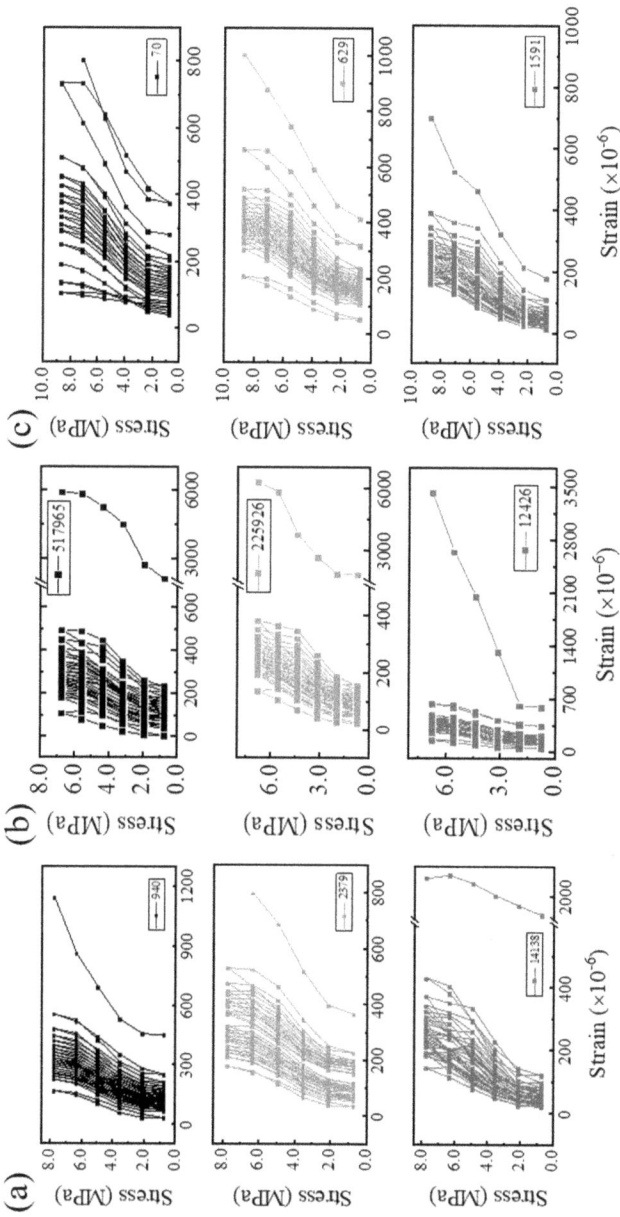

Figure 5.5 Flexural fatigue stress–strain hysteresis curves of UHPC reinforced with 0.5 vol% SSWs at maximum stress level of (a) 0.7; (b) 0.8; (c) 0.9

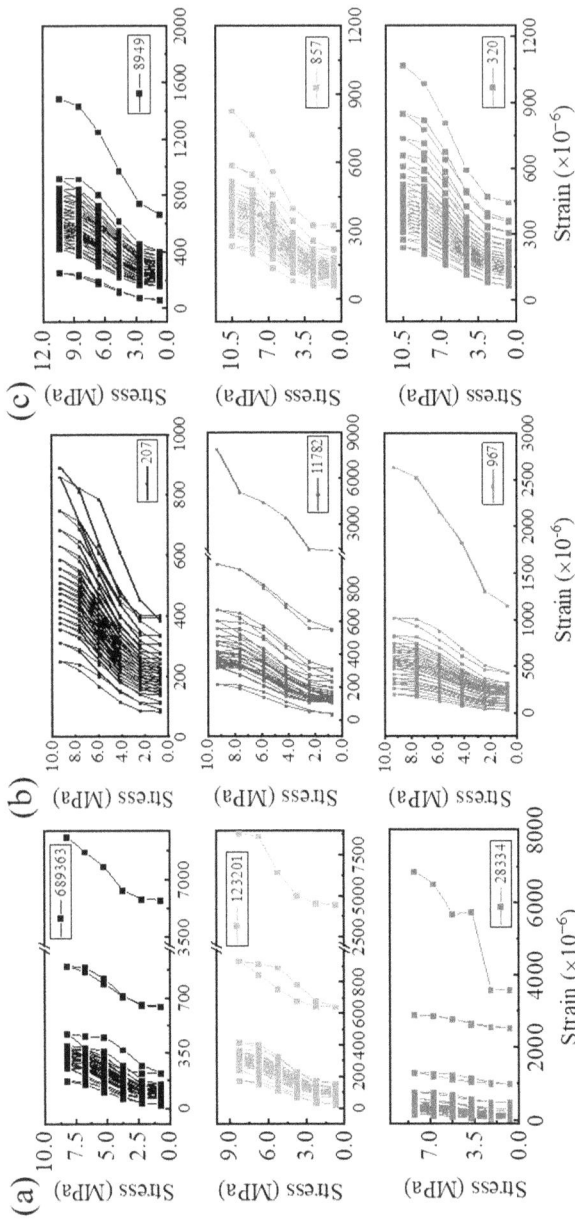

Figure 5.6 Flexural fatigue stress–strain hysteresis curves of UHPC reinforced with 1.0 vol% SSWs at maximum stress level of (a) 0.7; (b) 0.8; (c) 0.9

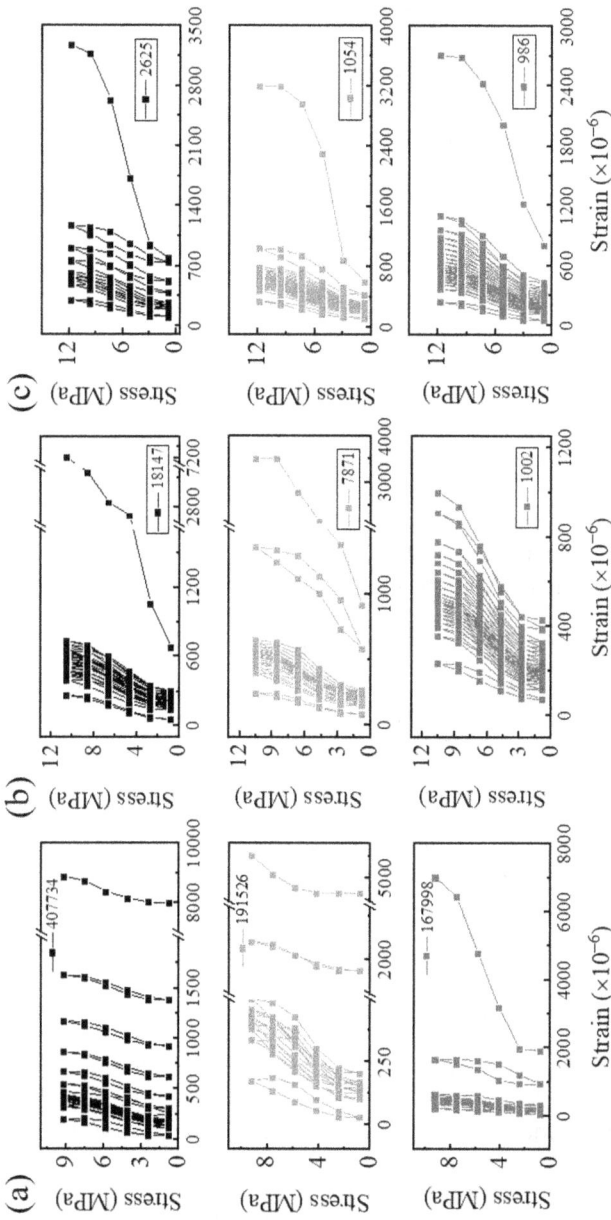

Figure 5.7 Flexural fatigue stress–strain hysteresis curves of UHPC reinforced with 1.5 vol% SSWs at maximum stress level of (a) 0.7; (b) 0.8; (c) 0.9

vol% SSWs has only a relatively weak inhibition effect on crack initiation. When the maximum stress level is 0.9, the slope of stress–strain hysteresis curves for UHPC without SSWs at the initial loading stage is lower than that when the stress level is 0.7 and 0.8, coming from the increasing initial damage, and then, the strain continues to increase rapidly until the specimens fail. However, it is found that the stress–strain hysteresis curves of UHPC reinforced with 0.5 vol% SSWs uniformly move to the right with the increase of fatigue cycles, showing the beneficial effect of SSWs on limiting crack development.

Figures 5.6 and 5.7 show that the flexural fatigue stress–strain hysteresis curves of UHPC reinforced with 1.0 vol%/1.5 vol% SSWs have the change of trend from sparse to dense to sparse and then to failure in addition to specimens with fatigue life of 28334, 967, and 167998. It is worth mentioning that specimens of UHPC reinforced with 0.5 vol% SSWs with fatigue life of 517965, 225926, 15459, and 629 also possess stress–strain fatigue curves with these same characteristics. The emergence of the first sparse stage results from the compactness effect of SSWs on the matrix microstructure and the inhibition effect of SSWs on crack initiation. At the maximum stress levels of 0.7 and 0.8, the proportions of fatigue life for UHPC reinforced with 1.5 vol% SSWs occupied by the first sparse and the last sparse stage are larger than that for UHPC reinforced with 1.0 vol% SSWs, illustrating that the compactness of the matrix microstructure and the bridging effect of SSWs increase with increasing SSWs content. When the maximum stress level is 0.9, the slope of the stress–strain hysteresis curves before failure decreases significantly with increasing SSWs content, indicating that the number of SSWs being pulled off increases. Meanwhile, compared with UHPC without SSWs and UHPC reinforced with 0.5 vol% SSWs, the tendency of stress–strain hysteresis curves to move to the right for UHPC reinforced with 1.0 vol%/1.5 vol% SSWs slows down at the maximum stress level of 0.9. According to this analysis, the stress–strain hysteresis curves reflect the development of deformation in real time and also show the role of SSWs in improving the flexural fatigue performance of UHPC.

5.2.2.2 Maximum flexural fatigue strain

The relationships between maximum flexural fatigue strain and number of cycles to fatigue life ratio (N/N_f) are demonstrated in Figure 5.8. As shown in Figure 5.8(a), the maximum fatigue strain for most specimens of UHPC without SSWs is divided into two stages, representing the sustained development and rapid aggregation of microcracks, respectively. The lack of a microcrack initiation stage is a manifestation of brittleness. In addition, there is a lower probability that the strain gauge will obtain intact damage development with increasing maximum stress level, also resulting from the

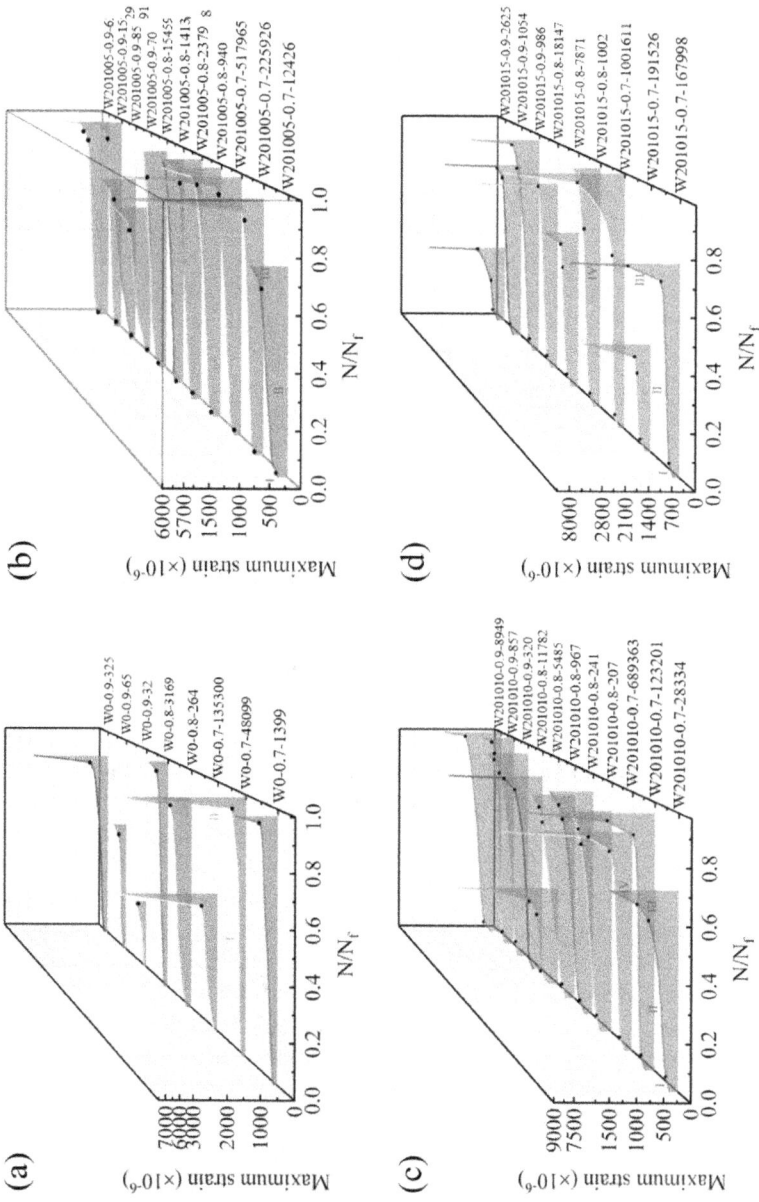

Figure 5.8 Maximum fatigue strain–number of cycles curves of SSWs-engineered multifunctional UHPC (a) UHPC without SSWs; (b) UHPC reinforced with 0.5 vol.% SSWs; (c) UHPC reinforced with 1.0 vol.% SSWs; (d) UHPC reinforced with 1.5 vol.% SSWs

brittleness characteristics of UHPC without SSWs. This phenomenon is not observed in Figure 5.8(b)–(d). Figure 5.8(b) shows that the development of maximum fatigue strain for most specimens of UHPC reinforced with 0.5 vol% SSWs goes through three stages: the microcrack initiation stage, the propagation stage, and the failure stage. It is observed that the strain growth rate in the second stage tends to increase with increasing fatigue cycles and maximum stress level, which is the effect of microcracks increasing. In the failure stage, the strain growth rate is still high, indicating that the effects of SSWs bridging and being pulled off are limited.

Figure 5.8(c) and (d) show that the maximum fatigue strain of UHPC reinforced with 1.0 vol%/1.5 vol% SSWs possesses four stage development trends with an increasing number of cycles, corresponding to the microcrack initiation stage, the crack stable development stage, the crack unstable development stage, and the failure stage, respectively. It can be clearly seen that the unstable development stage of maximum strain for UHPC reinforced with 1.5 vol% SSWs is longer than that for UHPC reinforced with 1.0 vol% SSWs due to the increasing numbers of bridging SSWs. It should be noted that although the complete monitoring of specimen flexural-tensile deformation under fatigue flexure has not been realized through strain gauges, the maximum strain can still be used to compare the effect of SSWs content on the flexural-tensile deformation. At the maximum stress levels of 0.7, 0.8, and 0.9, the monitored maximum strain for UHPC without SSWs is 6612, 2288, and 3121, respectively. The increase of maximum stress level leads to the initial damage increasing, thus decreasing the value of maximum strain. The monitored maximum fatigue strain for UHPC reinforced with 0.5 vol% SSWs is 6270, 4755, and 1001, corresponding to maximum stress levels of 0.7, 0.8, and 0.9, respectively. Adding 1.0 vol% SSWs increases the maximum fatigue strain by 45.1%, 245.3%, and −110.0% at the same three maximum stress levels, reaching 9597, 7901, and 1486, respectively. The increasing maximum strain at the maximum stress level of 0.7 results from the modification effect of SSWs on the UHPC matrix compactness, and the increase at the maximum stress level of 0.8 is the result of a synergistic effect of SSWs on the matrix microstructure and inhibiting effect on crack development. The maximum strain of UHPC reinforced with 1.5 vol% SSWs at the maximum stress levels of 0.7, 0.8, and 0.9 is 8863, 7201, and 3265, respectively, which is 34.0%, 214.7%, and 4.6% higher than that of UHPC without SSWs. This result indicates that the high content of SSWs can not only modify the microstructure of UHPC to improve deformation ability at a low maximum stress level but also effectively bridge cracks to achieve large flexural-tensile deformation at a high maximum stress level. The change of trend in maximum strain increment indicates that SSWs at a high content not only have a better hindering effect on crack propagation at a high maximum stress level but are also better at improving the microstructural compactness of UHPC. This conclusion is consistent with that obtained from Table 5.1.

5.2.2.3 Residual flexural fatigue strain

The residual flexural fatigue strain of SSWs-engineered multifunctional UHPC against the increase of fatigue cycles is presented in Figure 5.9. It can be seen from Figure 5.9(a) that the residual fatigue strain of UHPC without SSWs only contains two stages at the maximum stress level of 0.9, reflecting that microcracks are emerging in large numbers and developing rapidly. However, the development of residual strain for UHPC without SSWs can be divided into three stages at the maximum stress levels of 0.7 and 0.8, and the slope of the three stages gradually increases with the increase of fatigue cycles, representing the acceleration of damage development. The three-stage variation characteristics of residual strain for UHPC without SSWs at low maximum stress levels are due to the reduction of initial microcracks. Except for some special specimens, the residual strain accumulation development of SSWs-engineered multifunctional UHPC has three stages, deriving from the initiation, propagation, and convergence of microcracks, respectively.

Through comparison and analysis, it can be found that the proportion of the first stage of residual strain in fatigue life for SSWs-engineered

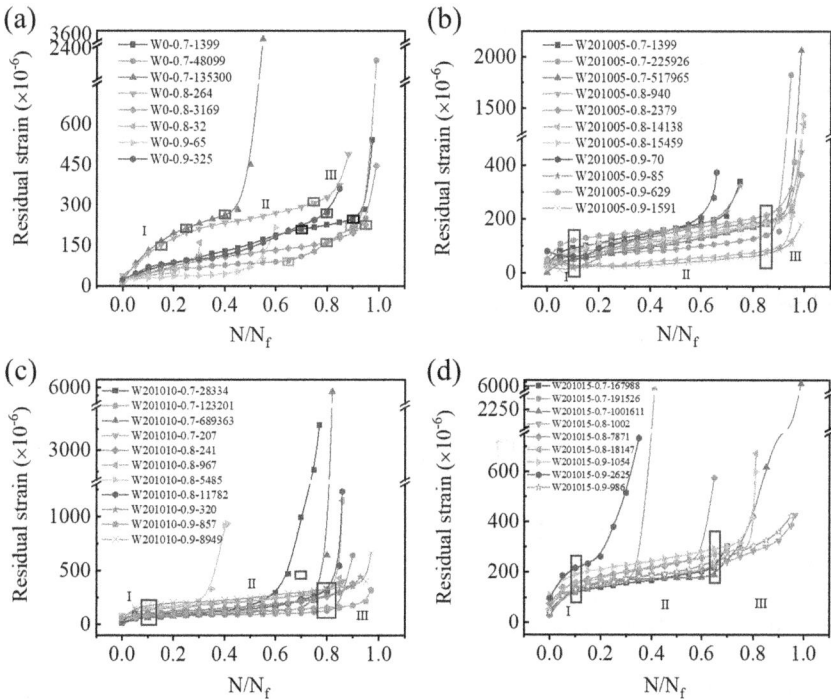

Figure 5.9 Residual flexural fatigue strain of SSWs-engineered multifunctional UHPC. (a) UHPC without SSWs; (b) UHPC reinforced with 0.5 vol% SSWs; (c) UHPC reinforced with 1.0 vol% SSWs; (d) UHPC reinforced with 1.5 vol% SSWs

multifunctional UHPC is lower than that for UHPC without SSWs, showing the limiting effect of SSWs on crack initiation. Meanwhile, the development slope of the second stage for residual strain is smaller than that of the first stage for specimens of SSWs-engineered multifunctional UHPC, and this is contrary to the development law of UHPC without SSWs. This results from the hindering effect of SSWs on microcrack propagation. Furthermore, Figure 5.9 also shows that with the increase of SSWs content, the accumulation slope of the third stage for residual strain shows a decreasing trend, benefiting from the increasing number of SSWs being pulled off.

5.2.2.4 Flexural fatigue damage

Since UHPC has good uniformity, its damage and residual strain under flexural fatigue load also conforms to Weibull distribution, as demonstrated in the following Eq. (5.14):

$$D = 1 - \exp\left[-\left(\frac{\varepsilon_{min}}{a} \right)^m \right] \tag{5.14}$$

where ε_{min} is the residual strain demonstrated in Figure 5.9, and a and m are the scale and shape parameters determined by the constitutive performance of SSWs-engineered multifunctional UHPC. Hence, as demonstrated in Eq. (5.15), the constitutive model of concrete under static flexural load can be used to calculate the values of a and m [29]:

$$\sigma = E\varepsilon \exp\left[-\left(\frac{\varepsilon}{a} \right)^m \right] \tag{5.15}$$

where σ and ε are the stress and strain of SSWs-engineered multifunctional UHPC under static flexure (in Figure 2.6(a)). Taking the derivative with respect to the strain in Eq. (5.15), the result is shown as follows:

$$\frac{d\sigma}{d\varepsilon} = E\left[1 - m\left(\frac{\varepsilon}{a} \right)^m \right] \exp\left[1 - \left(\frac{\varepsilon}{a} \right)^m \right] \tag{5.16}$$

According to the boundary conditions at the peak strain under static flexure, it can be known that when $\varepsilon = \varepsilon_{max}$, $\sigma = \sigma_{max}$ and $d\sigma/d\varepsilon = 0$ are established, i.e. the following Eqs (5.17) and (5.18) are obtained (the values of ε_{max} and σ_{max} are shown in Figure 2.6(a)):

$$\left(\frac{\varepsilon_{max}}{a} \right)^m = \frac{1}{m} \tag{5.17}$$

$$\frac{\sigma_{max}}{E\varepsilon_{max}} = \exp\left[-\left(\frac{\varepsilon_{max}}{a}\right)^m\right] \qquad (5.18)$$

Taking the natural logarithm of both sides of Eq. (5.18) twice, the following Eq. (5.19) can be acquired:

$$\ln\left[\ln\left(\frac{E\varepsilon_{max}}{\sigma_{max}}\right)\right] = m\ln\left(\frac{\varepsilon_{max}}{a}\right) \qquad (5.19)$$

Equation (5.20) is obtained by combining Eq. (5.17) and Eq. (5.19):

$$\ln\left(\frac{E\varepsilon_{max}}{\sigma_{max}}\right) = \frac{1}{m} \qquad (5.20)$$

The values of parameters m and a can be calculated from the following Eqs (5.21) and (5.22):

$$m = \frac{1}{\ln\left(\dfrac{E\varepsilon_{max}}{\sigma_{max}}\right)} \qquad (5.21)$$

$$a = \frac{\varepsilon_{max}}{\left(1/m\right)^{1/m}} \qquad (5.22)$$

Substituting the expression of parameter a into Eq. (5.14), the flexural fatigue damage of SSWs-engineered multifunctional UHPC is as follows:

$$D = 1 - \exp\left[-\frac{1}{m}\left(\frac{\varepsilon_{min}}{\varepsilon_{max}}\right)^m\right] \qquad (5.23)$$

Based on the parameters displayed in Table 5.1 and Eq. (5.23), the damage of representative specimens for SSWs-engineered multifunctional UHPC under flexural fatigue load is calculated and demonstrated in Figures 5.10–5.13. Eq. (5.23) shows that the damage is proportional to residual strain when the static peak strain for a specimen is constant. With the appearance and increase of microcracks under flexural fatigue load, the residual strain and the corresponding damage appear and show a steady increase trend, which is the crack initiation stage. When microcracks converge to a certain extent, even a small number of fatigue cycles leads to the occurrence of large residual strain and damage, represented by a rapid increase in residual strain and damage with the increase of fatigue cycles, signifying the crack stable development stage. However, with the further expansion of microcracks, the aggregation speed is accelerated, leading to an abrupt development in corresponding residual strain/damage and cracks progressing to the unstable stage. As macrocracks form and the specimen fails, a jump in damage happens. Therefore, the development stage of cracks can be described qualitatively by an increasing trend of damage with an increase in fatigue cycles. The damage at different crack development stages has been marked by lines in Figures 5.10–5.13. It can be found that the damage of composites accumulates with the increase of fatigue cycles. What needs illustration is that some specimens continues to bear load after the failure of the strain gauge. Therefore, the maximum value of damage for these specimens appears before the maximum fatigue life is reached. Figures 5.10 and 5.11 show that the damage development of UHPC without SSWs and that reinforced with 0.5 vol% SSWs at the maximum stress levels of 0.7 and 0.8 is divided into three stages, standing for the initiation of cracks, propagation of cracks, and failure of specimens, respectively. However, the proportion of each stage in the whole fatigue life has no obvious character-istics. The fatigue development of UHPC reinforced with 0.5 vol% SSWs at the maximum stress level of 0.9 and of UHPC reinforced with 1.0 vol%/1.5 vol% SSWs at the maximum stress levels of 0.7, 0.8, and 0.9 includes the following four stages: the initiation stage of cracks, the stable development stage of cracks, the unstable development stage of cracks, and the failure

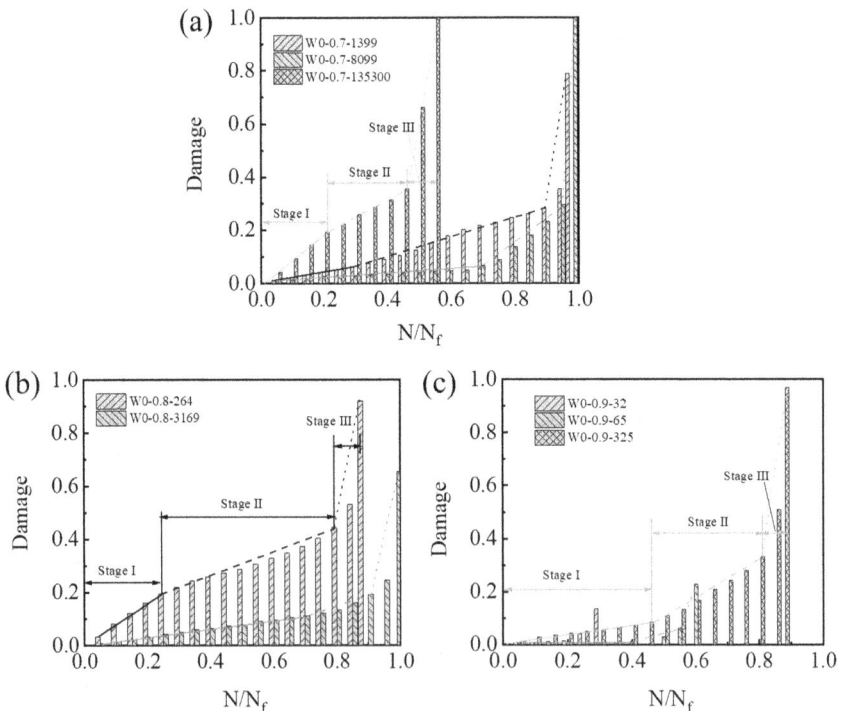

Figure 5.10 Flexural fatigue damage of UHPC without SSWs at maximum stress level of (a) 0.7; (b) 0.8; (c) 0.9

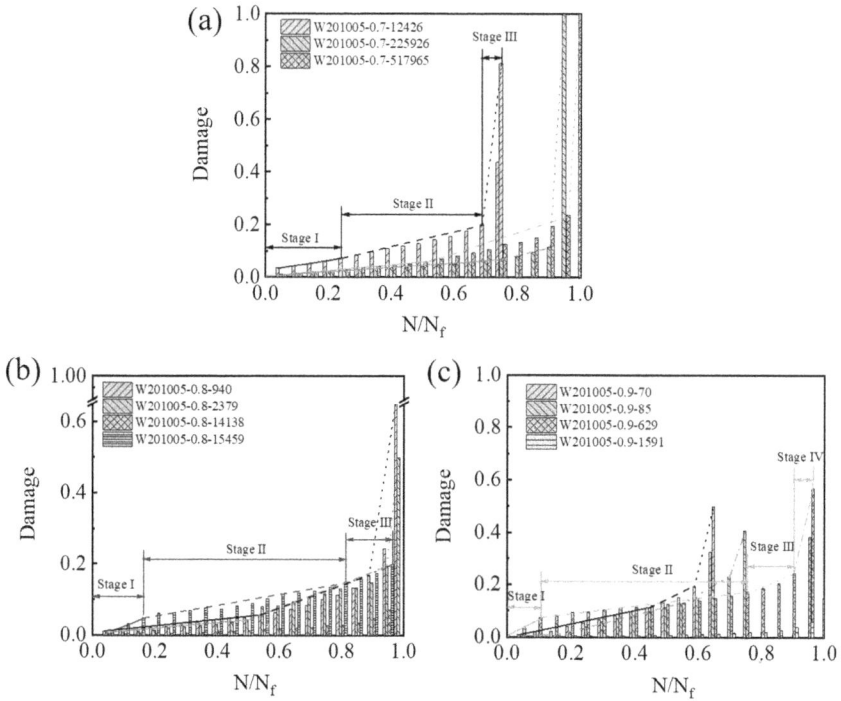

Figure 5.11 Flexural fatigue damage of UHPC reinforced with 0.5 vol% SSWs at maximum stress level of (a) 0.7; (b) 0.8; (c) 0.9

stage of specimens. The unstable development stage of cracks results from the bridging effect of SSWs. Comparative analysis shows that the first stage of crack development for SSWs-engineered multifunctional UHPC is shorter than that for UHPC without SSWs, resulting from the inhibiting effect of SSWs on crack initiation. The slope of the stable crack development stage for UHPC without SSWs and that reinforced with 0.5 vol% SSWs at the maximum stress levels of 0.7 and 0.8 is larger than that of the first stage, while this phenomenon has the opposite trend for UHPC reinforced with 0.5 vol% SSWs at the maximum stress level of 0.9 and UHPC reinforced with 1.0 vol%/1.5 vol% SSWs at the maximum stress levels of 0.7, 0.8, and 0.9. This illustrates that the damage of UHPC without SSWs develops rapidly after the appearance of microcracks, but the propagation and aggregation of microcracks are inhibited by adding SSWs. Meanwhile, the slope of the stable crack development stage decreases with increasing SSWs content, representing the enhanced inhibiting effect of SSWs. Furthermore, the slope of the unstable crack development stage for UHPC reinforced with 1.5 vol% SSWs is lower than that for UHPC reinforced with 1.0 vol% SSWs, resulting from the increase in the numbers of SSWs acting as a bridge. Compared

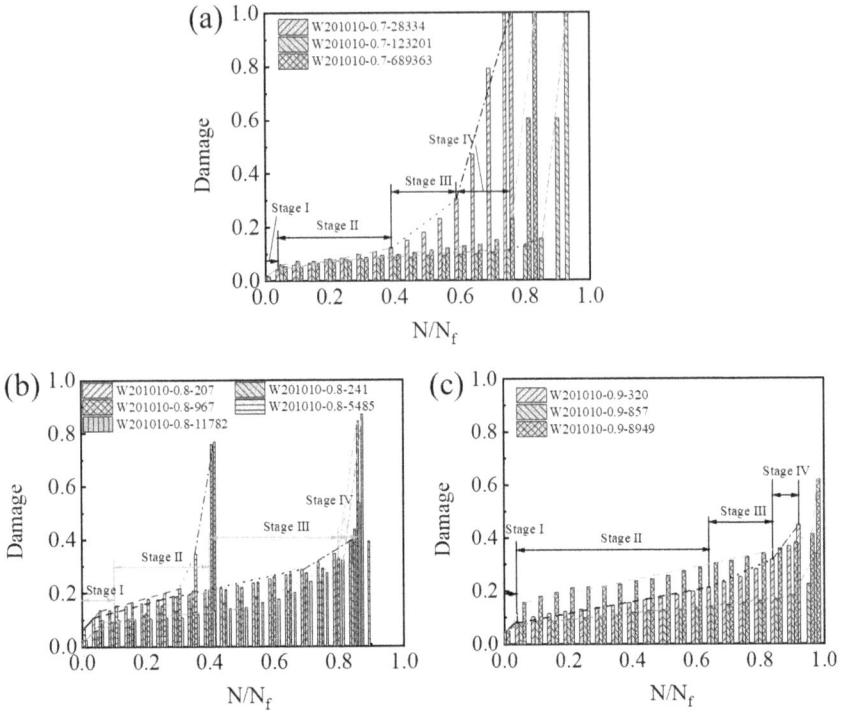

Figure 5.12 Flexural fatigue damage of UHPC reinforced with 1.0 vol% SSWs at maximum stress level of (a) 0.7; (b) 0.8; (c) 0.9

with UHPC reinforced with 0.5 vol%/1.0 vol% SSWs, the slope of the failure stage for UHPC reinforced with 1.5 vol% SSWs significantly decreases, indicating that the specimen has a tendency to multiple cracking. In addition, the damage value of the initiation stage of cracks for UHPC reinforced with 1.5 vol% SSWs is comparable to that of UHPC without SSWs, but higher than that of UHPC reinforced with 0.5 vol%/1.0 vol% SSWs, resulting from the increase of macrodefects caused by the decrease of the composites' workability. The failure damage of representative specimens at the maximum stress level of 0.9 is 0.96798, 0.5666, 0.61929, and 0.65965 for UHPC without SSWs and UHPC reinforced with 0.5 vol%/1.0 vol%/1.5 vol% SSWs, respectively. This means that the addition of 0.5 vol%, 1.0 vol%, and 1.5 vol% SSWs reduces the failure damage of UHPC by 41.5%, 36.0%, and 31.9%, respectively, a typical expression of the transformation from microscopic cracks to macroscopic cracks. When the maximum stress level is 0.8, the failure damage of representative specimens for UHPC without SSWs and UHPC reinforced with 0.5 vol%/1.0 vol%/1.5 vol% SSWs is 0.92076, 0.99993, 0.89545, and 0.70658, respectively, which is a decrease of −8.6%, 2.7%, and 23.3% due to the addition of SSWs. This result again

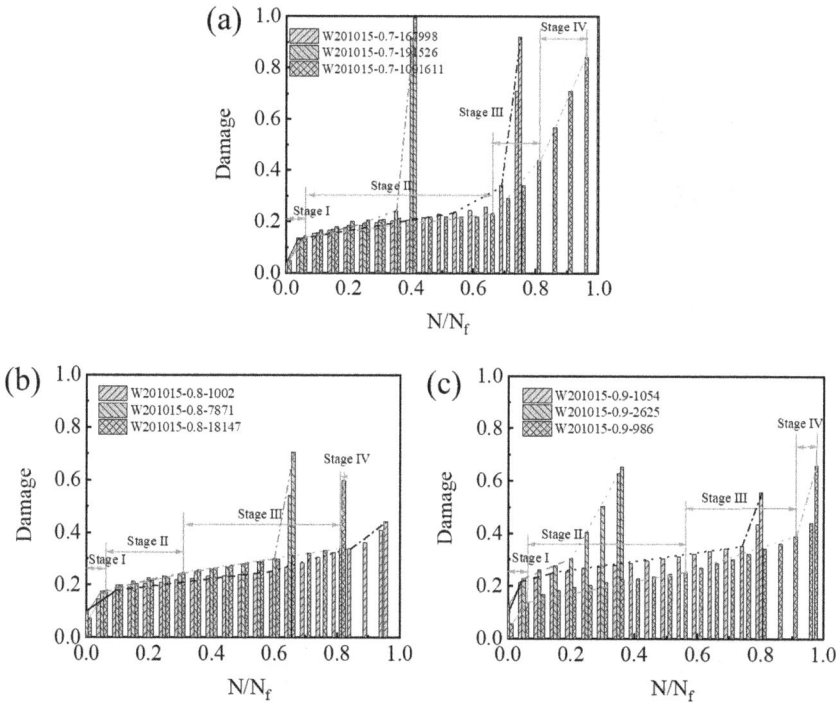

Figure 5.13 Flexural fatigue damage of UHPC reinforced with 1.5 vol% SSWs at maximum stress level of (a) 0.7; (b) 0.8; (c) 0.9

verifies that incorporating 0.5 vol% SSWs is more conducive to modifying the fatigue performance of UHPC at a high maximum stress level compared with that at a low maximum stress level, and is consistent with the conclusion that the flexural fatigue hysteresis stress–strain curves uniformly develop to the right.

5.2.3 Flexural failure state

Figures 5.14–5.17 show the failure surface for SSWs-engineered multifunctional UHPC after flexural fatigue load. Meanwhile, it can be seen from the appearance status of specimens that the addition of SSWs significantly reduces the number of air pores. It is observed during the test that there is no warning before the failure of specimens of UHPC without SSWs and UHPC reinforced with 0.5 vol% SSWs, while microcracks can be seen for specimens of UHPC reinforced with 1.0 vol%/1.5 vol% SSWs before failure. Meanwhile, the bending degree of main cracks increases with increasing SSWs content, resulting from the inhibiting effect of SSWs on the initiation and propagation of cracks. These effects also lead to main cracks gradually

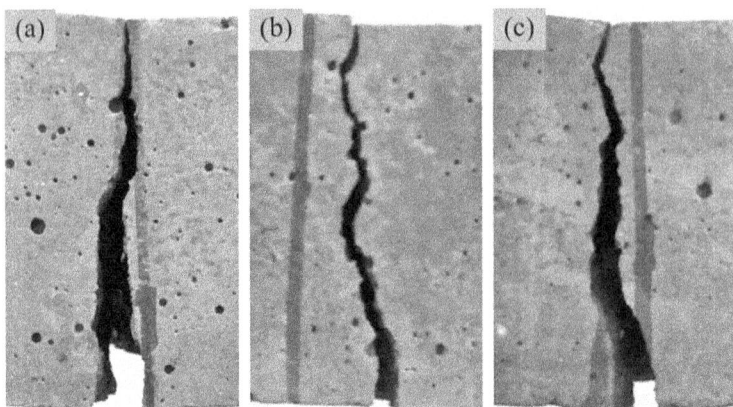

Figure 5.14 Flexural fatigue failure state of UHPC without SSWs. (a) S_{max}=0.7, N_f=125300; (b) S_{max}=0.8, N_f=3169; (c) S_{max}=0.9, N_f=325

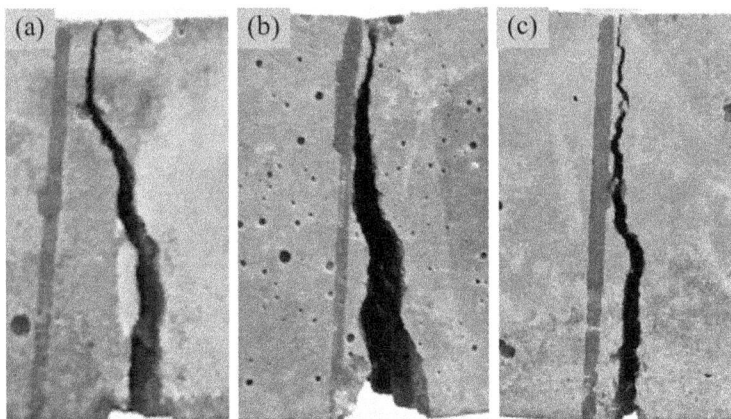

Figure 5.15 Flexural fatigue failure state of UHPC reinforced with 0.5 vol% SSWs. (a) S_{max}=0.7, N_f=12426; (b) S_{max}=0.8, Nf=940 (c) S_{max}=0.9, N_f=1591

tapering from the bottom to the top of specimens. Furthermore, SSWs bridging and being pulled off have been clearly observed at the location of main cracks for specimens of SSWs-engineered multufuntional UHPC. As displayed in Figures 5.16 and 5.17, the bridging and pulling off of SSWs are more significant at a high maximum stress level than at a low maximum stress level. This is consistent with the results showing that the improvement rate in fatigue life of composites caused by the incorporation of SSWs is more significant at a high maximum stress level.

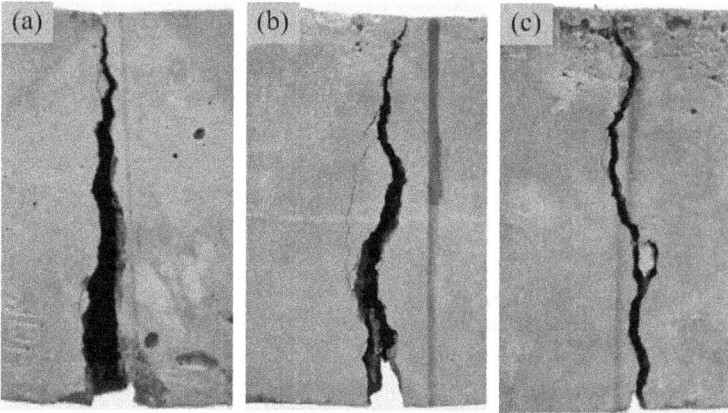

Figure 5.16 Flexural fatigue failure state of UHPC reinforced with 1.0 vol% SSWs. (a) S_{max}=0.7, N_f=123201; (b) S_{max}=0.8, N_f=967; (c) S_{max}=0.9, N_f=320

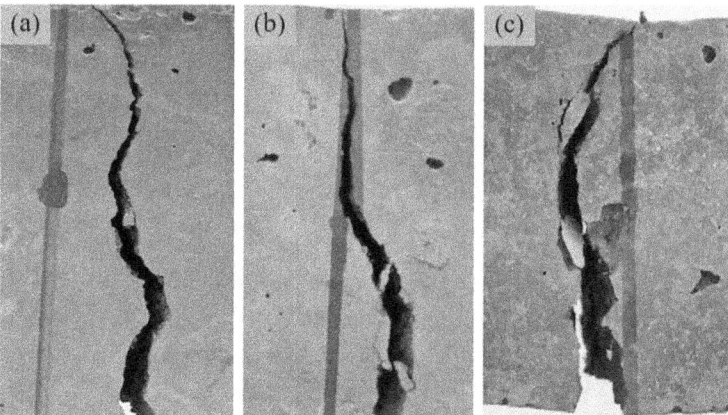

Figure 5.17 Flexural fatigue failure state of UHPC reinforced with 1.5 vol% SSWs. (a) S_{max}=0.7, N_f=191526; (b) S_{max}=0.8, N_f=1002; (c) S_{max}=0.9, N_f=2625

5.3 COMPRESSIVE FATIGUE PROPERTIES OF SSWS-ENGINEERED MULTIFUNCTIONAL UHPC

5.3.1 Compressive fatigue life and S-N curves

5.3.1.1 Compressive fatigue life

During the compressive fatigue test, the sinusoidal signal type of load with a frequency of 5 Hz was also adopted. It is worthy of note that a preload was applied to specimens of SSWs-engineered UHPC with sizes of 40 mm × 40 mm × 80 mm at a rate of 1.2 mm/min until the corresponding mean load P_{medium} was reached. The maximum load P_{max} was 75%, 80%, 85%, and 90% of the static uniaxial compressive peak load, i.e. the maximum stress level S_{max} was equal to 0.75, 0.80, 0.85 and 0.90, respectively. The minimum load P_{min} was set at 10% of the static uniaxial compressive peak load, i.e. the minimum stress level S_{min} was 0.1.

The fatigue life, i.e. number of fatigue cycles, of SSWs-engineered multifunctional UHPC at different maximum stress levels are listed in Table 5.7. It is worth noting that there were 12 specimens for each mix proportion, but the fatigue life of some of the specimens was not obtained due to operational error or instrument failure, especially at low maximum stress level, which did not affect the regularity of the test results. Table 5.7 shows that the fatigue life of composites decreases with the increase of maximum stress level, as expected. As the maximum stress level increases from 0.80 to 0.90, the average fatigue life of UHPC without SSWs and UHPC reinforced with 0.5 vol%/1.0 vol%/1.5 vol% SSWs decreases by 27.6 times, 64.2 times, 14.4 times, and 30.4 times, respectively. At the same maximum stress levels of 0.80, 0.85, and 0.90, the average fatigue life of UHPC reinforced with 0.5 vol% SSWs is 252.0%, 76.0%, and 54.6% higher, respectively, than that of UHPC without SSWs. That is to say, the incorporation of only 0.5 vol% SSWs can significantly improve the fatigue life of UHPC, confirming the effective distribution of SSWs in UHPC at a low dosage of 0.5 vol%. Meanwhile, the enhancement effect becomes more pronounced as the maximum stress level decreases. Overall, due to the micron diameter and high specific surface area of SSWs, the microstructure of UHPC with 0.5 vol% SSWs is enhanced, and the original flaws are reduced (as shown in Figure 5.18). This leads to a significantly improved fatigue life of the composite at a low maximum stress level. At a higher maximum stress level, the bridging effect of SSWs plays a more active role in enhancing the fatigue life of UHPC.

Table 5.7 also demonstrates that compared with UHPC without SSWs, the average compressive fatigue life of UHPC reinforced with 1.0 vol% SSWs is changed by −14.2%, −7.2%, 42.8%, and 71.9% at the maximum stress level of 0.75, 0.80, 0.85, and 0.90, respectively. It is also found that the average compressive fatigue life of UHPC reinforced with 1.5 vol% SSWs is reduced by 58.5%, 42.6%, 87.1%, and 47.7% at the maximum stress levels of 0.75, 0.80, 0.85, and 0.90, respectively. This means that when the dosage of SSWs is 1.0 vol%, it can improve the compressive fatigue life of UHPC at a high maximum stress level due to the bridging effect. However, the original flaws of hardened UHPC increase with increasing SSWs content, which can be attributed to the increase of superplasticizer content. This phenomenon can be observed through microstructural observation of SSWs-engineered UHPC, as shown in Figure 5.18. The conclusion is consistent with that obtained by Cachim et al. for steel fiber reinforced concrete [30].

Table 5.7 Uniaxial compressive fatigue life of SSWs-engineered multifunctional UHPC

SSWs content (vol%)	Specimen number	Stress level	Maximum stress (MPa)	Stress amplitude (MPa)	Fatigue life (number of load cycles)	Average fatigue life
0	1	0.75	81.6	35.4	1233400	1233400
	2	0.75	81.6	35.4	–	
	3	0.75	81.6	35.4	–	
	4	0.80	87.0	38.1	212331	152088
	5	0.80	87.0	38.1	139384	
	6	0.80	87.0	38.1	–	
	7	0.85	92.4	40.8	45593	27569
	8	0.85	92.4	40.8	30336	
	9	0.85	92.4	40.8	6778	
	10	0.90	97.5	43.3	9160	5318
	11	0.90	97.5	43.3	769	
	12	0.90	97.5	43.3	16	
0.5	1	0.75	93.8	40.7	>2000000	>2000000
	2	0.75	93.8	40.7	–	
	3	0.75	93.8	40.7	–	
	4	0.80	100.0	43.8	423204	535951
	5	0.80	100.0	43.8	811231	
	6	0.80	100.0	43.8	252000	
	7	0.85	106.3	46.9	85381	48515
	8	0.85	106.3	46.9	33147	
	9	0.85	106.3	46.9	27017	
	10	0.90	112.5	50	11109	8223
	11	0.90	112.5	50	10046	
	12	0.90	112.5	50	3518	
1.0	1	0.75	95.6	41.4	1650259	1058442
	2	0.75	95.6	41.4	1466625	
	3	0.75	95.6	41.4	–	
	4	0.80	102.0	44.6	292401	141022
	5	0.80	102.0	44.6	88863	
	6	0.80	102.0	44.6	41814	
	7	0.85	108.4	47.8	54886	39384
	8	0.85	108.4	47.8	49503	
	9	0.85	108.4	47.8	13764	
	10	0.90	114.8	51	10486	9142
	11	0.90	114.8	51	8852	
	12	0.90	114.8	51	8089	
1.5	1	0.75	108.8	47.2	864318	512078
	2	0.75	108.8	47.2	385605	
	3	0.75	108.8	47.2	–	
	4	0.75	108.8	47.2	286311	
	5	0.80	116.0	50.8	89263	87350
	6	0.80	116.0	50.8	85437	
	7	0.85	123.3	54.4	6886	3554
	8	0.85	123.3	54.4	2768	
	9	0.85	123.3	54.4	1008	
	10	0.90	130.5	58.0	4660	2779
	11	0.90	130.5	58.0	1948	
	12	0.90	130.5	58.0	1728	

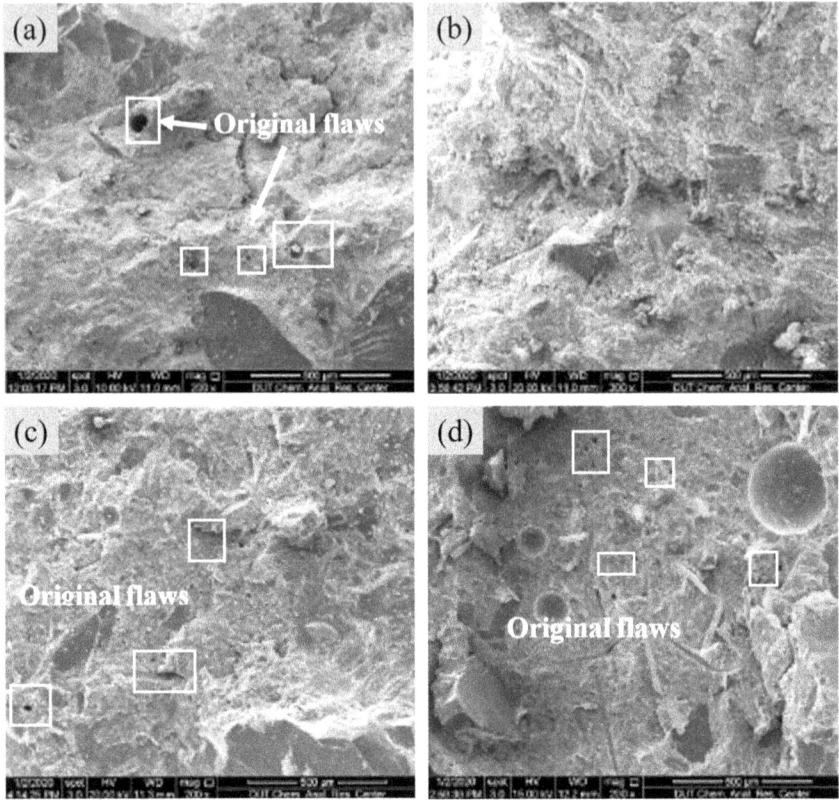

Figure 5.18 Original flaws of SSWs-engineered multifunctional UHPC specimens for compressive fatigue test. (a) UHPC without SSWs; (b) UHPC reinforced with 0.5 vol% SSWs; (c) UHPC reinforced with 1.0 vol% SSWs; (d) UHPC reinforced with 1.5 vol% SSWs

The difference between the maximum and the minimum fatigue life of SSWs-engineered multifunctional UHPC at the same maximum stress level is lower than that in previous published research [30, 31]. The enhancement effect on the compressive fatigue life of UHPC caused by the incorporation of 0.5 vol% SSWs is significantly higher than that caused by steel fiber at the stress level of 0.85 [30]. Meanwhile, at the same maximum stress level of 0.85 and 0.90, the fatigue life of UHPC reinforced with 0.5 vol%/1.0 vol% SSWs is considerably higher than that of PVA fiber reinforced concrete (with a content of 2 vol%) [31, 32].

5.3.1.2 S-N curves and P-S-N curves of compressive fatigue life

Because the maximum stress level S_{max} is the only parameter considered in the investigation of compressive fatigue, Eq. (5.4) was used to describe the relationship between maximum stress level and compressive fatigue life of SSWs-engineered multifunctional UHPC.

The S-N fatigue equations of SSWs-engineered multifunctional UHPC are demonstrated in Figure 5.19 and Table 5.8. The results show that the correlation coefficients between fatigue equations and experimental results for composites are larger than 0.92, indicating that the relationship between the logarithm of maximum stress level and the logarithm of fatigue life is linear for composites. When the content of SSWs is less than 1.0 vol%, the slope of the S-N fatigue equation first decreases and then increases. The decreasing slope represents increasing resistance to new microcrack initiation and propagation. Meanwhile, it is found that the intercept increases as the SSWs dosage increases from 0 to 1.0 vol%, manifesting enhancement of the bridging crack effect of SSWs. It is noteworthy that the slope of the fatigue equation of UHPC reinforced with 1.5 vol% SSWs is smaller than that of UHPC without SSWs due to large amounts of SSWs.

It can be seen from Table 5.8 that the compressive fatigue life of SSWs-engineered multifunctional UHPC has certain scatter characteristics. Meanwhile, as shown in Eq. (5.5), the compressive fatigue life of SSWs-engineered UHPC conforms to a two-parameter Weibull distribution with the location parameter considered as zero [24, 25, 33–35]. Taking the

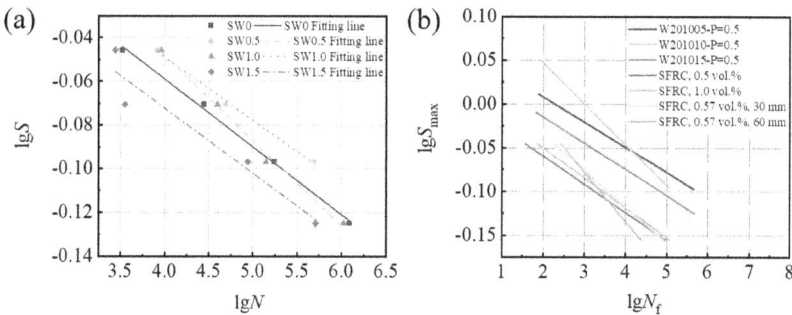

Figure 5.19 S-N and P-S-N curves for compressive fatigue life of SSWs-engineered multifunctional UHPC. (a) S-N curves; (b) P-S-N curves and references. SW0, SW0.5, SW1.0, and SW1.5 represent W0, W201005, W201010, and W201015, respectively

Table 5.8 S-N Fatigue equations for compressive fatigue life of SSWs-engineered multifunctional UHPC

Group	$\log_{10}C_1$	C_2	R^2
W0	0.06479	−0.03095	0.98745
W201005	0.06523	−0.02860	0.99915
W201010	0.10695	−0.03882	0.87819
W201015	0.04689	−0.02980	0.95018

natural logarithm of both sides of Eq. (5.5) twice, Eq. (5.24) is obtained as follows:

$$\ln\left[\ln\left(1/P\right)\right] = b\ln N_f - b\ln\alpha \qquad (5.24)$$

Then, the fatigue life of composites in Table 5.8 at a certain stress level is arranged in ascending order. The empirical survival probability of each fatigue life can be obtained by the following Eq. (5.25):

$$P = 1 - \frac{i}{n+i} \qquad (5.25)$$

where i is the sequence of fatigue life; n is the specimen numbers of composites at a specified stress level. Plotting $\ln N_f$ against $\ln\left[\ln\left(1/P\right)\right]$, the values of distribution parameters can be gained through linear fitting. The fitting results of scale parameter α and shape parameter b are listed in Table 5.9.

Table 5.9 shows that there is a good linear correlation between $\ln N_f$ and $\ln\left[\ln\left(1/P\right)\right]$, reflecting that the two-parameter Weibull distribution gives an accurate description of fatigue life of composites. Therefore, the fatigue failure probability P' of SSWs-engineered multifunctional UHPC can be expressed by the following Eq. (5.26):

$$P' = 1 - \exp\left[-\left(\frac{\bar{N}}{\alpha}\right)^b\right] \qquad (5.26)$$

Table 5.9 Distribution parameters of SSWs-engineered multifunctional UHPC at different stress levels

Group	Distribution parameters	Stress levels			
		0.75	0.80	0.85	0.90
W0	b	–	–	0.76056	0.24584
	α	–	–	37046	2761
	R^2	–	–	0.93883	0.99665
W201005	b	–	1.33336	1.16095	1.15097
	α	–	609934	66621	10675
	R^2	–	0.98228	0.83034	0.86497
W201010	b	–	0.78810	0.94667	5.80550
	α	–	177079	52590	9618
	R^2	–	0.96157	0.85665	0.94220
W201015	b	1.30492	–	0.81932	1.29029
	α	634638	–	4519	3490
	R^2	0.89596	–	0.99851	0.78595

where $\bar{N}$ is the equivalent fatigue life at failure probability of P'. Combining the scale and shape parameters at a specified maximum stress level listed in Table 5.9, the equivalent fatigue life of composites with a certain failure probability can be calculated by the following Eq. (5.27):

$$\bar{N} = \alpha \left| \ln\left(1 - P'\right) \right|^{1/b} \tag{5.27}$$

Assuming the failure probability of 50%, the corresponding P-S-N fatigue equations of composites are displayed in Table 5.10.

The correlation coefficients shown in Table 5.10 show that the P-S-N compressive fatigue equations of composites have higher accuracy than those obtained by Lee et al. and Cachim et al. [30, 36]. The slope of compressive fatigue equation of UHPC reinforced with 0.5 vol% SSWs is the smallest, indicating the highest compactness of the concrete matrix. Meanwhile, the intercept of the compressive fatigue equation of UHPC reinforced with 1.0 vol% SSWs is the largest, resulting from the bridging effect of SSWs. The P-S-N compressive fatigue equations are plotted in Figure 5.19(b). Substituting $N_f = 2 \times 10^6$ into the fatigue equations in Table 5.10, the fatigue limit strength can be calculated for SSWs-engineered UHPC and steel fiber reinforced concrete (SFRC). It can be seen from Table 5.10 that the compressive fatigue limit strengths of UHPC reinforced with 0.5 vol%/1.0 vol%/1.5 vol% SSWs under failure probability of 50% are 0.7660 f_c, 0.7008 f_c, and 0.7182 f_c, respectively, which are much higher than those for conventional concrete at the same failure probability (ranging from 57% to 67% of static strength) [37] and for SFRC [30]. The fatigue limit strength

Table 5.10 P-S-N equations for compressive fatigue life of SSWs-engineered multifunctional UHPC

| Group | P-S-N Fatigue equation | | | Fatigue limit strength |
	$\log_{10}C_1$	C_2	R^2	
W201005	0.06533	−0.02873	0.99820 (P′ = 0.5)	0.7660 f_c
W201010	0.14002	−0.04672	0.99485 (P′ = 0.5)	0.7008 f_c
W201015	0.04496	−0.02995	0.83037 (P′ = 0.5)	0.7182 f_c
Steel fiber reinforced concrete (SFRC, 0.5 vol%)	0.00497	−0.03232	0.6332 [36]	0.6328 f_c
SFRC, 1.0 vol%	0.01756	−0.03406	0.2371 [36]	0.6350 f_c
SFRC, 0.57 vol% and length of 30 mm	0.04798	−0.04139	0.781 [30]	0.6126 f_c
SFRC, 0.57 vol% and length of 60 mm	0.0918	−0.05660	0.572 [30]	0.6420 f_c

Note: f_c is the static uniaxial compressive strength

of UHPC reinforced with 1.0 vol%/1.5 vol% SSWs of about 0.7 f_c confirms the reliability of the results. In addition, the fatigue limit strength of SSWs-engineered UHPC is larger than the specified value in the standard of ACI 318-08.

5.3.2 Compressive fatigue strain and energy dissipation capacity

5.3.2.1 Maximum fatigue strain

It is worth noting that during the experimental process, not all the strains for the whole test procedure can be measured, as some strain gauges were damaged by crack propagation. Meanwhile, when the maximum stress level is 0.75, the acquired strains are not sufficient to support analysis. Hence, the following analysis is conducted on specimens at the maximum stress levels of 0.80, 0.85, and 0.90. The maximum fatigue strain–time curves of SSWs-engineered multifunctional UHPC are drawn based on the time at 5%, 10%, 20%, 40%, 60%, 80%, and 100% of fatigue times before the failure cycle, as shown in Figsures 5.20–5.23. Figure 5.20 shows that three stages

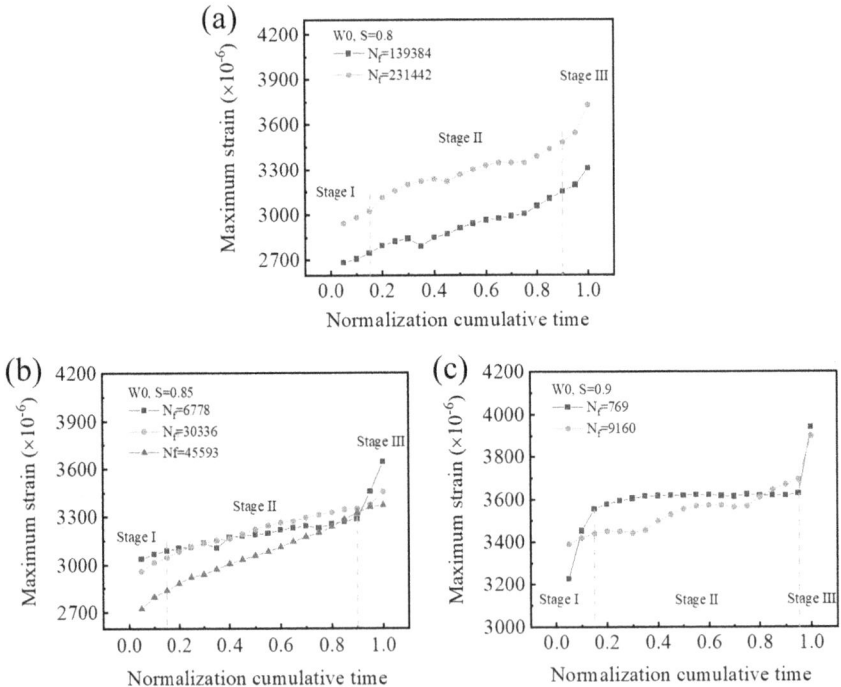

Figure 5.20 Maximum compressive fatigue strain–time curves of UHPC without SSWs. (a) S_{max}=0.8; (b) S_{max}=0.85; (c) S_{max}=0.9

of maximum compressive fatigue strain–time curves for UHPC without SSWs can be observed, similar to the results obtained in previous research [31, 38]. Stage I, accounting for 15% of the fatigue life, represents the cyclic creep and crack latent stage, and it is closely related to the compactness of the microstructure. Stage II is the crack initiation and progressive propagation stage, and it accounts for about 80% of the fatigue life. Stage III, about 5% of the fatigue life, is the instability failure stage. It can be clearly seen that the strain growth rate in stage II increases with the decrease of maximum stress level. This illustrates that the damage rapidly develops at a low maximum stress level but is not extensive before the instantaneous failure of specimens of UHPC without SSWs at a high maximum stress level. This is the typical manifestation of brittle materials.

Figures 5.21–5.23 show that the maximum compressive fatigue strain–time curves of SSWs-engineered multifunctional UHPC can be divided into four stages: the fatigue creep and crack latent stage, the microcrack initiation stage, the crack stable propagation stage, and the instability failure stage. When the maximum stress level is 0.90, the average maximum strain of UHPC at 100% cycles of fatigue life is improved by 12.0%,

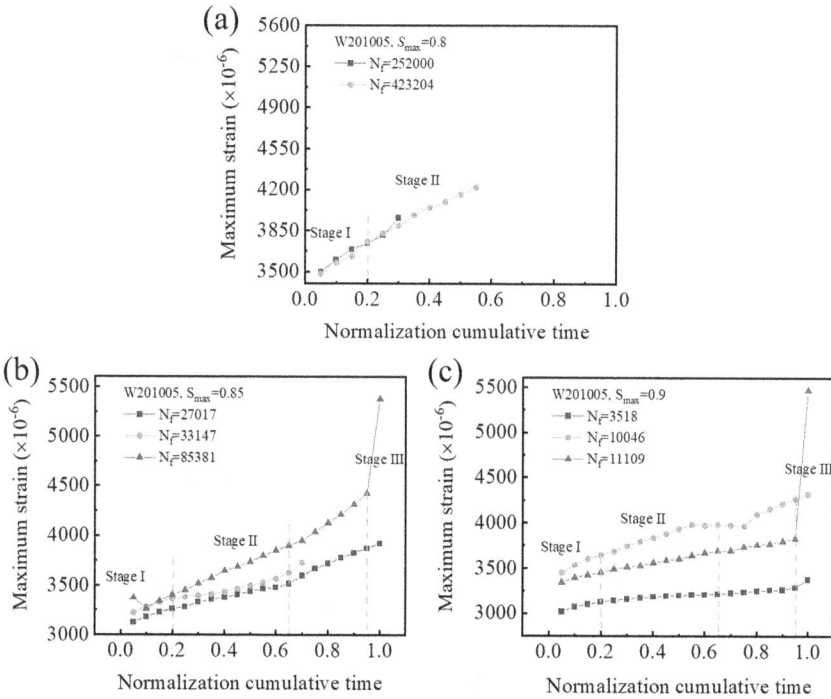

Figure 5.21 Maximum compressive fatigue strain–time curves of UHPC reinforced with 0.5 vol% SSWs. (a) S_{max}=0.8; (b) S_{max}=0.85; (c) S_{max}=0.9

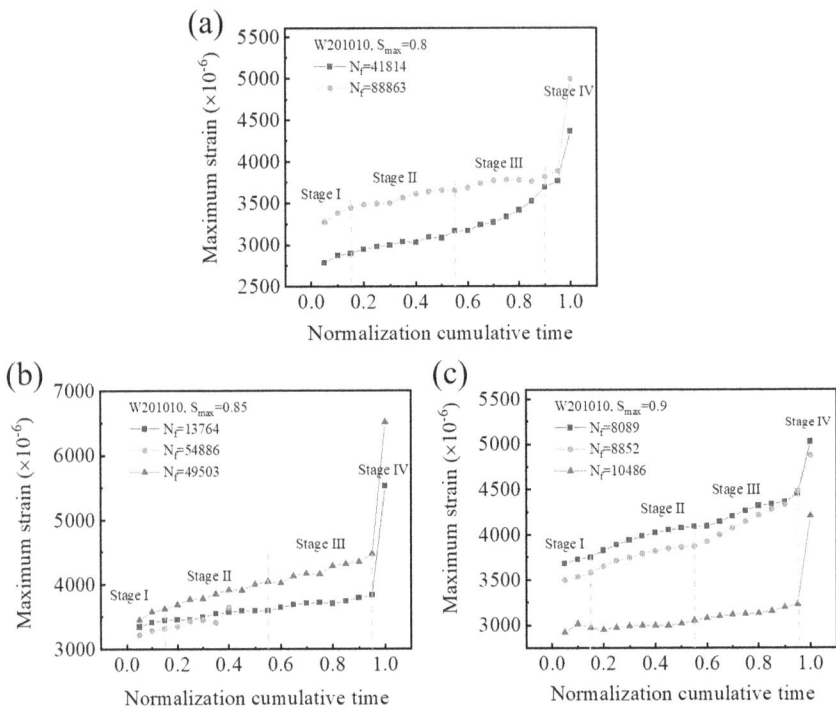

Figure 5.22 Maximum compressive fatigue strain–time curves of UHPC reinforced with 1.0 vol% SSWs. (a) S_{max}=0.8; (b) S_{max}=0.85; (c) S_{max}=0.9

20.0%, and 48.8%, respectively, due to the addition of 0.5 vol%, 1.0 vol%, and 1.5 vol% SSWs. The corresponding increments are 54.1%, 72.4%, and 73.7% for the maximum stress level of 0.85. The creep and crack latent stage, i.e. stage I, accounts for 20% of the fatigue life for UHPC reinforced with 0.5 vol% SSWs, resulting from the improvement of microstructural compactness. For UHPC reinforced with 1.0 vol%/1.5 vol% SSWs, the percentage of stage I is the same as that of UHPC without SSWs due to the increase of original flaws. Due to the widely distributed reinforcing network of SSWs, the initiation of cracks is limited. SSWs can transfer crack tip stress and cause stress to be redistributed. Hence, stage II accounts for 40–45% of compressive fatigue life for UHPC incorporating 0.5 vol%/1.0 vol%/1.5 vol% SSWs. In stage III, the development of cracks is inhibited by the bridging effect of SSWs. This stage comprises about 35% of the compressive fatigue life of composites. The fatigue failure stage, i.e. stage IV, accounts for 5% of the compressive fatigue life for UHPC reinforced with 0.5 vol%/1.0 vol% SSWs and 10% for UHPC reinforced with 1.5 vol% SSWs.

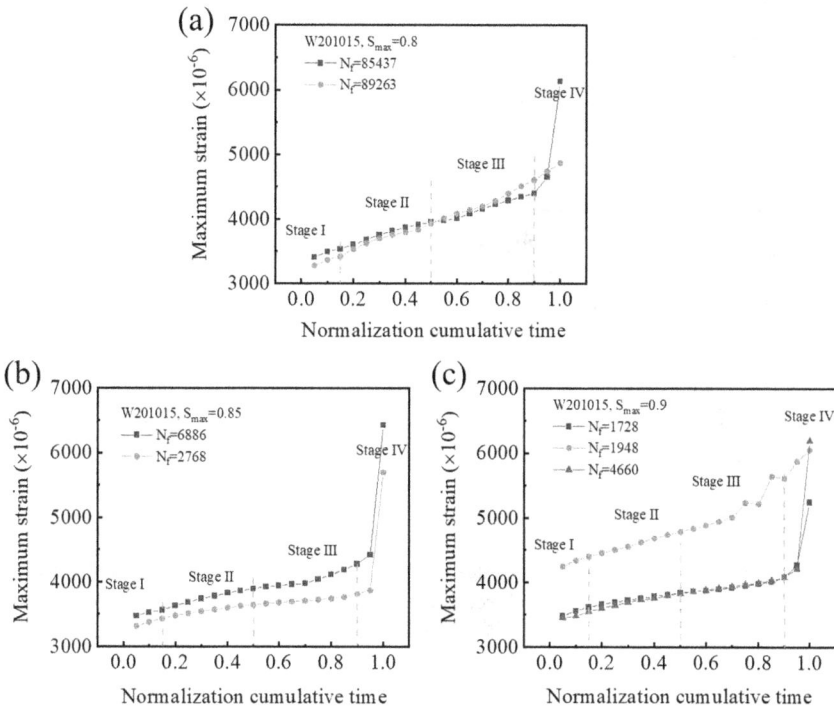

Figure 5.23 Maximum compressive fatigue strain–time curves of UHPC reinforced with 1.5 vol% SSWs. (a) S_{max}=0.8; (b) S_{max}=0.85; (c) S_{max}=0.9

5.3.2.2 Fatigue strain–stress curves and energy dissipation capacity

The representative strain–stress curves of UHPC without SSWs and UHPC reinforced with 0.5 vol% SSWs for the whole compressive fatigue test process are displayed in Figure 5.24. It can be seen that the compressive fatigue strain–stress curve of UHPC without SSWs presents a change rule from sparse to intensive with the increase of fatigue cycles, while that of UHPC reinforced with 0.5 vol% SSWs has characteristics from sparse to intensive and then to sparse again. SSWs in UHPC can inhibit the initiation and propagation of cracks and can then bridge cracks until they are pulled off. This effect finally leads to increased damage of composites, especially when the specimens are close to failure, and this is also the reason why the second sparse stage appears on the strain–stress curve of UHPC reinforced with 0.5 vol% SSWs. When SSWs are incorporated into UHPC at a volume fraction of 1.0 vol% and 1.5 vol%, the strain–stress curves also present three stage characteristics, from sparse to intensive and then to sparse.

Experimental results show that the compressive fatigue strain–stress hysteresis curves of composites are mainly influenced by the incorporation of SSWs and the maximum stress level. Therefore, the strain–stress hysteresis curves of the representative specimens are plotted on the basis of 10%, 20%, 40%, 60%, 80%, and 100% cycles of compressive fatigue life before failure, as shown in Figures 5.25 to 5.28. Similarly to Figure 5.25, the compressive fatigue strain–stress hysteresis curves of UHPC without

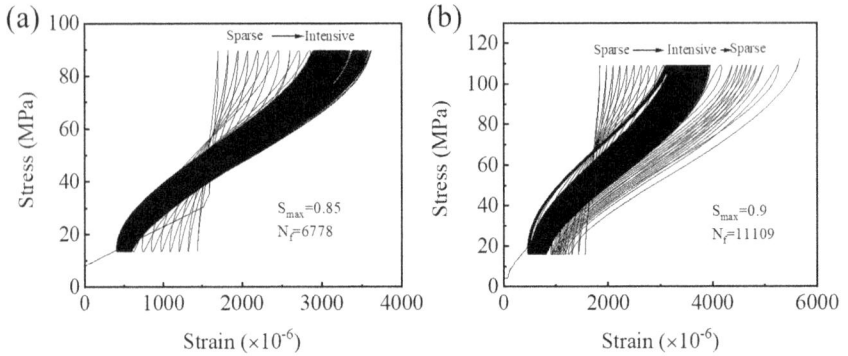

Figure 5.24 Representative compressive fatigue stress–strain curves of UHPC without SSWs and UHPC reinforced with 0.5 vol% SSWs. (a) UHPC without SSWs (b) UHPC reinforced with 0.5 vol% SSWs

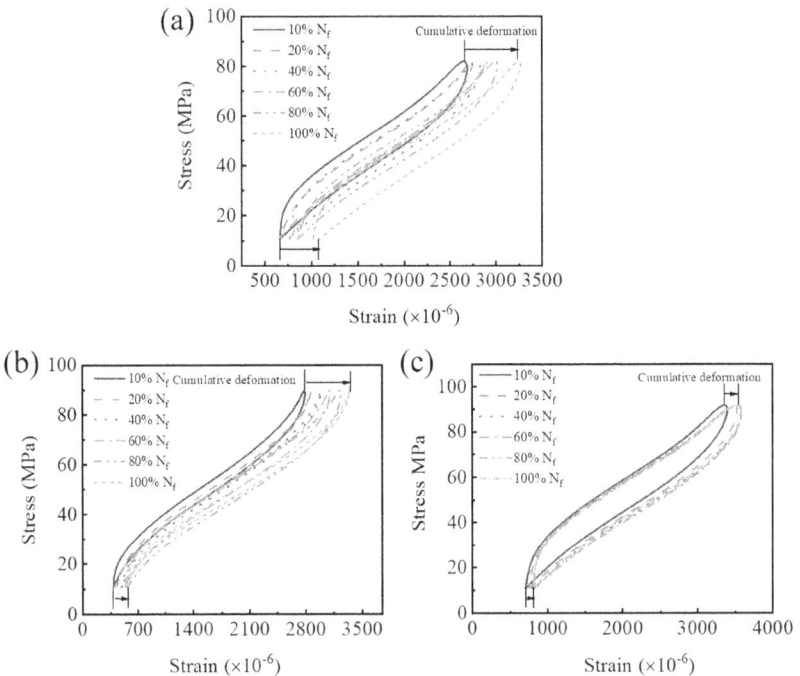

Figure 5.25 Representative compressive strain–stress hysteretic curves of UHPC withoutSSWs. (a) S_{max}=0.8,N_f=139384; (b) S_{max}=0.85,N_f=45593; (c) S_{max}=0.9,N_f=769

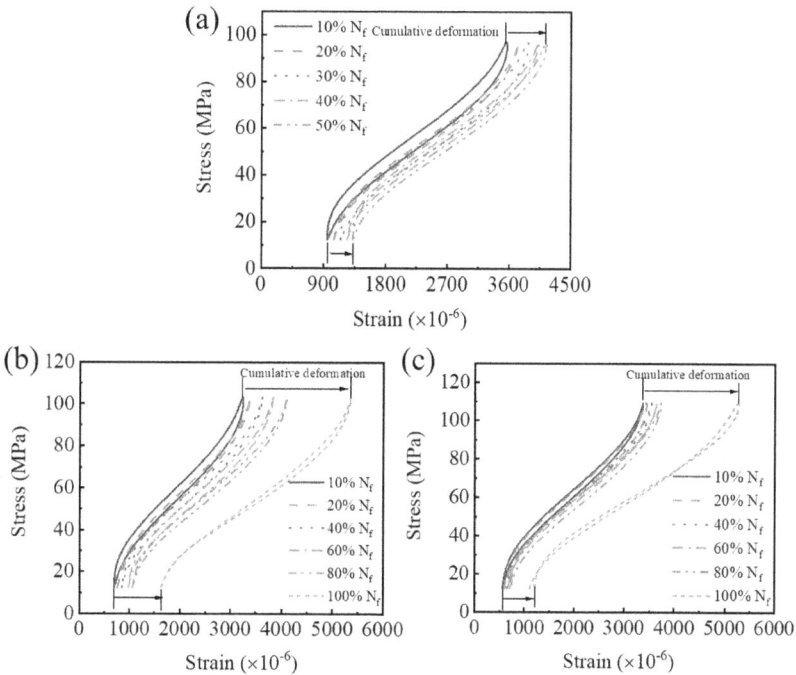

Figure 5.26 Representative compressive fatigue strain–stress hysteretic curves of UHPC reinforced with 0.5 vol% SSWs. (a) S_{max}=0.8, N_f=423204 (strain is not fully collected); (b) S_{max}=0.85, N_f=85381; (c) S_{max}=0.9, N_f=11109

SSWs show a dense state, while those for UHPC with SSWs have the change law of dense–sparse as the cycles increase. The results also show that the strains of composites corresponding to the maximum stress level (S_{max}) and the minimum stress level (S_{min}) increase with the number of cycles, in good agreement with results obtained for conventional concrete [39]. Figure 5.25 shows that with the increase in fatigue cycles, the compressive fatigue strain–stress hysteresis curves of UHPC without SSWs gradually move to the right and slope toward the x-axis, indicating the accumulation of deformation and damage. It also shows that the deformation and damage accumulation of UHPC without SSWs increase with decreasing maximum stress level. Especially at the maximum stress level of 0.9, the specimen possesses minimal accumulation damage as the failure occurs.

Figures 5.26 to 5.28 show that compared with UHPC without SSWs, the compressive fatigue strain–stress hysteresis curves of UHPC reinforced with 0.5 vol%/1.0 vol%/1.5 vol% SSWs have a greater magnitude of movement to the right at the same maximum stress level, reflecting the increase of energy dissipation capacity. Meanwhile, it is found that the loading stage of compressive fatigue strain–stress hysteresis curves for UHPC without SSWs and UHPC reinforced with 0.5 vol%/1.0 vol%/1.5 vol% SSWs all display an inverse S-shape. The unloading stage of strain–stress hysteresis

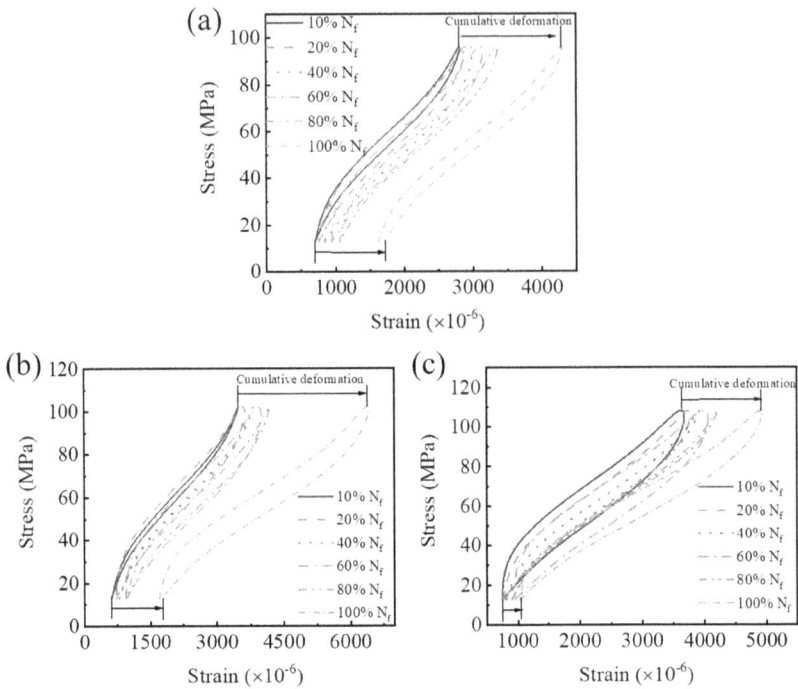

Figure 5.27 Representative compressive fatigue strain–stress hysteretic curves of UHPC reinforced with 1.0 vol% SSWs. (a) S_{max}=0.8, N_f=41814; (b) S_{max}=0.85, N_f=49503; (c) S_{max}=0.9, N_f=8089

curves for UHPC reinforced with 0.5 vol% SSWs also appears as an inverse S-shape, while that for UHPC without SSWs and UHPC reinforced with 1.0 vol%/1.5 vol% SSWs is closer to a straight line, illustrating that UHPC reinforced with 0.5 vol% SSWs has a highly compact structure. The distance between the compressive fatigue strain–stress hysteresis curve at 80% cycles of fatigue life and that at 100% cycles of fatigue life shows a tendency to increase with increasing SSWs dosage, showing the combination effect of SSWs bridging and being pulled off. It is worth noting that the area enclosed by the reloading and unloading curves for each strain–stress cycle is significantly affected by original flaws, and it can also be used to qualitatively describe the structure compactness of composites. Comparative analysis shows that the enclosed area of compressive fatigue strain–stress curves of each cycle for UHPC reinforced with 0.5 vol% SSWs in Figure 5.26 is markedly smaller than for other composites, again indicating a highly compact structure. The enclosed area of envelope curves from 10% to 100% cycles of fatigue life is used to characterize the energy dissipation capacity of composites in this chapter, as shown in Figure 5.29. Due to strain gauge damage, the energy dissipation value of UHPC reinforced with 0.5 vol% SSWs at the maximum stress level of 0.8 cannot be measured completely. Incorporating 1.0 vol% and 1.5 vol% SSWs enables the energy

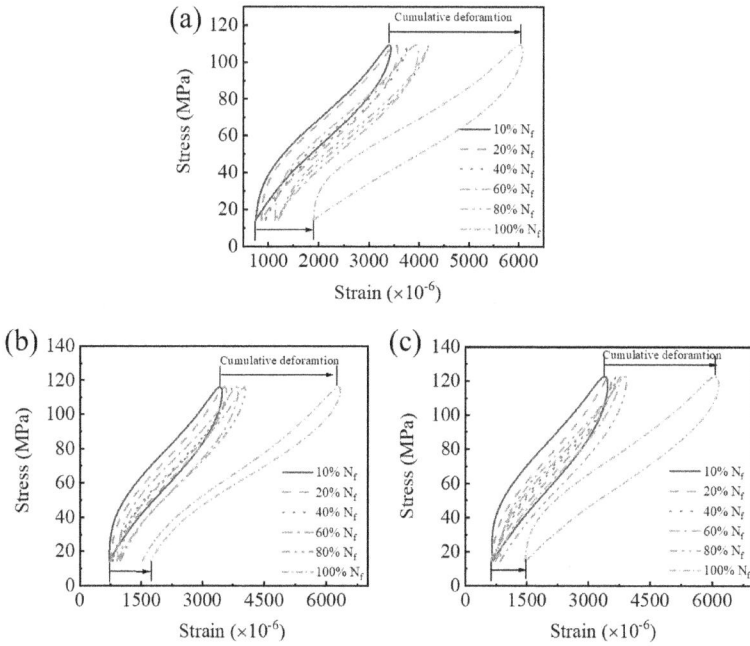

Figure 5.28 Representative compressive fatigue strain–stress hysteretic curves of UHPC reinforced with 1.5 vol% SSWs. (a) S_{max}=0.8, N_f =85437; (b) S_{max}=0.85, N_f =6886; (c) S_{max}=0.9, N_f=4660

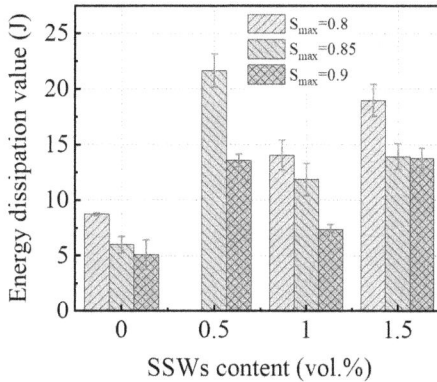

Figure 5.29 Compressive fatigue energy dissipation value of SSWs-engineered multifunctional UHPC at different maximum stress levels

dissipation value of UHPC at the maximum stress level of 0.8 to increase by 61.5% and 117.9%, respectively. It can also be found that at the maximum stress level of 0.85 and 0.90, UHPC reinforced with 0.5 vol% SSWs displays the largest energy dissipation capacity. At the maximum stress level of 0.85, the energy dissipation value of UHPC reinforced with 0.5 vol%/1.0

vol%/1.5 vol% SSWs is 262.3%, 98.5%, and 133.1% higher, respectively, than that of UHPC without SSWs. Meanwhile, the energy dissipation value of SSWs-engineered multifunctional UHPC is increased by 167.8%, 45.3%, and 171.5% compared with that of UHPC without SSWs at the maximum stress level of 0.90. The high energy dissipation value of UHPC reinforced with 0.5 vol% SSWs is mainly attributed to the highly compact structure, and the value of UHPC reinforced with 1.5 vol% SSWs can be attributed to the effect of SSWs bridging and being pulled off.

5.3.3 Compressive fatigue deformation and damage

5.3.3.1 Residual strain

The cumulative development of residual strain under compressive fatigue load reflects the cyclic creep strain and irreversible fatigue deformation caused by the appearance of fatigue cracks simultaneously. Based on the 10%, 20%, 40%, 60%, 80%, and 100% cycles of fatigue life, the residual strain of composites is plotted in Figures 5.30 to 5.33. Figure 5.30 shows that the compressive fatigue residual strains of UHPC without SSWs are divided into three stages with increasing fatigue cycles at different maximum stress

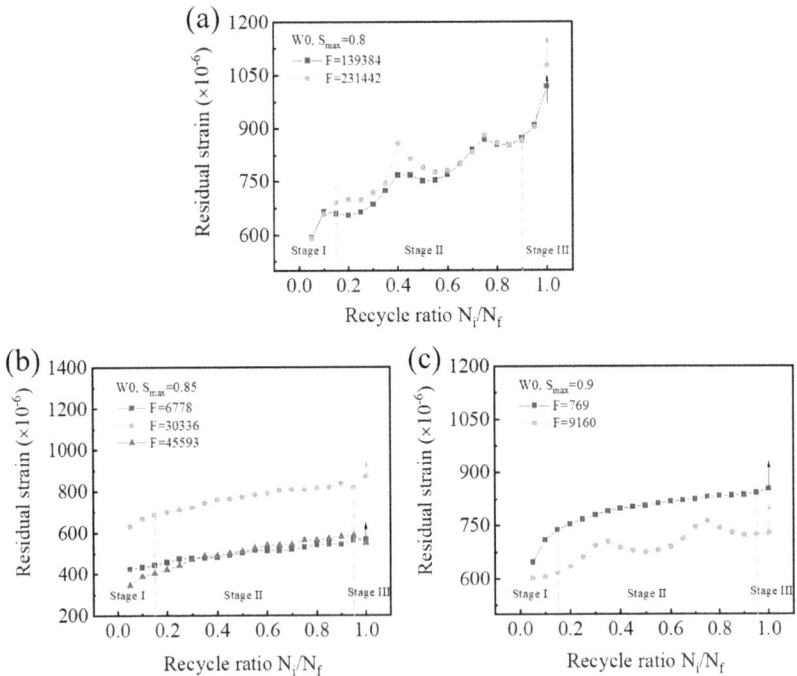

Figure 5.30 Compressive fatigue residual strain of UHPC without SSWs. (a) S_{max}=0.8; (b) S_{max}=0.85; (c) S_{max}=0.9

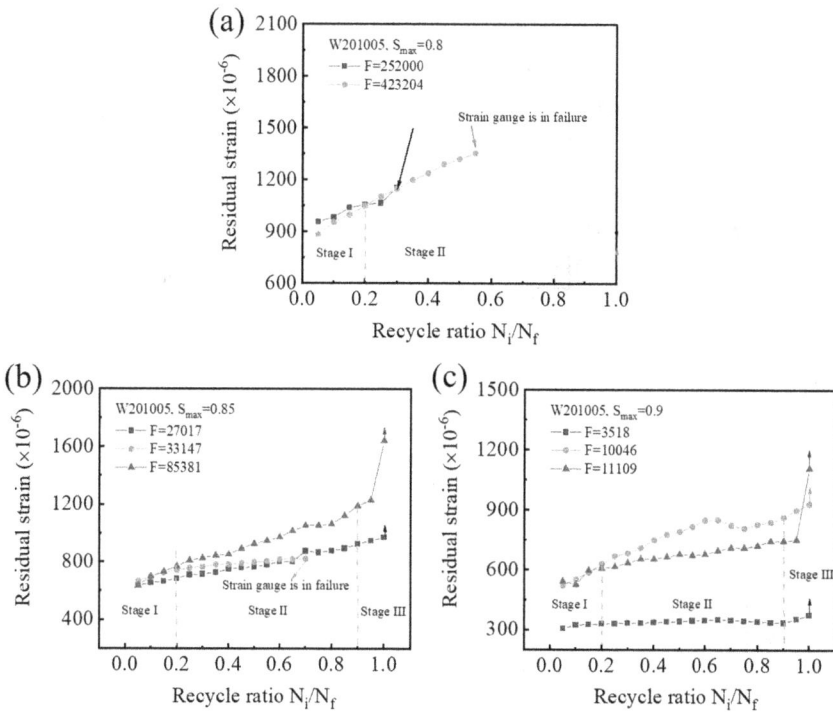

Figure 5.31 Compressive fatigue residual strain of UHPC reinforced with 0.5 vol% SSWs. (a) S_{max}=0.8; (b) S_{max}=0.85; (c) S_{max}=0.9

levels. The compressive fatigue residual strain of UHPC without SSWs in stage I initially develops rapidly, and then, the deformation accumulation rate decreases gradually, mainly resulting from the creep deformation caused by original flaws. This stage accounts for about 15% of the fatigue life. Stage II, accounting for about 80% of the fatigue life, is the stable development stage of compressive fatigue cumulative deformation, and the deformation accumulation rate shows an increasing trend with decreasing maximum stress level. When the mechanical properties of composites begin to decay, the corresponding cycle ratio is defined as the macroscopic fatigue damage threshold. Hence, it can be concluded that the macroscopic compressive fatigue damage threshold for UHPC without SSWs is 15% cycle ratio in this chapter. The compressive fatigue residual strain of UHPC without SSWs rapidly increases in stage III due to the unstable propagation and convergence of fatigue cracks. Stage III comprises only about 5% of the fatigue life for UHPC without SSWs.

As shown in Figures 5.31 to 5.33, four stages are observed in the compressive fatigue residual strain curves of SSWs-engineered multifunctional UHPC, which is different from that of UHPC without SSWs. The

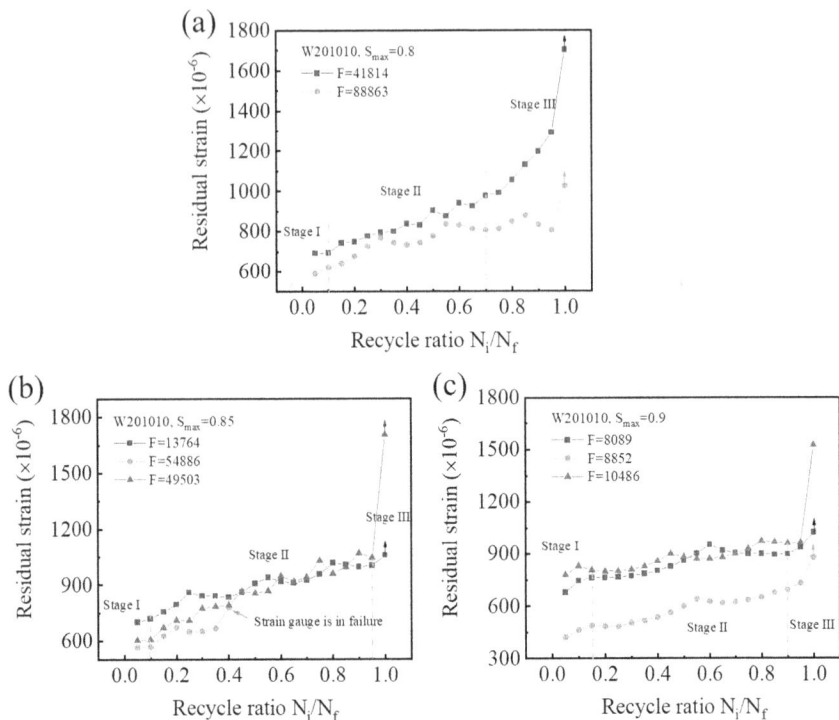

Figure 5.32 Compressive fatigue residual strain of UHPC reinforced with 1.0 vol% SSWs. (a) S_{max}=0.8; (b) S_{max}=0.85; (c) S_{max}=0.9

cumulative damage in stage I first develops rapidly and then slows down due to cyclic creep. Stage I accounts for 20% of the compressive fatigue life for UHPC reinforced with 0.5 vol% SSWs, while it accounts for 15% of the compressive fatigue life of UHPC reinforced with 1.0 vol%/1.5 vol% SSWs. This is due to the highly compact structure of UHPC reinforced with 0.5 vol% SSWs. Stage II and stage III on the compressive fatigue residual strain curves of composites are the crack initiation and stable developing stages, respectively. It is found that the rate of increase of compressive fatigue residual strain of stage III is higher than that of stage II. The turning point of residual strain between stage II and stage III happens at about the cycle ratio of 60%. In stage II, SSWs can effectively inhibit the initiation of microcracks, leading to a small rate of increase of compressive fatigue residual strain. In stage III, cracks enter the propagation stage, and the corresponding residual strain growth rate is accelerated. SSWs can hinder the opening of cracks through their bridging effect. The compressive fatigue residual strain presents a sharply increasing trend in stage IV, resulting from the convergence of major cracks. In this process, SSWs are gradually pulled

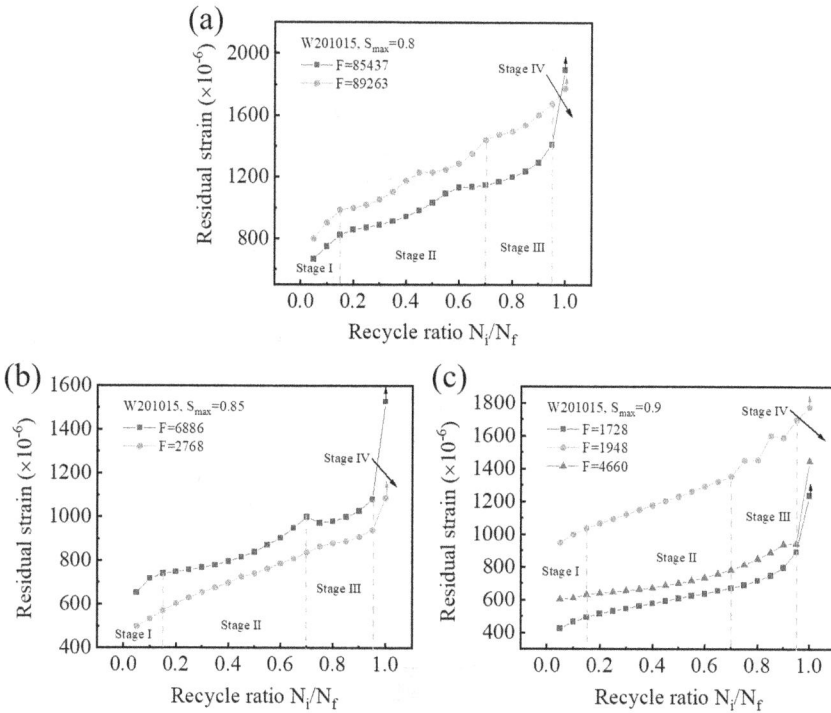

Figure 5.33 Compressive fatigue residual strain of UHPC reinforced with 1.5 vol% SSWs. (a) S_{max}=0.8; (b) S_{max}=0.85; (c) S_{max}=0.9

off, such that this stage for UHPC reinforced with 1.5 vol% SSWs accounts for 10% of the compressive fatigue life, while for UHPC reinforced with 0.5 vol%/1.0 vol% SSWs, it comprises about 5% of the compressive fatigue life. Figures 5.30 to 5.33 also show that at 100% cycles of compressive fatigue life before failure, the addition of SSWs markedly increases the residual strain of UHPC under different maximum stress levels. The increase of compressive fatigue residual strain also characterizes the enhancement of the energy dissipation capacity of composites. At the maximum stress level of 0.85 and 0.90, the average compressive fatigue residual strain corresponding to 80% cycles of compressive fatigue life for UHPC reinforced with 0.5 vol% SSWs is 28.7% and 87.1% higher, respectively, than that of UHPC without SSWs. At 100% cycles of compressive fatigue life, the average residual strain of UHPC reinforced with 0.5 vol%/1.0 vol%/1.5 vol% SSWs is 28.7%, 20.2%, and 69.5%, respectively, higher than that of UHPC without SSWs at the maximum stress level of 0.90. Meanwhile, the corresponding increments are 87.1%, 51.4%, and 87.2%, respectively, at the maximum stress level of 0.85.

5.3.3.2 Strain amplitude

The relationships between compressive fatigue strain amplitude and normalized fatigue lives of SSWs-engineered multifunctional UHPC at different stress levels are displayed in Figures 5.34 to 5.37. The increase rate of strain amplitude can be used to express the propagation of cracks and damage evolution. Figure 5.34 shows that the strain amplitude curves of UHPC without SSWs at the maximum stress levels of 0.85 and 0.90 all include two stages. The strain amplitude in stage I shows little change with the increase of fatigue cycles, and this stage is about 95% of the fatigue life. Stage II is only about 5% of the fatigue lives. As the maximum stress level is 0.8, the compressive fatigue strain amplitude of UHPC without SSWs increases linearly with the increase of fatigue cycles. After that, the specimens of UHPC without SSWs suddenly fail, indicating that UHPC without SSWs shows typical brittleness characteristics whatever the maximum stress level.

Unlike UHPC without SSWs, the compressive fatigue strain amplitude curves of SSWs-engineered multifunctional UHPC can be divided into three stages: stage I, stage II, and stage III, accounting for 60%, 30–35%, and 5–10% cycles of fatigue life, respectively. At stage I, widely distributed

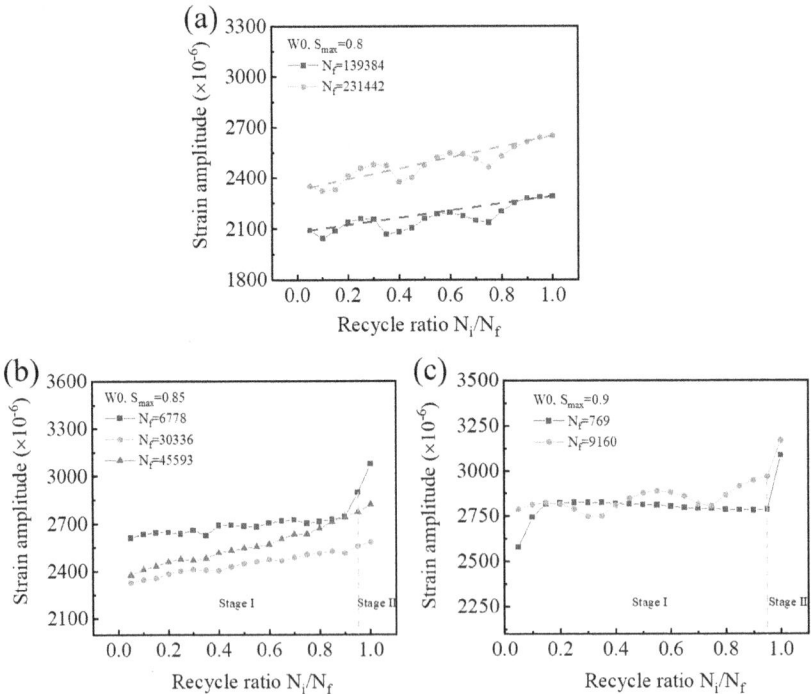

Figure 5.34 Compressive fatigue strain amplitude of UHPC without SSWs. (a) S_{max}=0.8; (b) S_{max}=0.85; (c) S_{max}=0.9

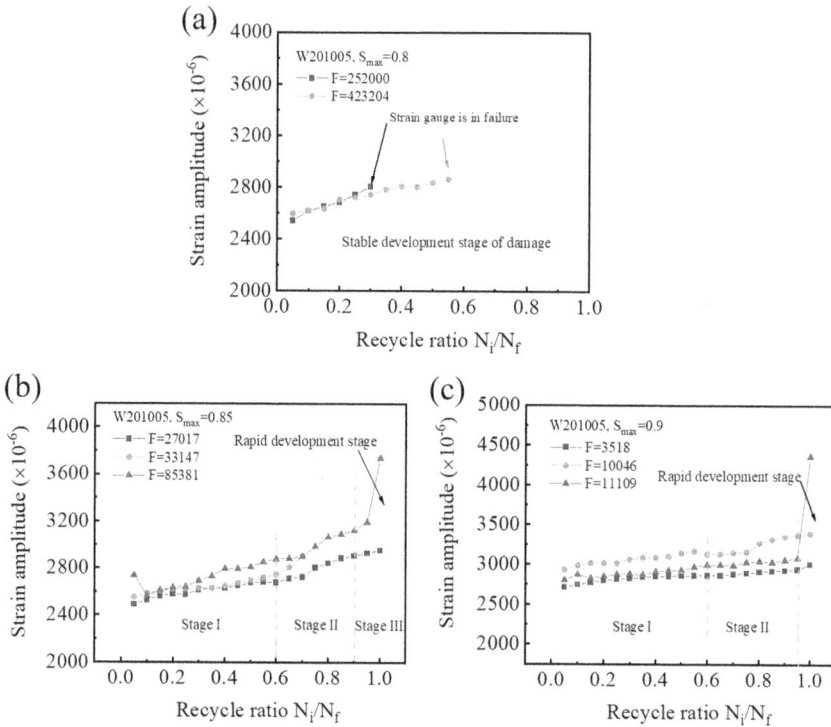

Figure 5.35 Compressive fatigue strain amplitude of UHPC reinforced with 0.5 vol% SSWs. (a) S_{max}=0.8; (b) S_{max}=0.85; (c) S_{max}=0.9

SSWs can effectively inhibit the generation of cracks, and the strain amplitude slightly changes with the increase of compressive fatigue cycles. Microcracks enter into stable propagation in stage II. SSWs can hinder the propagation of cracks through bridging, deflection, and the debonding effect [40]. It can be observed that the increasing rate of strain amplitude in stage II is larger than that in stage I. Stage III is the failure stage, accompanied by a sufficient number of unstable cracks. The strain amplitude varies at an increasing speed, and SSWs are gradually pulled off until failure occurs.

5.3.3.3 Fatigue modulus and damage evolution

Compressive fatigue modulus is defined as the ratio of stress to strain amplitudes, reflecting the change of elastic modulus with fatigue cycles. In this chapter, the compressive fatigue modulus is normalized with the initial static elastic modulus obtained in Figure 2.12. The relationships between normalized compressive fatigue modulus and normalized fatigue cycles of

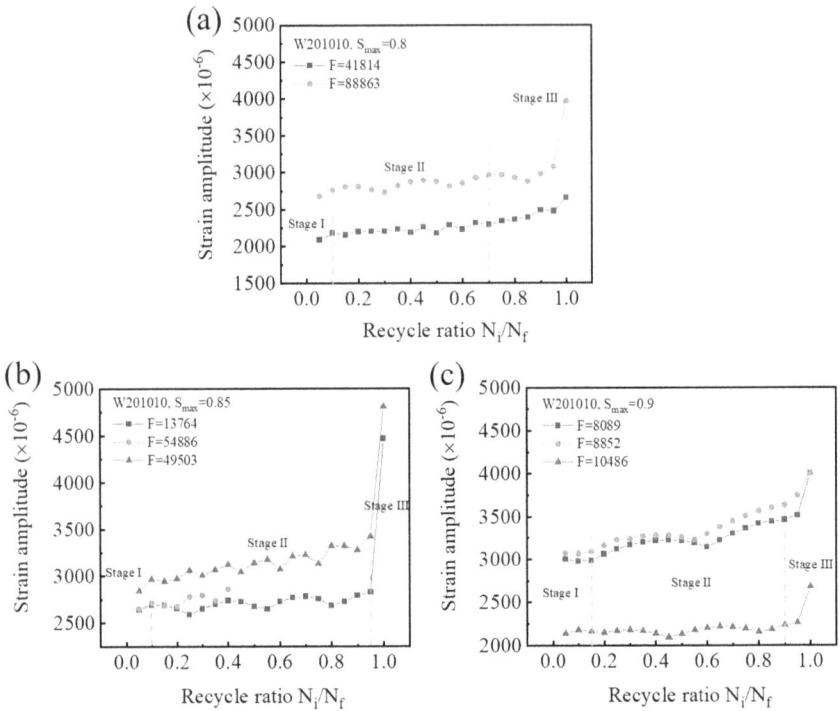

Figure 5.36 Compressive fatigue strain amplitude of UHPC reinforced with 1.0 vol%
SSWs .(a) S_{max}=0.8; (b) S_{max}=0.85; (c) S_{max}=0.9

composites can be observed in Figures 5.38 to 5.41. Because the stress range
is constant at the same maximum stress level, the normalized fatigue modulus is directly related to the deformation (i.e. strain amplitude). Hence, the
same two and three stages can be found in normalized compressive fatigue
modulus curves for UHPC without SSWs and UHPC reinforced with SSWs,
respectively, as shown in Figures 5.38(a) to 5.41(a). It can be observed that
the compressive fatigue modulus displays a decreasing trend with the number of cycles, resulting from the increase of deformation. Figure 5.38(a)
shows that the compressive fatigue modulus of UHPC without SSWs is less
affected by the maximum stress level. It first presents a slow linear decrease
with fatigue cycles and then sharply falls after 95% of the fatigue life. Up to
failure, the compressive fatigue modulus of UHPC without SSWs is about
83.7% of the initial fatigue modulus for the 0.85 stress level or 87.4% for
the 0.90 stress level. Figures 5.39(a) to 5.41(a) demonstrate that for SSWs-
engineered multifunctional UHPC, the decrease rate of compressive fatigue
modulus in stage II is larger than that in stage I. This is because the compressive fatigue modulus change in stage I is mainly caused by fatigue creep and
crack initiation, and that in stage II results from stable crack propagation.

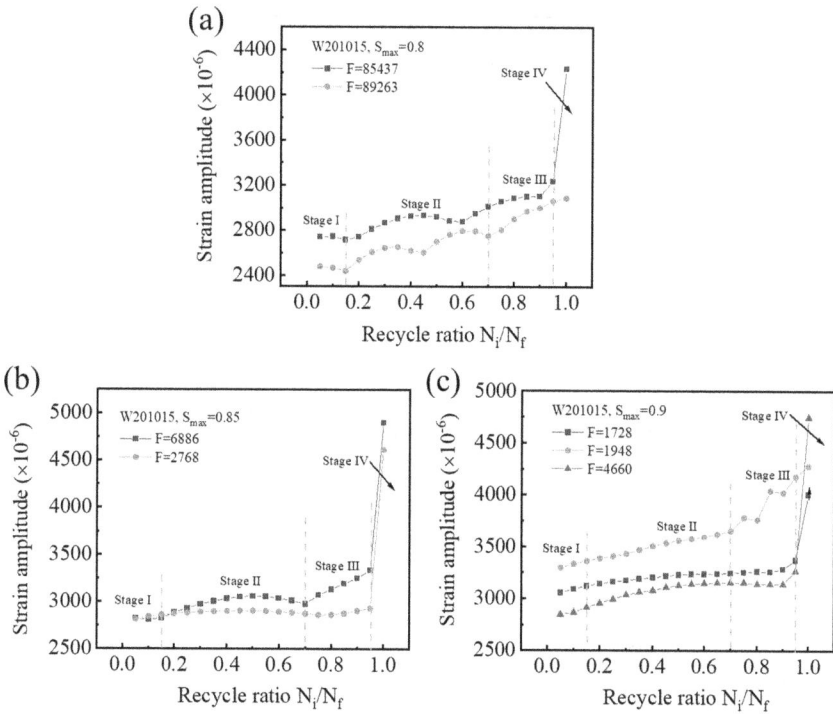

Figure 5.37 Compressive fatigue strain amplitude of UHPC reinforced with 1.5 vol% SSWs. (a) S_{max}=0.8; (b) S_{max}=0.85; (c) S_{max}=0.9

Due to the large amounts of SSWs, the percentage occupied by stage III (the crack unstable propagation and convergence stage) for UHPC reinforced with 1.5 vol% SSWs increases to 10%. The influence of maximum stress level on the compressive fatigue modulus of UHPC reinforced with 1.0 vol%/1.5 vol% SSWs is limited, while the compressive fatigue modulus of UHPC reinforced with 0.5 vol% SSWs decreases with decreasing maximum stress level. At the maximum stress level of 0.90, the compressive fatigue modulus of UHPC reinforced with 0.5 vol%/1.0 vol%/1.5 vol% SSWs is reduced to 76.8%, 77.1%, and 70.2% of initial fatigue modulus up to failure, respectively. The decrease of compressive fatigue modulus also represents the improvement of energy dissipation capacity.

Compressive fatigue damage is an important parameter to characterize the deformation ability and toughness of composites. Based on one-dimensional linear elastic law, the following Eq. (5.28) can be established for damage composites:

$$\varepsilon_f = \frac{\sigma_f}{E_0} \tag{5.28}$$

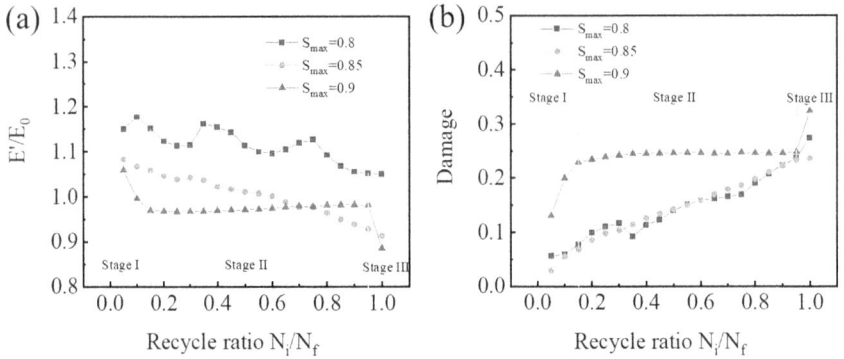

Figure 5.38 Compressive fatigue modulus and damage of UHPC without SSWs. (a) Fatigue modulus; (b) Fatigue damage

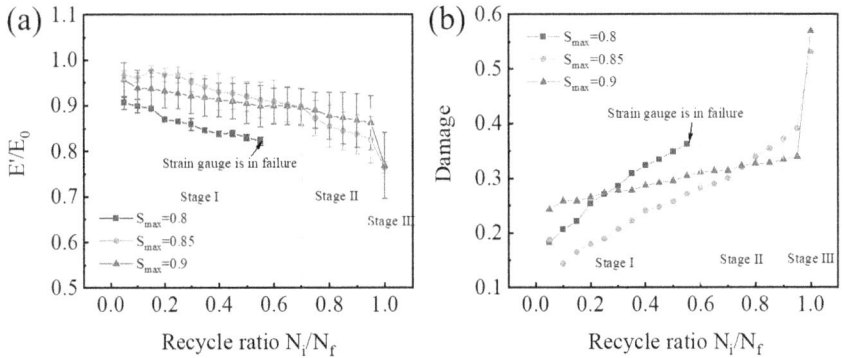

Figure 5.39 Compressive fatigue modulus and damage of UHPC reinforced with 0.5 vol% SSWs. (a) Fatigue modulus; (b) Fatigue damage

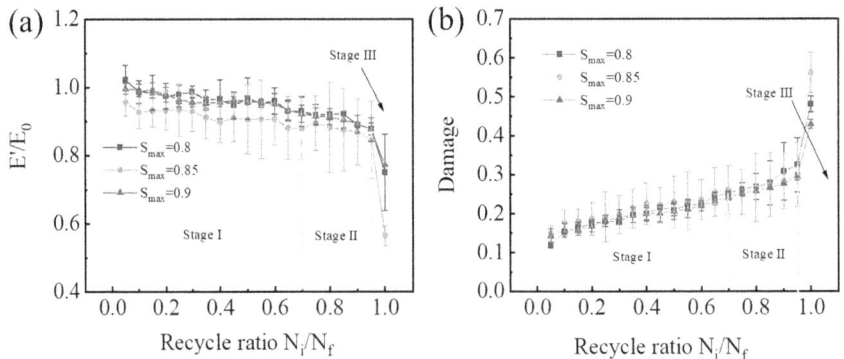

Figure 5.40 Compressive fatigue modulus and damage of UHPC reinforced with 1.0 vol% SSWs. (a) Fatigue modulus; (b) Fatigue damage

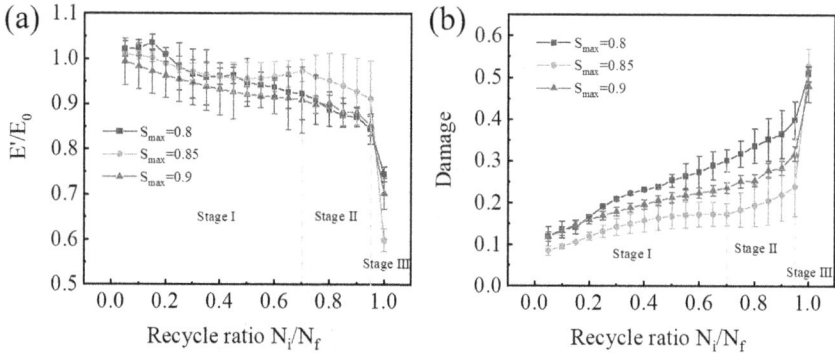

Figure 5.41 Compressive fatigue modulus and damage of UHPC reinforced with 1.5 vol%
SSWs. (a) Fatigue modulus; (b) Fatigue damage

where σ_f is the nominal stress, and E_0 is the initial elastic modulus. For non-damaged composites, the linear elastic relationship can be expressed by Eq. (5.29), as follows:

$$\varepsilon_0 = \frac{\sigma}{E'} \tag{5.29}$$

where σ is the effective stress, and E' is the fatigue modulus. On the basis of the damage variable (D) definition proposed by Rabotnov [41], the following relation presented by Eq. (5.30) should be satisfied:

$$\sigma_f = \frac{\sigma}{1-D} \tag{5.30}$$

According to the strain equivalence hypothesis [42], the strain produced by nominal stress σ_f on the damaged composites is equal to that produced by effective stress σ on the non-damaged composites, as shown in Eq. (5.31):

$$\varepsilon_f = \varepsilon_0 \tag{5.31}$$

By combining Eqs (5.28)–(5.31), Eq. (5.32) can be obtained:

$$D = 1 - \frac{E'}{E_0} \tag{5.32}$$

Considering the influence of irreversible deformation during the compressive fatigue test, Eq. (5.32) is modified, and Eq. (5.33) is established:

$$D = 1 - \left(\frac{\varepsilon_{max} - \varepsilon_r}{\varepsilon_{max}}\right)\frac{E'}{E_0} \tag{5.33}$$

where ε_{max} is the maximum compressive fatigue strain; ε_r is the compressive fatigue residual strain. The fatigue damage of SSWs-engineered multifunctional UHPC is plotted in Figures 5.38(b) to 5.41(b).

Figure 5.38(b) shows that due to cyclic creep, the compressive fatigue damage of UHPC without SSWs quickly increases, followed by a slower trend until 95% of fatigue lives at the maximum stress level of 0.9, while the damage increases linearly for 0.8 and 0.85 stress level. As the cycle ratio increases from 0.95 to 1.0, the compressive fatigue damage of UHPC without SSWs abruptly rises to 0.31, 0.26, and 0.30 for the maximum stress levels of 0.80, 0.85, and 0.90, respectively. Figures 5.39(b)–5.41(b) show that the damage development during the whole fatigue test procedure can be divided into three stages, accounting for 60%, 30–35%, and 5–10% of fatigue life, respectively. The damage in stage I mainly comes from the coupling action of fatigue creep and flaw initiation, and that in stage II results from the slow and progressive growth of microcracks, such that the damage growth rate in stage II is higher than that in stage I. When the unstable microcracks form macrocracks, the compressive fatigue damage moves into stage III, manifesting a sharply increasing trend. With the increase in content of bridging SSWs, stage III is prolonged. It can also be observed that the compressive fatigue damage of UHPC reinforced with 0.5 vol%/1.5 vol% SSWs increases with the decrease of maximum stress level at the same fatigue cycle ratio, while that of UHPC without SSWs and UHPC reinforced with 1.0 vol% SSWs has no obvious change rule with the maximum stress level. At 100% cycles of fatigue life, the compressive fatigue damage of UHPC reinforced with 0.5 vol%/1.0 vol%/1.5 vol% SSWs is 0.36, 0.43, and 0.48, respectively, for the 0.9 maximum stress level, increased by 20.0%, 43.0%, and 60.3% compared with that of UHPC without SSWs. When the maximum stress level is 0.85, the compressive fatigue damage of composites is 66.3%, 110.5%, and 98.1% higher than that of UHPC without SSWs due to the addition of 0.5 vol%, 1.0 vol%, and 1.5 vol% SSWs.

5.3.4 Compressive fatigue failure state

The compressive fatigue failure state of SSWs-engineered multifunctional UHPC is presented in Figures 5.42 to 5.45. It can be seen by the naked eye that the addition of 0.5 vol% SSWs significantly reduces the amount of macropores and improves the structural compactness, justifying the increased fatigue life. Meanwhile, the addition of SSWs markedly reduces the fatigue breakage of UHPC specimens. At the same maximum stress level, the specimens of UHPC reinforced with 0.5 vol% SSWs have higher integrity than other composites. During the experiment, the specimens of UHPC without SSWs broke into pieces immediately after failure, while there were some small chips falling off specimens of UHPC reinforced with SSWs before failure, and the specimens retained their integrity due to the bridging interaction between the SSWs and the UHPC matrix. The degree

Figure 5.42 Compressive fatigue failure state of W0. (a) S_{max}=0.8, N_f=139384; (b) S_{max}=0.85, N_f=45593; (c) S_{max}=0.9, N_f=769

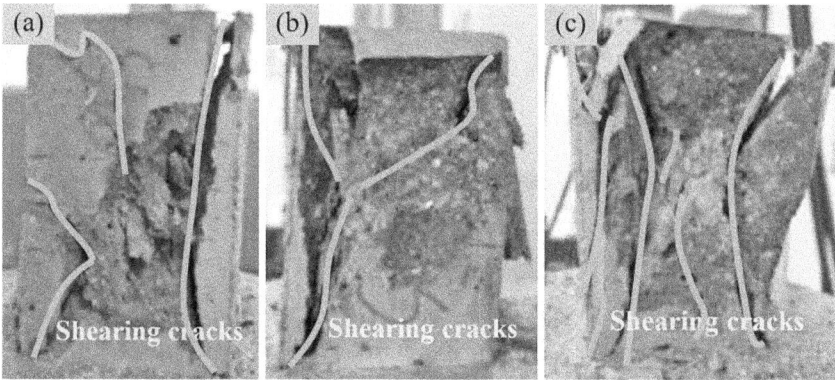

Figure 5.43 Compressive fatigue failure state of UHPC reinforced with 0.5 vol% SSWs. (a) S_{max}=0.8, N_f=423204; (b) S_{max}=0.85, N_f=85381; (c) S_{max}=0.9, N_f=11109

of breakage and number of cracks of composites increase with increasing maximum stress level. The compressive fatigue failure specimens of UHPC without SSWs present a typical inverted cone shape, caused by the hoop effect. However, the influence of the hoop effect on the compressive fatigue failure state of SSWs-engineered multifunctional UHPC is weakened due to the constraining effect of the SSWs' network on lateral deformation. Meanwhile, the compressive fatigue cracks of UHPC without SSWs are close to the vertical splitting-tensile type, while those of SSWs-engineered multifunctional UHPC are the inclined shearing type.

The compressive fatigue failure surface of representative specimens of UHPC without SSWs and UHPC reinforced with 1.5 vol% SSWs is shown in Figure 5.46. It can be seen that the failure specimen of UHPC reinforced with 1.5 vol% SSWs has more cracks compared with that of UHPC without

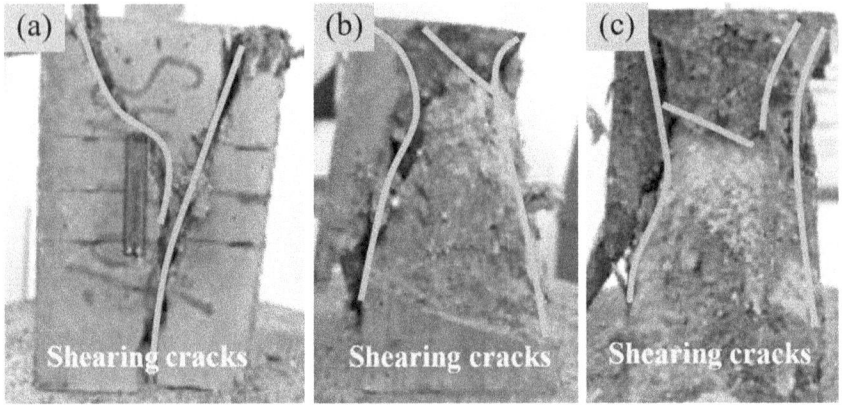

Figure 5.44 Compressive fatigue failure state of UHPC reinforced with 1.0 vol% SSWs. (a) S_{max}=0.8, N_f=41814; (b) S_{max}=0.85, N_f=49503; (c) S_{max}=0.9, N_f=8089

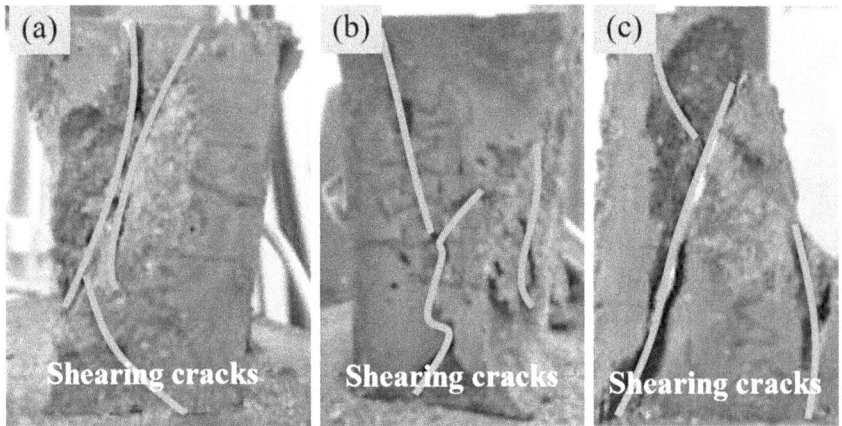

Figure 5.45 Compressive fatigue failure state of UHPC reinforced with 1.5 vol% SSWs. (a) S_{max}=0.8, N_f=85437; (b) S_{max}=0.85, N_f=6886; (c) S_{max}=0.9, N_f=4660

SSWs, explaining the low fatigue life of UHPC reinforced with 1.5 vol% SSWs but high energy dissipation capacity. The formation of cracks for UHPC without SSWs under compressive fatigue load includes three stages. In stage I, vertical cracks (on the hoop surface) appear on both ends of specimens due to local compression, and then, there are no other visible cracks with the increase of fatigue cycles. This can be attributed to the increased propagation resistance of vertical cracks, the high loading rate, and the short energy accumulation process. Under continuous fatigue load, an oblique crack is generated in the middle of specimens and then gradually expands to both ends, signifying stage II of crack initiation and propagation. When the number of cycles continues to increase and exceeds 95%

of the total lives, the original fatigue cracks gradually merge into a single critical diagonal crack. The compressive fatigue failure of UHPC without SSWs happens suddenly through the friction zone (the area covered by the yellow dotted line) in Figure 5.46(a). The friction zone in Figure 5.46 presents a rough state and white color resulting from the dislocation sliding of composites on the fracture surface of specimens.

The formation of compressive fatigue failure cracks for UHPC reinforced with 1.5 vol% SSWs can be divided into four stages. The cracks in stage I are mainly oblique cracks on the side, while the cracks inside the specimens are in the latent stage. In stage II, there are many microcracks initiated from the original flaws, but they are inhibited by SSWs. Due to the high loading and unloading rate, the energy dissipation is mainly used to produce microcracks. With the increase of compressive fatigue cycles and the repeated accumulation of energy in stage III, the microcracks continue to grow, and SSWs can hinder and bridge the cracks, as shown in Figure 5.46(b). When the connected or extended cracks reach a critical value, the cracks begin to grow unsteadily, and the specimens quickly break down. In this stage, the SSWs are pulled off, and the friction zone is generated in failure specimens (the area inside the yellow dotted line in Figure 5.46(b)). Comparative analysis shows that the area of the friction zone in UHPC reinforced with 1.5 vol% SSWs is larger than that in UHPC without SSWs, and there are many visible cracks in the friction zone of UHPC reinforced with 1.5 vol% SSWs due to the inhibiting effect of SSWs on the initiation and propagation of cracks.

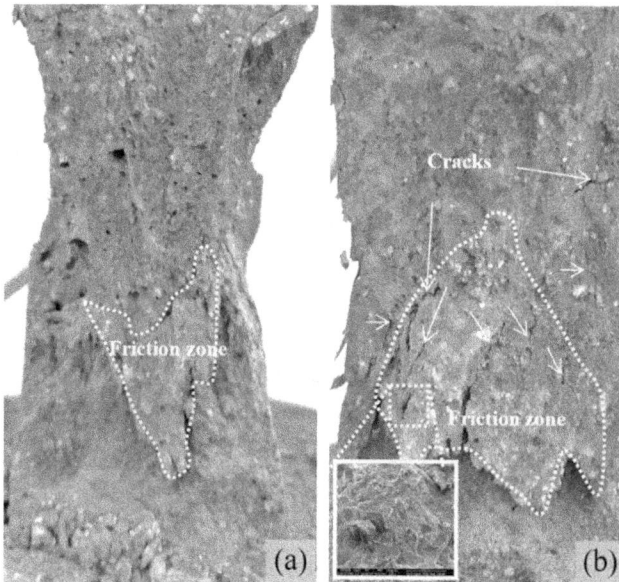

Figure 5.46 Compressive fatigue failure surface of SSWs-engineered multifunctional UHPC. (a) SW0, S_{max}=0.8, N_f=139384; (b) SW1.5, S_{max}=0.85, N_f=6886

5.4 ENHANCEMENT MECHANISMS OF SSWS ON FATIGUE PROPERTIES OF UHPC

5.4.1 Pore structure

The cumulative pore volume curves of SSWs-engineered multifunctional UHPC are displayed in Figure 5.47(a), and they can be divided into four regions according to the growth patterns. The pore with a diameter larger than 100 μm in region I is a large capillary pore. When the diameter ranges from 1 μm to 100 μm, the pore is a medium capillary pore. The pore diameter corresponding to region III and region IV is 50–1000 nm and 5–50 nm, and these pores are a small capillary pore and a gel pore, respectively. The inclination of cumulative pore volume curves toward the x-axis represents a refinement of pore structure. As shown in Figure 5.47(a), adding 0.5 vol% SSWs has no positive effect on the pore refinement, and incorporating 1.0 vol% SSWs can only refine pores with diameters ranging from 5 to 50 nm. When the dosage of SSWs is 1.5 vol%, pores with diameters of 5–1000 nm have been significantly refined. The

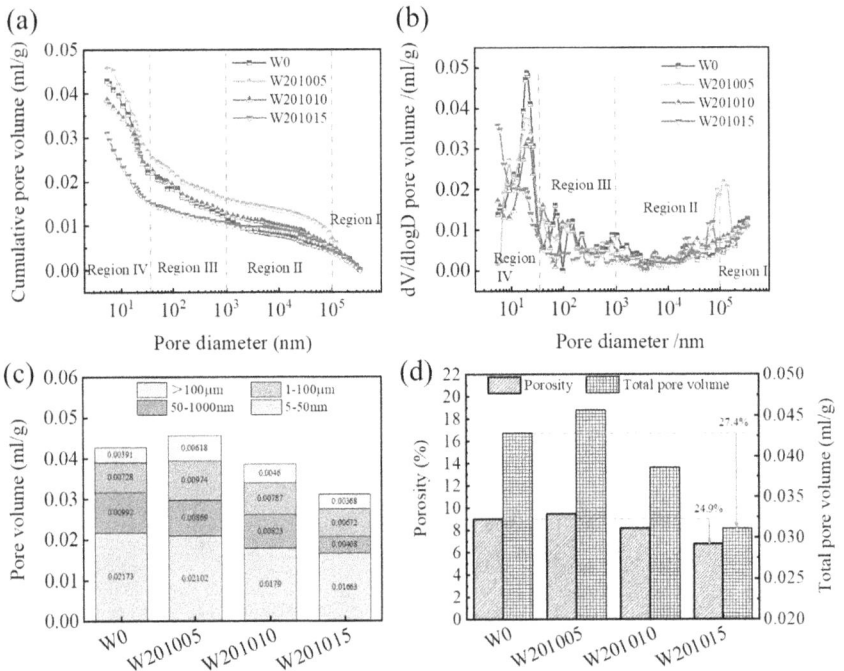

Figure 5.47 Pore structure and porosity of SSWs-engineered multifunctional UHPC. (a) Cumulative pore volume; (b) Pore size distribution; (c) Pore volume fraction; (d) Porosity and total pore volume

pore size distribution curves of SSWs-engineered multifunctional UHPC contain one remarkable wave crest in region IV, as demonstrated in Figure 5.47(b), and this wave crest represents the most probable pore size of concrete. The values of the most probable pore size are listed in Table 5.11. The most probable pore size of UHPC reinforced with 1.5 vol% SSWs is 72.5% lower than of UHPC without SSWs, characterizing the refinement of pore structure.

Figure 5.47(c) shows the pore volume in different regions of SSWs-engineered multifunctional UHPC. It can be discovered that the volume of pores with a diameter larger than 100 µm for UHPC reinforced with 0.5 vol%/1.0 vol% SSWs is 57.8% and 17.6% larger, respectively, than that for UHPC without SSWs. Meanwhile, the addition of 0.5 vol% and 1.0 vol% SSWs also increases the volume of pores with diameter ranging from 1 µm to 100 µm by 33.8% and 8.1%, respectively, due to the air entraining effect of SSWs and the workability reduction of UHPC. However, the volume of pores with diameter larger than 100 µm and between 1 and 100 µm is decreased by 5.9% and 7.7%, respectively, due to the addition of 1.5 vol% SSWs. This can be attributed to the compact microstructure induced by the large specific surface area of SSWs and the high pozzolanic activity of silica fume simultaneously. The compact microstructure–inducing and formation effect of 1.5 vol% SSWs is dominant and makes up for the negative effect of air entraining and workability reduction, leading to the improvement of microstructure homogeneity. This is also what causes the volume of pores with diameters of 50–1000 nm and 5–50 nm for UHPC reinforced with 1.5 vol% SSWs to be reduced by 58.9% and 23.5%, respectively, compared with that for UHPC without SSWs. Especially noteworthy is that the pore volume corresponding to a diameter of 50–1000 nm and 5–50 nm for UHPC reinforced with 0.5 vol%/1.0 vol% SSWs is also 12.4%/3.3% and 17.0%/23.5%, respectively, lower than that for UHPC without SSWs, confirming that SSWs have the function of improving the microstructural compactness of UHPC.

Figure 5.47(d) displays the porosity and total pore volume of SSWs-engineered multifunctional UHPC. The results show that the porosity and

Table 5.11 Pore structure characteristic parameters of SSWs-engineered multifunctional UHPC

Group	Average pore diameter (nm)	Median pore diameter (nm)	Most probable pore diameter (nm)	Characteristic length (nm)	Tortuosity
W0	26.94	46.59	19.94	262067.2	2.7455
W201005	30.53	84.16	18.90	109797.4	4.2265
W201010	30.33	97.18	18.91	35767.3	11.7216
W201015	19.72	31.99	5.48	24230.2	14.2110

total pore volume of UHPC are reduced due to the addition of SSWs at a content larger than 1.0 vol%. A reduction rate of 24.9% and 27.4%, respectively, can be achieved as 1.5 vol% SSWs is incorporated. As discussed earlier, adding SSWs introduces air into concrete and reduces the workability of concrete, leading to an increase of large and medium capillary pore volume especially. However, SSWs and silica fume can work together to enhance the microstructural compactness; meanwhile, this function is gradually strengthened with the increase of SSWs content. This is why the porosity and total pore volume of UHPC first increase and then decrease with increasing SSWs content. However, despite this observation, it was shown that the flexural fatigue life of UHPC reinforced with 0.5 vol% SSWs is higher than that of UHPC without SSWs. This illustrates that in addition to the modification effect of SSWs on the microstructure of UHPC, the inhibiting effect of SSWs on crack initiation and propagation also plays a great role in modifying the fatigue performance of concrete.

As listed in Table 5.11, the average pore diameter and median pore diameter of SSWs-engineered multifunctional UHPC share the same change of trend and porosity as SSWs content increases. Incorporating 1.5 vol% SSWs decreases the average pore diameter and median pore diameter by 36.6% and 31.3%, respectively, resulting from the enhancing effect of SSWs on microstructural compactness. The characteristic length of pores decreases and the tortuosity of pores increases with increasing SSWs content, and this represents refinement of the pore structure. The addition of 0.5 vol%, 1.0 vol%, and 1.5 vol% SSWs reduces the characteristic length of pores by 58.1%, 86.4%, and 90.8%, respectively. Meanwhile, the tortuosity of pores is increased by 53.9%, 326.9%, and 417.6%, respectively. The change of pore structure of SSWs-engineered multifunctional UHPC indicates that the refinement of pore structure is beneficial to improve the fatigue damage resistance of concrete.

5.4.2 Hydration product characteristics

Figure 5.48 shows the molar ratio of CaO to SiO_2 (Ca/Si) of C-S-H gels at different locations of composites. It can be found that the average Ca/Si ratio of C-S-H gels in UHPC without SSWs is 1.1461, and this value decreases to 1.1116 and 1.0338 for the location of the SSWs' surface and the gap between SSWs, respectively, in the specimen of UHPC reinforced with 1.5 vol% SSWs. Although the values of the Ca/Si ratio at each test point in Figure 5.48 fluctuate up and down, there is no significant increase compared with that for the UHPC matrix. This indicates that no weak interfacial transition zone between the SSWs and the UHPC matrix is found in 1.5 vol% SSWs reinforced UHPC to any extent. Weakening or eliminating the influence of the interface transition zone is also the reason why the addition of SSWs can improve the fatigue life of UHPC.

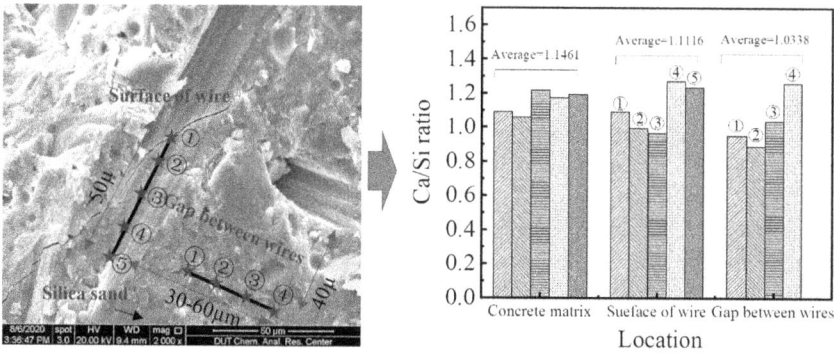

Figure 5.48 Ca/Si ratios at different locations of SSWs-engineered multifunctional UHPC

5.4.3 Scanning electron microscopy (SEM) images of specimens after flexural fatigue test

The microstructure and failure cracks in the tension zone of SSWs-engineered multifunctional UHPC specimens after flexural fatigue load with the maximum stress level of 0.9 are displayed in Figure 5.49. As shown in Figure 5.49(a), cracks in specimens of UHPC without SSWs are continuous and cross-cutting; meanwhile, obvious original flaws and a loose gel aggregation structure are observed in the SEM image, whereas Figure 5.49(b) demonstrates that the addition of 1.5 vol% SSWs can markedly reduce the original flaws of UHPC and effectively block the connection of cracks by deflecting, bridging, and being pulled off.

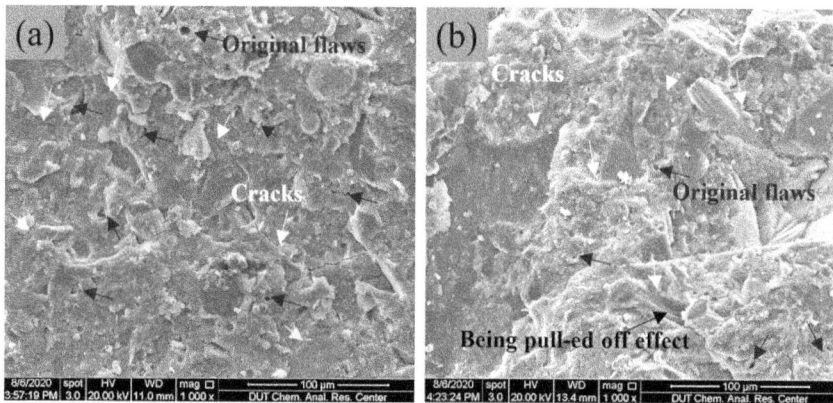

Figure 5.49 Microstructure and failure cracks of SSWs-engineered multifunctional UHPC after flexural fatigue load with maximum stress level of 0.9. (a) UHPC without SSWs at the maximum stress level of 0.9 and fatigue life of 32; (b) 1.5 vol% SSWs reinforced UHPC at the maximum stress level of 0.9 and fatigue life of 986

The cracks in the tension zone of failure specimens for SSWs-engineered multifunctional UHPC subjected to flexural fatigue load with the maximum stress level of 0.7 are displayed in Figure 5.50. It can be seen that cracks in UHPC without SSWs are long and even penetrate fine aggregates, while the development of cracks in SSWs-engineered multifunctional UHPC is suppressed by SSWs. This is consistent with the crack development characteristics of composites under flexural fatigue load of 0.9 stress level. Furthermore, the long and wide cracks have been transformed into emission cracks surrounding SSWs. By comparing Figure 5.50(b)–(d), it is clearly seen that the cracks become narrow as the SSWs content increases. Figure 5.50(d) demonstrates that the macrocracks are

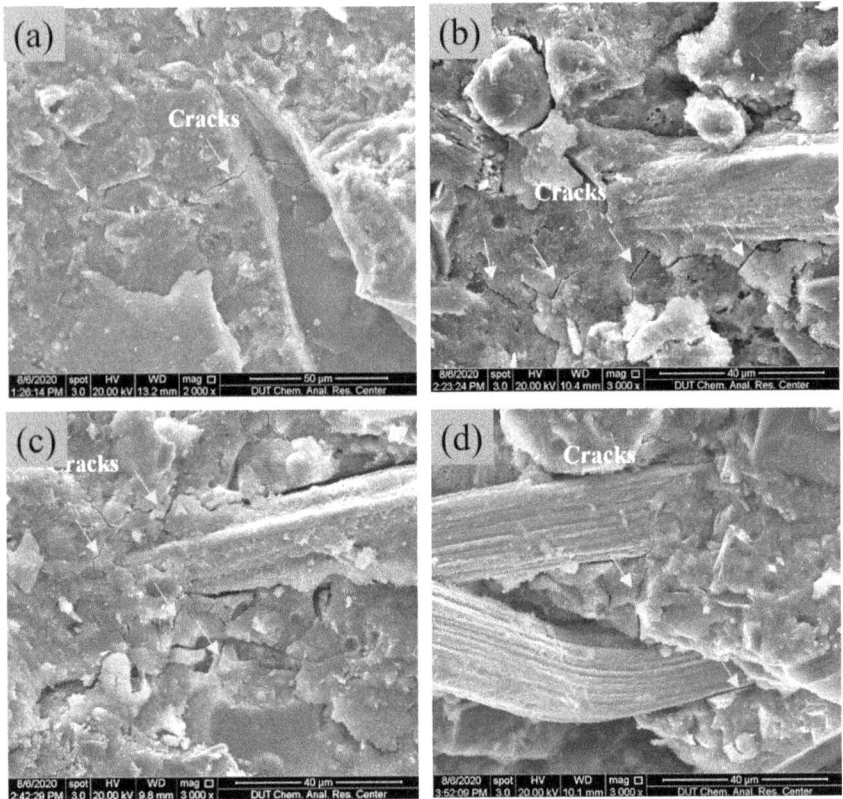

Figure 5.50 Cracks in failure specimens of SSWs-engineered multifunctional UHPC after flexural fatigue load with maximum stress level of 0.7. (a) UHPC without SSWs at the maximum stress level of 0.7 and fatigue life of 135300; (b) 0.5 vol% SSWs reinforced UHPC at the maximum stress level of 0.7 and fatigue life of 517965; (c) 1.0 vol% SSWs reinforced UHPC at the maximum stress level of 0.7 and fatigue life of 689363; (d) 1.5 vol% SSWs reinforced UHPC at the maximum stress level of 0.7 and fatigue life of 1001611

significantly reduced by adjacent SSWs. These phenomena prove the modification effect of SSWs on the flexural fatigue performance of UHPC, mainly attributed to the following three aspects. Firstly, the addition of SSWs enhances the microstructural compactness of the concrete matrix, leading to the reduction of initial cracks as the fatigue load is first applied. This effect is particularly significant for composites subjected to fatigue load with a high maximum stress level, and it is also one of the reasons that make the flexural fatigue life enhancement effect of 1.5 vol% SSWs far superior to that of 0.5 vol%/1.0 vol% SSWs at the maximum stress level of 0.9. Secondly, SSWs can effectively inhibit the initiation and propagation of microcracks by transferring crack tip stress to prolong the flexural fatigue life of UHPC, and this effect is dominant when the applied flexural fatigue load is at a low maximum stress level. Thirdly, SSWs can bridge macrocracks until they are pulled off to continuously improve the flexural fatigue life of UHPC.

5.4.4 SEM images of specimens after compressive fatigue test

SEM images of compressive fatigue failure specimens of SSWs-engineered multifunctional UHPC are shown in Figure 5.51. Figure 5.51(a) and (b) show that there are penetrating long and wide cracks in compressive fatigue failure specimens of UHPC without SSWs. However, the compressive fatigue cracks of UHPC reinforced with 0.5 vol% SSWs mainly exist radially at the root of SSWs, as displayed in Figure 5.51(c)–(e). This indicates that the presence of SSWs can convert a single main crack into multiple cracks to improve the compressive fatigue life and energy dissipation capacity of composites. This phenomenon has not been found in previous research on fiber reinforced concrete [30, 31, 36]. Meanwhile, it can be observed that compressive fatigue cracks in failure specimens of UHPC reinforced with 0.5 vol% SSWs mainly exist in the UHPC matrix, which can be attributed to the compact microstructure and high cycle numbers. Fatigue cracks on the interface between quartz sand and the UHPC matrix occur in compressive fatigue failure specimens of UHPC without SSWs and UHPC reinforced with 1.0 vol%/1.5 vol% SSWs. In addition, SSWs in UHPC can effectively inhibit crack propagation, deflect cracks, bridge cracks, and be pulled off, leading to the addition of SSWs in volume fractions of 1.0 vol% and 1.5 vol% improving the compressive fatigue energy dissipation capacity of UHPC.

It is worth noting that the dense interface between the SSWs and the UHPC matrix results in large compressive fatigue energy being consumed during the debonding process, as confirmed by Figure 5.52, showing that parts of the UHPC matrix are pulled out during the process of compressive fatigue failure. The increase of interface bond strength is positively correlated with the increase of compressive fatigue life for composites.

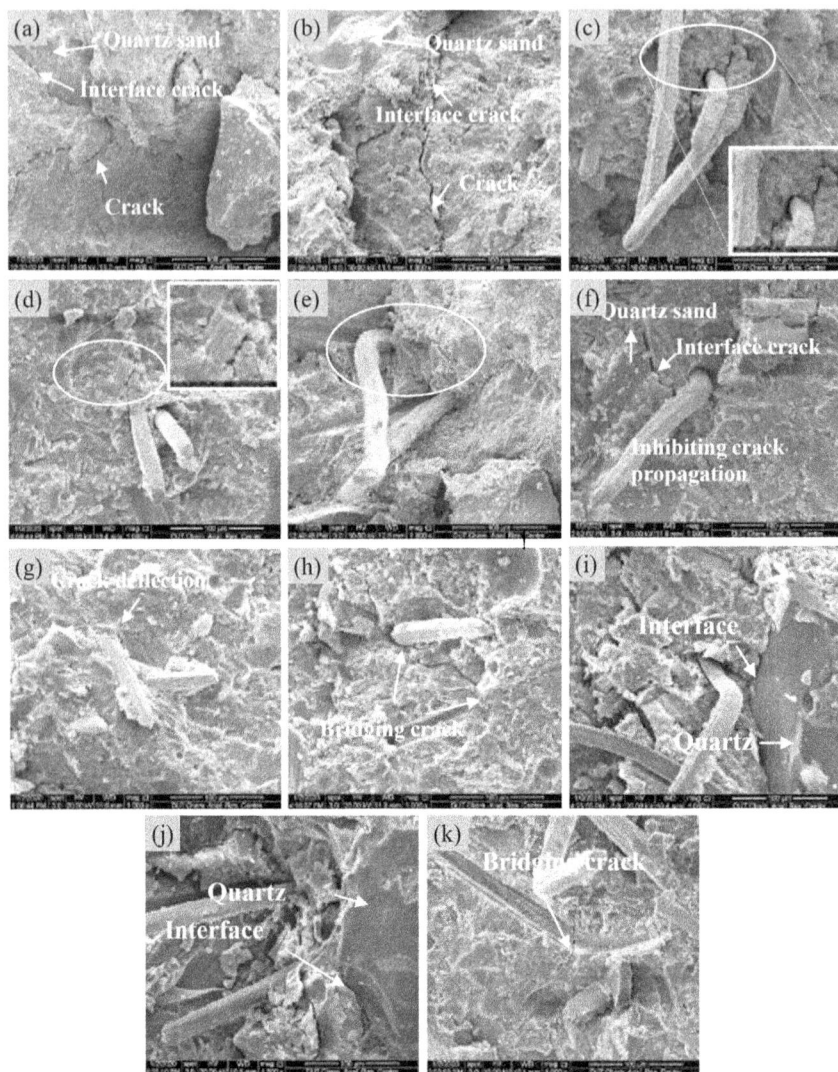

Figure 5.51 SEM images of compressive fatigue failure specimens of SSWs-engineered multifunctional UHPC. (a) W0, S_{max}=0.8; (b) W0, S_{max}=0.9; (c) W201005, S_{max}=0.8; (d) W201005, S_{max}=0.85; (e) W201005, S_{max}=0.9; (f) W201010, S_{max}=0.8; (g) W201010, S_{max}=0.85; (h) W201010, S_{max}=0.9; (i) W201015, S_{max}=0.8; (j) W201015, S_{max}=0.85; (k) W201015, S_{max}=0.9

According to these analyses, the enhancing mechanisms of SSWs on the fatigue performance of UHPC mainly include the following two aspects.

On the one hand, the incorporation of SSWs can increase the microstructural compactness of UHPC and enhance the interfacial transition zone, which has been proved by the decrease of porosity, total pore volume, and

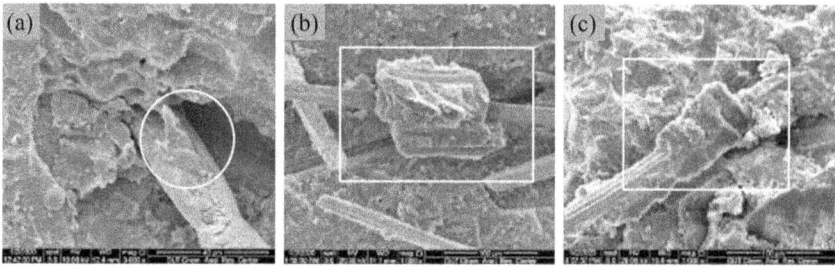

Figure 5.52 Interface failure between SSWs and UHPC matrix after compressive fatigue test. (a) W201005, S_{max}=0.9; (b) W201010, S_{max}=0.85; (c) W201015, S_{max}=0.85

Ca/Si ratio. Because the first step of preparation process is to stir silica fume, water, water reducer, and SSWs together, silica fume easily concentrates on the surface of SSWs, benefiting from the large specific surface area of SSWs, as shown in Table 5.12. After cement is added, Ca^{2+} approaches the surface of SSWs under the influence of water migration. Then, Ca^{2+} reacts with silica fume to generate C-S-H gels on the surface of the wires because of the high pozzolanic activity of silica fume. This is why the Ca/Si ratio on the surface of the wires is lower than that of the UHPC matrix. Furthermore, it is worth mentioning that the reaction process of Ca^{2+} with silica fume consumes water, leading to a decrease in the water-to-binder ratio around SSWs. This phenomenon is conducive to improving the microstructural compactness and C-S-H gel polymerization around SSWs and can be verified by the decrease of the Ca/Si ratio at the gap between SSWs. Therefore, a microstructure enhancement zone is formed with SSWs as the core, resulting from the large specific surface area of SSWs and the high pozzolanic activity of silica fume simultaneously, and this is completely different from the weak interface transition zone between steel fibers and concrete matrix [24, 25]. As listed in Table 5.12, the presence of the microstructure enhancement zone caused by SSWs helps to improve the compactness of cement outer hydration products. In particular, when the content of SSWs is high, the existence of a large number of microstructure enhancement zones has the potential to reduce the adverse effects of the aggregate interface transition zone [35]. That is to say, the structural homogeneity of UHPC is improved by the incorporation of SSWs, which plays an important role in improving fatigue life.

On the other hand, the addition of SSWs prolongs the fatigue life of UHPC by exerting crack resistance and the toughening function of fibers. At the stage of unloading, the excellent thermal conductivity of SSWs can effectively transmit hydration heat to reduce original cracks caused by the hydration heat gradient [43]. At the initial loading stage, SSWs can effectively inhibit the initiation of microcracks due to the increase of microstructural compactness. With increasing fatigue cycles, main cracks appear in

Table 5.12 The microstructure enhancement zone formed around SSWs

Materials	Before hydration	After hydration
W0		
W201005		
W201010/ W201015		
Symbols	● Cement ⟩ SSW ● Fly ash ● Silica fume	⟩ Interface products ● Inner products Outer products

specimens of UHPC, and damage propagates quickly along the main crack. For the specimens of SSWs-engineered multifunctional UHPC, the widely distributed SSWs can effectively transfer crack tip stress during the medium loading stage, leading to the appearance of multiple cracks and radial cracks centered on SSWs. The significance and curvature of main cracks decrease and increase, respectively, with increasing SSWs' dosage. At the stage of final loading, specimens of UHPC without SSWs show a state of instantaneous complete destruction. Due to the bridging effect of SSWs, specimens of SSWs-engineered multifunctional UHPC can be continuously loaded under the condition of multiple microcracks until the SSWs are pulled off.

5.5 SUMMARY

This chapter introduced the flexural and compressive fatigue performance of SSWs-engineered multifunctional UHPC, including fatigue life, stress–strain hysteretic relationships, and fatigue damage. The modification mechanisms of SSWs based on observation of the microstructure and analyses of the characteristics of hydration products were presented. The main conclusions are as follows:

1) 0.5 vol% SSWs can effectively increase the average flexural fatigue life of UHPC at different maximum stress levels. The flexural fatigue life of SSWs-engineered multifunctional UHPC obeys a Weibull distribution, and the method of moments and method of maximum likelihood are used to calculate scale and shape parameters. The ratio of flexural fatigue limit strength to static flexural strength for UHPC can be increased by 7.8% due to the addition of 1.5 vol% SSWs. The flexural fatigue stress–strain hysteresis curves of SSWs-engineered multifunctional UHPC change from sparse to dense to sparse and then, to failure with the increase of fatigue cycles, and the slope of stress–strain curves before failure significantly decreases with increasing SSWs content. The maximum flexural fatigue strain and fatigue damage of SSWs-engineered multifunctional UHPC are 34.0% higher and 41.5% lower, respectively, than those of UHPC without SSWs.

2) At the maximum stress levels of 0.80, 0.85, and 0.90, 0.5 vol% SSWs enables the compressive fatigue life of UHPC to increase by 252.0%, 76.0%, and 54.6%, respectively, compared with that of UHPC without SSWs. At the failure probability of 50%, the compressive fatigue limit strength of composites reaches 76.6% of the static uniaxial compressive strength. The compressive fatigue energy dissipation of UHPC reinforced with 0.5 vol% SSWs is 262.3% and 167.8% higher than that of UHPC without SSWs at the average fatigue lives of 48515 and 8223, respectively. The compressive fatigue deformation ability of composites increases with increasing SSWs content, and the fatigue damage before failure is increased by 98.1%, due to SSWs bridging, debonding, and being pulled off.

3) The microstructure enhancement zone formed by the synergistic effect of SSWs and silica fume improves the homogeneity of UHPC, represented by the decrease of porosity, total pore volume, and Ca/Si ratio of C-S-H gels. Meanwhile, SSWs exerts a crack resistance and toughening function to reduce original cracks and inhibit the initiation and propagation of microcracks, shown by the long link-up microcracks in fatigue failure specimens having been transformed into emission cracks centered on SSWs.

REFERENCES

1. X. Wang, Q. Zheng, S. Dong, A. Ashour, B. Han. Interfacial characteristics of nano-engineered concrete composites, *Construction and Building Materials.* 259 (2020) 119803.
2. T. Hsu. Fatigue of plain concrete, *Journal of the American Concrete Institute.* 78(4) (1981) 292–305.
3. B. Han, L. Zhang, S. Zeng, S. Dong, X. Yu, R. Yang, J. Ou. Nano-core effect in nano-engineered cementitous composites, *Composites Part A: Applied Science and Manufacturing.* 95 (2017) 100–109.
4. L. Li, Q. Zheng, B. Han, J. Ou. Fatigue behaviors of graphene reinforcing concrete composites under compression, *International Journal of Fatigue.* 151 (2021) 106354.
5. X. Wang, S. Dong, A. Ashour, S. Ding, B. Han. Bond behaviors between nano-engineered concrete and steel bars, *Construction and Building Materials.* 299 (2021) 124261.
6. W. Sebastian, T. Keller. Ductility of civil engineering structures incorporating fibre reinforced polymers (FRPs), *Construction and Building Materials.* 49 (2013) 913–914.
7. F. Zhang, L. Poh, M. Zhang. Resistance of cement-based materials against high-velocity small caliber deformable projectile impact, *International Journal of Impact Engineering.* 144 (2020) 103629.
8. W. Sebastian, S. Webb, H.S. Nagree. Orthogonal distribution and dynamic amplification characteristics of partially prefabricated timber-concrete composites, *Engineering Structures.* 219 (2020) 110693.
9. F. Zhang, A.S. Shedbale, R. Zhong, L. Poh, M. Zhang. Ultra-high performance concrete subjected to high-velocity projectile impact: Implementation of K&C model with consideration of failure surfaces and dynamic increase factors, *International Journal of Impact Engineering.* 155 (2021) 103907.
10. S. Dong, X. Wang, A. Ashour, B. Han, J. Ou. Enhancement and underlying mechanisms of stainless steel wires to fatigue properties of concrete under flexure, *Cement and Concrete Composites.* 126 (2022) 104372.
11. S. Dong, Y. Wang, A. Ashour, B. Han, J. Ou. Uniaxial compressive fatigue behavior of ultra-high performance concrete reinforced with super-fine stainless wires, *International Journal of Fatigue.* 142 (2021) 105959.
12. J. Wang, B. Han, Z. Li, X. Yu, X. Dong. Effect investigation of nanofillers on C-S-H gel structure with Si NMR spectra, *Journal of Materials in Civil Engineering.* 31(1) (2019) 04018352.
13. S.R. Kasu, S. Deb, N. Mitra, A.R. Muppireddy, S.R. Kusam. Influence of aggregate size on flexural fatigue response of concrete, *Construction and Building Materials.* 229 (2019) 116922.
14. K. Yeon, Y. Choi, K.K. Kim, K. Kim, J. Yeon. Flexural fatigue life analysis of unsaturated polyester-methyl methacrylate polymer concrete, *Construction and Building Materials.* 140 (2017) 336–343.
15. J.D. Ríos, H. Cifuentes, S. Blasón, M. López-Aenlle, A. Martínez-De La Concha. Flexural fatigue behaviour of a heated ultra-high-performance fibre-reinforced concrete, *Construction and Building Materials.* 276(4) (2021) 122209.

16. N. Saoudi, B. Bezzazi. Flexural fatigue failure of concrete reinforced with smooth and mixing hooked-end steel fibers, *Cogent Engineering*. 6(1) (2019) 1594508.

17. D. Meng, Y. Zhang, C. Lee. Flexural fatigue behaviour of steel reinforced PVA-ECC beams, *Construction and Building Materials*. 221 (2019) 384–398.

18. X. Zhu, X. Chen, S. Liu, S. Li, W. Xuan, Y. Chen. Experimental study on flexural fatigue performance of rubberised concrete for pavement, *International Journal of Pavement Engineering*. 21(9) (2018) 1–12.

19. B.S. Saini, S.P. Singh. Flexural fatigue life analysis of self compacting concrete containing 100% coarse recycled concrete aggregates, *Construction and Building Materials*. 253 (2020) 119176.

20. R. Tepfers. Fatigue of plain concrete subject to stress reversals, fatigue of concrete structures, *American Concrete Institute*. 75 (1982) 195–216.

21. A.S. Vesic, S.K. Saxena. Analysis of structural behaviour of road test rigid pavements, highway research record (No.291), 48th annual meeting of highway research board, national research council. 291 (1969) 156–158.

22. J. Kim, Y. Kim. Experimental study of the fatigue behavior of high strength concrete, *Cement and Concrete Research*. 26(10) (1996) 1513–1523.

23. P.H. Wirsching, J.T.P. Yao. Statistical method in structural fatigue, *ASCE Journal of Structural Engineering*. 96(6) (1970) 1201–1219.

24. B.H. Oh. Fatigue analysis of plain concrete in flexure, *ASCE Journal of Structural Engineering*. 112(2) (1986) 273–288.

25. B.H. Oh. Fatigue life distributions of concrete for various stress levels, *ACI Materials Journal*. 88(2) (1991) 122–128.

26. J. Li, Q. Zhang. A iterative method for the maximum likelihood estimate of the two parameters in the two-parameters Weibull distribution, *Journal of Mechanical Strength*. 3 (1994) 67–68. (In Chinese).

27. S. Goel, S.P. Singh, P. Singh. Flexural fatigue strength and failure probability of self compacting fibre reinforced concrete beams, *Engineering Structures*. 40 (2012) 131–140.

28. S. Bawa, S.P. Singh. Flexural fatigue strength prediction of hybrid fibre-reinforced self-compacting concrete, *Construction Materials*. 173(5) (2019) 1–40.

29. S. Dong, B. Han, X. Yu, J. Ou. Constitutive model and reinforcing mechanisms of uniaxial compressive property for reactive powder concrete with super-fine stainless wire, *Composites Part B: Engineering*. 166 (2019) 298–309.

30. P.B. Cachima, J.A. Figueiras, P. Pereira. Fatigue behavior of fiber-reinforced concrete in compression, *Cement and Concrete Composites*. 24(2) (2002) 211–217.

31. Q. Li, B. Huang, S. Xu, B. Zhou, R.C. Yu. Compressive fatigue damage and failure mechanism of fiber reinforced cementitious material with high ductility, *Cement and Concrete Research*. 90 (2016) 174–183.

32. B.T. Huang, Q.H. Li, S.L. Xu, M. ASCE, B. Zhou. Frequency effect on the compressive fatigue behavior of ultrahigh toughness cementitious composites: Experimental study and probabilistic analysis, *Journal of Structural Engineering*. 143(8) (2017) 04017073.

33. Medeiros, X. Zhang, G. Ruiz, R.C. Yu, M. de Souza Lima Velasco. Effect of the loading frequency on the compressive fatigue behavior of plain and fiber reinforced concrete, *International Journal of Fatigue*. 70 (2015) 342–350.

34. V.C. Li. From micromechanics to structural engineering-the design of cementitious composites for civil engineering applications, *Journal of Structural Mechanics and Earthquake Engineering*. 10(2) (1993) 37–48.

35. H. Li, M. Zhang, J. Ou. Flexural fatigue performance of concrete containing nano-particles for pavement, *International Journal of Fatigue*. 29(7) (2007) 1292–1301.

36. M.K. Lee, B.I.G. Barr. An overview of the fatigue behavior of plain and fibre reinforced concrete, *Cement and Concrete Composites*. 26(4) (2004) 299–305.

37. J.O. Holmen. Fatigue of concrete by constant and variable amplitude loading, ACI Specification Publication. 75 (1982) 71–110.

38. H.A.W. Cornelissen. Fatigue failure of concrete in tension, *Heron*. 29(4) (1984) 1–68.

39. D.A. Hordijk, G.M. Wolsink, J. de Vries. Fracture and fatigue behavior of a high strength limestone concrete as compared to gravel concrete, *Heron*. 40 (1995) 125–146.

40. B. Han, S. Dong, J. Ou, C. Zhang, Y. Wan, X. Yu, S. Ding. Microstructure related mechanical behaviors of short-cut super-fine stainless wire reinforced reactive powder concrete, *Materials and Design*. 96 (2016) 16–26.

41. Y.N. Rabotnov. *Creep Rupture // Applied Mechanics*, Springer, 1969.

42. J. Lemaitre. Evaluation of dissipation and damage on metals submitted to dynamic loading, international conference of mechanical behavior of materials, 1971.

43. S. Dong, X. Wang, H. Xu, J. Wang, B. Han. Incorporating super-fine stainless wires to control thermal cracking of concrete structures caused by heat of hydration, *Construction and Building Materials*. 271 (2021) 121896.

Chapter 6

Electrical and self-sensing properties of stainless steel wires-engineered multifunctional ultra-high performance concrete

6.1 INTRODUCTION

With the development of modern infrastructures in terms of scale, complexity, functionality, and intelligence, concrete should not only have excellent mechanical, durability, and processing properties, necessary for structural materials, but should also have functional properties such as electrical, thermal, and electromagnetic properties as well as intelligent properties such as self-sensing, self-healing, and self-regulating properties [1–4]. Endowing concrete with high electrically conductive and intrinsic self-sensing characteristics based on electrical signals provides a new approach to realize real-time in-situ monitoring of the condition of infrastructure, and this is conducive to prolonging the service life of concrete structures and provide early warning of the degradation of structure safety [5–9]. Stainless steel wires (SSWs) with high electrical conductivity and high aspect ratio can be widely distributed in ultra-high performance concrete (UHPC) at a low content, thus developing electrically conductive and self-sensing UHPC.

In this chapter, the electrically conductive properties, including electrical resistivity measured by different test methods, and electrochemical impedance spectroscopy (EIS) and equivalent circuit of SSWs-engineered multifunctional UHPC are shown, and the electrical resistivity responses and sensing mechanisms of SSWs-engineered multifunctional UHPC under cyclic compressive, monotonic compressive, flexural, and fracture load are presented. The main contents of this chapter are shown in Figure 6.1.

6.2 ELECTRICALLY CONDUCTIVE PROPERTIES OF SSWS-ENGINEERED MULTIFUNCTIONAL UHPC

6.2.1 Electrical resistivity

6.2.1.1 Measured using the two-electrode alternating current (AC) method

It is worthwhile to note that SSWs with a diameter of 8 μm and 20 μm and a length of 10 mm were used in this chapter. Meanwhile, SSWs were added in

DOI: 10.1201/9781003276357-6

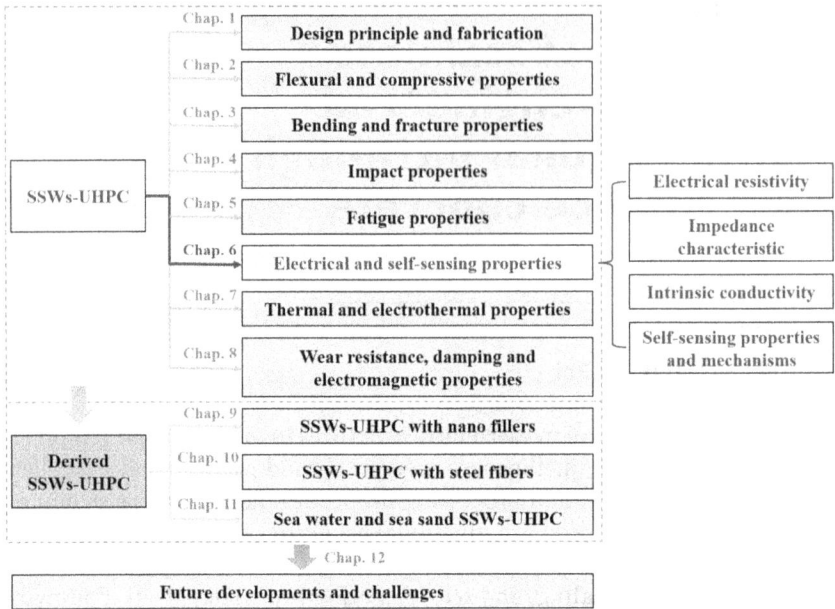

Figure 6.1 Main contents of Chapter 6

the amounts of 0, 0.5 vol%, 1.0 vol%, and 1.5 vol% of UHPC, and the corresponding specimens were marked as W0, W0810 (W081005, W081010, W081015), and W2010 (W201005, W201010, W201015). In addition, a series of specimens of SSWs-engineered multifunctional UHPC were cured in water at 20 ° until they reached certain curing ages (called water curing), and the other specimens were exposed to air after water curing for 14 days (called natural curing).

The electrical resistivity of SSWs-engineered multifunctional UHPC measured using the two-electrode AC method is shown in Figure 6.2. As can be seen from Figure 6.2, the incorporation of SSWs significantly lowers the electrical resistivity of UHPC, and the electrical resistivity values have been reduced from $9.4 \times 10^4 \Omega$ ·cm for UHPC without SSWs to $10.4 \times 10^2 \Omega$ ·cm and 19.8Ω ·cm for UHPC reinforced with 1.5 vol% SSWs with a diameter of 8 μm and 20 μm, respectively, after natural curing for 28 days. The electrical resistivity of SSWs-engineered UHPC increases with curing age, which can be attributed to the development of cementitious material hydration. Meanwhile, the electrical resistivity decreases with increasing SSWs content due to the greater connection between the adjacent SSWs. It is worth noting that the electrical resistivity of SSWs-engineered multifunctional UHPC after natural curing for 28 days decreases abruptly when the content of SSWs with a diameter of 8 μm and 20 μm is 1.5 vol% and 1.0 vol%, respectively. However, this phenomenon for UHPC reinforced

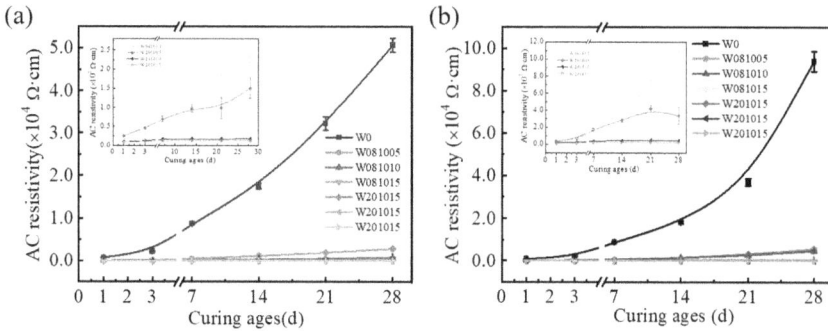

Figure 6.2 Electrical resistivity of SSWs-engineered multifunctional UHPC measured using the two-electrode AC method. (a) Water curing; (b) Natural curing

with SSWs at a diameter of 8 μm after water curing for 28 days occurs at a content of 1.0 vol%. The experimental results also show that the electrical resistivity of UHPC reinforced with SSWs at a diameter of 20 μm is far lower than that of UHPC reinforced with SSWs at a diameter of 8 μm, and the electrical resistivity of UHPC reinforced with 0.5 vol% SSWs at a diameter of 20 μm is already lower than that of UHPC reinforced with 1.5 vol% SSWs at a diameter of 8 μm.

Comparing Figure 6.2(a) to Figure 6.2(b), the electrical resistivity of SSWs-engineered multifunctional UHPC after water curing for 28 days is smaller than that of composites after natural curing for 28 days. The difference is particularly obvious for SSWs-engineered multifunctional UHPC at a low content, because ionic conduction plays an important role in the conductive pathway of these composites. Meanwhile, the two-electrode method is easily affected by the contact resistance between the electrode and the UHPC matrix [10]. Therefore, the moistening of the electrode surface caused by water curing partly contributes to the decrease of the electrical resistivity.

6.2.1.2 Measured using the two-electrode direct current (DC) method

The electrical resistivity of SSWs-engineered multifunctional UHPC measured using the two-electrode DC method is shown in Figure 6.3. Figure 6.3 shows that the electrical resistivity decreases with increasing SSWs content, and the electrical resistivity of UHPC reinforced with SSWs at a diameter of 8 μm is considerably higher than that of UHPC reinforced with SSWs at a diameter of 20 μm, which indicates that the dispersal uniformity of SSWs at a diameter of 20 μm is better than those at a diameter of 8 μm at the same content. Figure 6.3(a) shows that the electrical resistivity of UHPC reinforced with 0.5 vol%/1.0 vol%/1.5 vol% SSWs at a diameter of 8 μm after water curing for 28 days is 21.8 × $10^4\Omega$ ·cm, 3.9 × $10^4\Omega$ ·cm, and 5.8 × $10^2\Omega$ ·cm, respectively. This indicates that the percolation phenomenon has already appeared when the content of SSWs at a diameter of 8 µm is 1.5 vol%, and the conductive pathway has shifted from the UHPC matrix to the SSWs network. Figure 6.3(c) shows that the electrical resistivity of UHPC reinforced with SSWs at a diameter of 8 µm after natural curing for 28 days increases with curing age and is relatively higher than that after water curing for 28 days. Figure 6.3(b) and (d) show that the percolation phenomenon of UHPC reinforced with SSWs at a diameter of 20 µm already exists when the content of SSWs is only 0.5 vol%, and the electrical resistivity of UHPC reinforced with 0.5 vol%/1.0 vol%/1.5 vol% SSWs at a diameter of 20 µm after natural curing for 28 days is 4.5 × $10^2\Omega$ ·cm, 44.0Ω ·cm, and 22.0Ω ·cm, respectively. The test results also show that the electrical resistivity of UHPC reinforced with SSWs at a diameter of 20 µm is less affected by curing age, which can be attributed to the SSWs playing an important role in the conductive pathway of composites. The UHPC reinforced with 0.5 vol% SSWs at a diameter of 20 µm after natural curing for 28 days has much higher electrical resistivity than that after natural curing

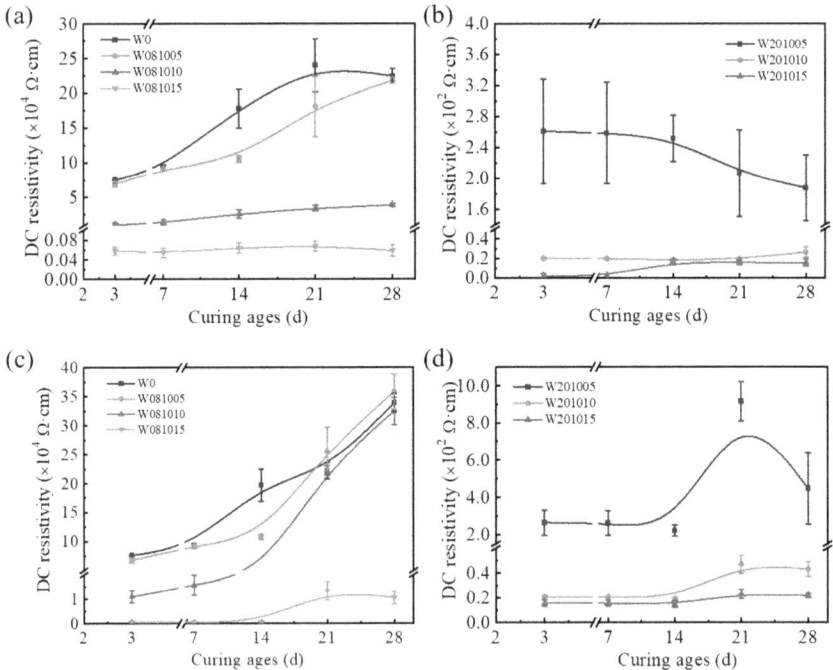

Figure 6.3 Electrical resistivity of SSWs-engineered multifunctional UHPC measured using the two-electrode DC method. (a) W0810, water curing; (b) W2010, water curing; (c) W0810, natural curing; (d) W2010, natural curing

for 21 days, because the overlapping of SSWs in these specimens is less, and the conductive pathway is susceptible to shrinkage crack caused by natural curing. Meanwhile, the electrical resistivity of UHPC reinforced with 0.5 vol%/1.0 vol%/1.5 vol% SSWs at a diameter of 20 μm after water curing for 14 days decreases with increasing curing age because of the effect of ionic conduction on the incompletely overlapping SSWs' conductive network.

It can be seen from Figures 6.2 and 6.3 that the percolation phenomenon of SSWs-engineered multifunctional UHPC is not subject to AC and DC voltage, but the percolation threshold of UHPC reinforced with SSWs at a diameter of 8 μm measured using AC voltage is affected by water content. The electrical resistivity of UHPC reinforced with SSWs at a diameter of 8 μm measured using the two-electrode DC method is considerably higher than that of composites measured using the two-electrode AC method, while there is little difference between the electrical resistivity of UHPC reinforced with SSWs at a diameter of 20 μm measured using these two methods. This is because the effect of polarization caused by DC voltage on electrical resistivity is closely related to the conductivity of composites, and it is more obvious when the conductive pathway of composites is dominated by the UHPC matrix.

6.2.1.3 Measured using the four-electrode DC method

The electrical resistivity of SSWs-engineered multifunctional UHPC after natural curing for certain ages measured using the four-electrode DC method is shown in Figure 6.4. Figure 6.4 shows that the electrical resistivity of SSWs-engineered multifunctional UHPC after natural curing for 14 days decreases slightly with the increase in curing age, except for UHPC without SSWs and with 1.0 vol% SSWs at a diameter of 20 μm. This can be attributed to the effect of shrinkage crack and pore solution concentration. The electrical resistivity of SSWs-engineered multifunctional UHPC decreases with increasing SSWs content, and the electrical resistivity of UHPC reinforced with 0.5 vol%/1.0 vol%/1.5 vol% SSWs at a diameter of 20 μm after natural curing for 28 days is 44.0Ω ·cm, 8.5Ω ·cm, and 4.7Ω ·cm, respectively. Meanwhile, the electrical resistivity of UHPC reinforced with SSWs at a diameter of 20 μm is far lower than that of UHPC reinforced with SSWs at a diameter of 8 μm. The percolation phenomenon has already appeared when the content of SSWs at a diameter of 8 μm and 20 μm are 1.5 vol% and 0.5 vol%, respectively. This is the same as the aforementioned results.

It can be seen from Figures 6.3 and 6.4 that the electrical resistivity of SSWs-engineered multifunctional UHPC measured using the two-electrode DC method is larger than that of composites measured using the four-electrode DC method, due to the contact resistance caused by the two-electrode method. The relationship between the test results measured using the four-electrode DC method and the two-electrode AC method varies with SSWs

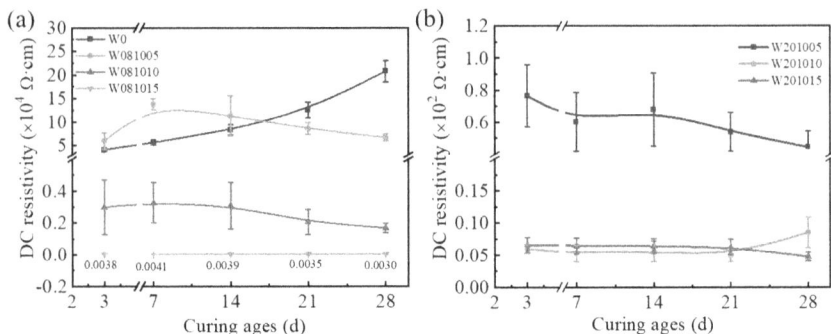

Figure 6.4 Electrical resistivity of SSWs-reinforced multifunctional UHPC measured using the four-electrode-DC method. (a) W0810; (b) W2010

content and diameter. The incorporation of SSWs at a diameter of 20 μm or 1.5 vol% SSWs at a diameter of 8 μm makes the electrical resistivity measured using the four-electrode DC method smaller than that of composites measured using the two-electrode AC method. This indicates that the effect of polarization on electrical resistivity is far less than that of contact resistance between the electrode and the UHPC matrix for high-conductivity composites. However, the experimental results for UHPC without SSWs and with 0.5 vol% SSWs at a diameter of 8 μm show the contrary trend, i.e. the effect of polarization caused by DC voltage is more obvious than that of contact resistance for composites with poor conductivity.

Figures 6.2–6.4 show that the resistivity variation trend of SSWs-engineered multifunctional UHPC caused by SSWs content and diameter is not subject to test method. In other words, the selection of a test method for electrical resistivity is determined by specimen dimensions, loading methods, working conditions, and test requirements together.

6.2.1.4 Electrical polarization phenomenon

The electrical polarization process of SSWs-engineered multifunctional UHPC after natural curing for certain ages is investigated using the two-electrode DC and the four-electrode DC method, as shown in Table 6.1 and Table 6.2, respectively. As shown in Table 6.1, the electrical resistivity measured using two-electrode DC increases with time during measurement, and it becomes stable after a period of time with the current on. Table 6.2 shows that the electrical resistivity of UHPC reinforced with SSWs at a diameter of 20 μm measured using four-electrode DC floats up and down because the value is too small. The polarization is obvious for specimens without SSWs or with a low content of SSWs at a diameter of 8 μm. This is because the conductivity of these composites is mainly dominated by ionic

Table 6.1 Change of electric resistivity of SSWs-engineered multifunctional UHPC measured using the two-electrode DC method with test time under polarization effect

Group	Age (d)	0	50	100	150	200	250	300	350	400	450	500
W0 ($\times 10^4 \Omega \cdot cm$)	3	5.8	7.2	7.4	7.5	7.5	7.5	7.5	–	–	–	–
	7	7.9	9.1	9.2	9.3	9.3	9.3	9.3	–	–	–	–
	14	13.8	16.7	17.1	17.5	17.7	17.7	17.7	–	–	–	–
	21	18.2	19.9	20.2	20.4	20.5	20.5	20.5	–	–	–	–
	28	28.3	31.5	31.9	32.2	32.2	32.3	32.3	–	–	–	–
W081005 ($\times 10^4 \Omega \cdot cm$)	3	3.9	6.1	6.5	6.7	6.8	6.8	6.8	6.8	–	–	–
	7	4.4	8.4	8.8	8.9	9.0	9.0	9.1	9.0	–	–	–
	14	7.1	9.7	10.1	10.2	10.3	10.4	10.5	10.5	–	–	–
	21	17.6	21.6	22.2	22.4	22.6	22.7	22.8	22.8	–	–	–
	28	26.8	33.2	34.0	34.4	34.6	34.8	34.9	34.9	–	–	–
W081010 ($\times 10^4 \Omega \cdot cm$)	3	0.3	0.7	0.8	0.9	0.9	0.9	0.9	0.9	1.0	1.0	1.0
	7	0.6	1.1	1.2	1.2	1.2	1.2	1.3	1.3	1.3	1.3	1.3
	14	1.5	1.8	2.0	2.1	2.2	2.2	2.3	2.3	2.4	2.4	2.4
	21	11.7	15.5	16.9	17.8	18.5	19.0	19.4	19.7	20.0	20.2	20.4
	28	22.3	27.2	28.7	29.7	30.4	30.9	31.2	31.5	31.7	31.9	32.1
W081015 ($\times 10^2 \Omega \cdot cm$)	3	5.5	5.6	5.7	5.7	5.7	5.8	5.8	5.8	5.8	5.8	–
	7	6.2	6.3	6.3	6.3	6.3	6.3	6.4	6.4	6.4	6.4	–
	14	6.0	6.4	6.4	6.4	6.4	6.5	6.5	6.5	6.5	6.5	–
	21	93.9	96.9	97.8	98.3	98.8	99.1	99.3	99.5	99.7	99.8	–
	28	63.7	77.2	77.8	78.1	78.3	78.3	78.3	78.3	78.3	78.3	–
W201005 ($\times 10^2 \Omega \cdot cm$)	3	2.3	2.6	2.6	2.6	2.6	2.6	–	–	–	–	–
	7	2.6	2.6	2.6	2.6	2.6	2.6	–	–	–	–	–
	14	2.5	2.5	2.5	2.5	2.5	2.5	–	–	–	–	–
	21	8.4	8.4	8.4	8.4	8.4	8.4	–	–	–	–	–
	28	6.6	6.7	6.7	6.7	6.7	6.7	–	–	–	–	–
W201010 ($\Omega \cdot cm$)	3	19.8	19.6	18.8	19.2	19.6	19.7	19.7	–	–	–	–
	7	19.6	19.6	19.6	19.6	19.6	19.6	19.6	–	–	–	–
	14	18.0	18.0	18.0	18.0	18.0	18.0	18.0	–	–	–	–
	21	50.0	50.1	50.1	50.1	50.1	50.2	50.2	–	–	–	–
	28	48.4	48.5	48.5	48.5	48.5	48.5	48.5	–	–	–	–
W201015 ($\Omega \cdot cm$)	3	18.3	18.3	18.2	18.1	18.3	18.3	18.3	–	–	–	–
	7	18.4	18.4	18.4	18.4	18.4	18.4	18.4	–	–	–	–
	14	17.4	17.4	17.4	17.4	17.4	17.4	17.4	–	–	–	–
	21	21.9	21.9	21.9	21.9	21.9	21.9	21.9	–	–	–	–
	28	21.6	21.6	21.6	21.6	21.6	21.6	21.6	–	–	–	–

Table 6.2 Change of electric resistivity of SSWs-engineered multifunctional UHPC measured using the four-electrode DC method with test time under polarization effect

Group	Age (d)	Time (s)										
		0	50	100	150	200	250	300	350	400	450	500
W0	3	3.9	5.0	5.1	5.2	5.2	5.2	6.2	5.2	5.2	5.1	5.1
($\times 10^4 \Omega \cdot cm$)	7	5.4	5.8	5.9	6.0	6.0	6.0	6.0	6.0	6.0	6.0	6.0
	14	5.6	7.7	8.5	9.0	9.5	9.9	10.2	10.4	10.7	10.9	11.1
	21	4.3	7.9	8.6	9.0	9.4	9.7	9.9	10.1	10.3	10.5	10.6
	28	10.3	13.4	14.5	15.5	16.4	17.1	17.8	18.3	18.9	19.3	19.6
W081005	3	5.8	5.8	5.9	5.9	6.0	6.0	6.0	6.0	6.1	6.1	6.2
($\times 10^4 \Omega \cdot cm$)	7	13.1	13.1	13.2	13.4	13.4	13.6	13.8	13.9	14.0	13.8	14.0
	14	11.1	11.2	11.3	11.4	11.6	11.6	11.6	11.6	11.5	11.7	11.8
	21	6.5	6.6	6.6	6.7	6.8	6.8	6.9	6.9	7.0	7.0	7.1
	28	5.7	5.9	6.1	6.2	6.4	6.5	6.5	6.6	6.7	6.8	6.8
W081010	3	0.16	0.20	0.22	0.23	0.23	0.24	0.24	0.24	0.24	–	–
($\times 10^4 \Omega \cdot cm$)	7	0.25	0.30	0.33	0.36	0.37	0.39	0.41	0.41	0.41	–	–
	14	0.23	0.23	0.23	0.26	0.24	0.26	0.26	0.26	0.24	–	–
	21	0.18	0.18	0.18	0.18	0.18	0.18	0.18	0.18	0.18	–	–
	28	0.15	0.15	0.15	0.15	0.15	0.15	0.15	0.15	0.15	–	–
W081015	3	30.8	30.9	31.0	31.1	31.0	31.1	31.1	31.1	–	–	–
($\Omega \cdot cm$)	7	30.8	34.9	22.7	28.1	21.4	32.2	31.4	32.2	–	–	–
	14	30.0	30.7	29.9	31.5	30.8	29.7	30.1	30.8	–	–	–
	21	30.0	30.6	30.0	28.7	29.4	27.7	30.2	30.5	–	–	–
	28	27.8	27.9	27.9	27.9	27.9	27.9	27.9	27.9	–	–	–
W201005	3	68.0	69.2	69.2	69.2	69.2	69.2	–	–	–	–	–
($\Omega \cdot cm$)	7	79.2	65.7	73.8	73.8	69.8	65.7	–	–	–	–	–
	14	73.6	70.9	68.9	67.6	79.0	76.3	–	–	–	–	–
	21	65.4	65.8	64.4	64.9	65.7	65.7	–	–	–	–	–
	28	41.1	41.0	41.1	41.1	41.1	41.1	–	–	–	–	–
W201010	3	2.6	2.2	3.5	3.8	4.2	4.0	–	–	–	–	–
($\Omega \cdot cm$)	7	4.3	4.2	4.2	4.1	4.2	4.2	–	–	–	–	–
	14	4.5	5.1	5.1	5.2	4.9	5.0	–	–	–	–	–
	21	2.8	2.7	3.1	3.2	3.6	3.5	–	–	–	–	–
	28	8.8	8.8	8.8	8.8	8.8	8.8	–	–	–	–	–
W201015	3	6.9	7.1	7.1	7.1	7.1	7.1	–	–	–	–	–
($\Omega \cdot cm$)	7	7.0	7.0	7.0	7.0	7.0	7.0	–	–	–	–	–
	14	7.0	7.1	7.0	7.0	7.1	7.0	–	–	–	–	–
	21	6.3	6.0	6.0	5.2	6.1	6.1	–	–	–	–	–
	28	4.6	4.6	4.6	4.6	4.6	4.6	–	–	–	–	–

conduction and double layer capacitance formed in the C-S-H gel surface and the interface between the SSWs and the UHPC matrix. As a result, the charging of the capacitors causes a linear increase in the resistivity [11]. The tendency to polarization decreases with increasing SSWs content and diameter, and the UHPC reinforced with SSWs at a diameter of 20 μm is not affected by polarization. Since a conductor cannot support a significant electric field, polarization will be weakened [12], i.e. the lower the electrical resistivity, the lower the polarization. The polarization change rule of SSWs-engineered multifunctional UHPC is consistent with that of electrical resistivity, and the experimental results of polarization are not affected by electrode layout.

The occurrence of the percolation phenomenon can also be observed by the change of polarization. The results show that the polarization is not obvious when the SSWs content is 1.5 vol% and 0.5 vol% for SSWs with a diameter of 8 μm and 20 μm, respectively, which represents the occurrence of percolation. This is the same as the results obtained from the change of electrical resistivity.

6.2.2 Impedance characteristic and intrinsic conductivity

6.2.2.1 Electrochemical impedance spectroscopy (EIS) and equivalent circuit

The EIS of SSWs-engineered multifunctional UHPC under different curing conditions can be drawn in a Nyquist plot, as shown in Figure 6.5. The frequency in each plot decreases from left to right. It can be seen from Figure 6.5 that the location, time parameter, and topological structure of EIS vary with SSWs content, diameter, and curing condition. The equivalent circuit, which is the most important and intuitive method for analyzing EIS, can connect EIS and the electrode process kinetics model. The equivalent element (e.g. resistance, capacitance, and inductance) and composite equivalent element (e.g. the series or parallel of resistance and capacitance), which have a definite physical meaning, can be used to represent the various stages of EIS during the drawing process of the equivalent circuit [13]. For example, the Nyquist plot of the parallel of resistance and capacitance is a semicircle in the first quadrant or a flattening semicircle because of the constant phase angle element carried by double layer capacitance in the electrode interface, and an oblique line in the first quadrant represents the diffusion impedance caused by charge diffusion. Based on the theoretical analysis of equivalent element and EIS, the corresponding equivalent circuits of SSWs-engineered UHPC are plotted in Figure 6.6.

Figure 6.5(a) shows that the EIS of UHPC without SSWs consists of three parts, and the corresponding equivalent circuit is shown in Figure 6.6(a). (1) The circular section can be represented by the parallel of resistance and

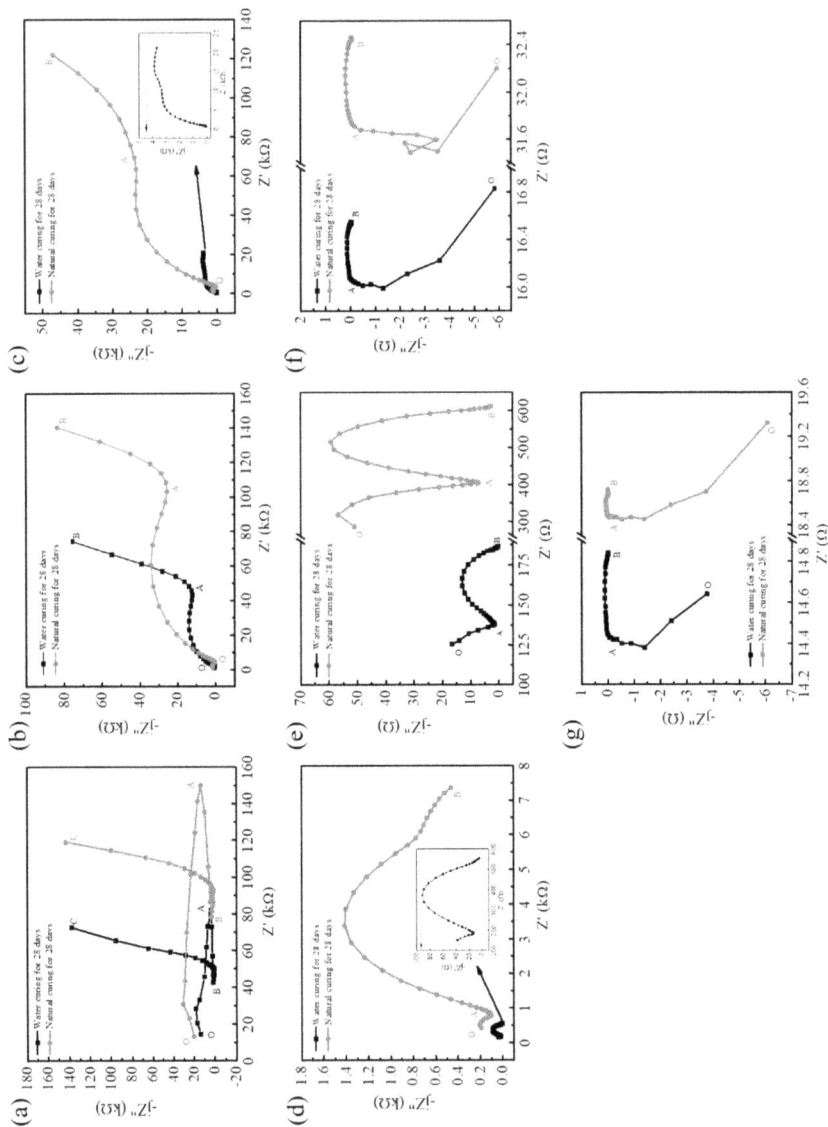

Figure 6.5 EIS of SSWs-engineered multifunctional UHPC. (a) W0; (b) W081005; (c) W081010; (d) W081015; (e) W201005; (f) W201010; (g) W201015

capacitance elements. The beginning of the EIS is marked as O, and it corresponds to pore solution resistance (R_s) in the equivalent circuit. The value of R_s is an inverse function of porosity and pore solution concentration. The arc diameter (OA section) in the high-frequency region contains the value of charge transfer resistance (R_{st}), which is closely related to the cement hydration product, water content, and ion concentration. The OA section of UHPC without SSWs after water curing for 28 days is smaller than that of composite after natural curing for 28 days. This means that the charge transfer resistance is lowered due to high water content and ion concentration. Meanwhile, the OA section also involves the resistance caused by C-S-H gel surface capacitance, which is represented by C_d in Figure 6.6(a). (2) The AB section shows that ion adsorption and desorption have occurred in the C-S-H gel surface in the process of voltage application, which is expressed as inductance (L) in the equivalent circuit. (3) The BC section in the low-frequency region shows that the real part and the imaginary part have a linear relationship, and the slope is closely related to the inner connected pore structure of SSWs-engineered UHPC. This electrochemical process is controlled by diffusion, and the corresponding impedance is called diffusion impedance (Z_w).

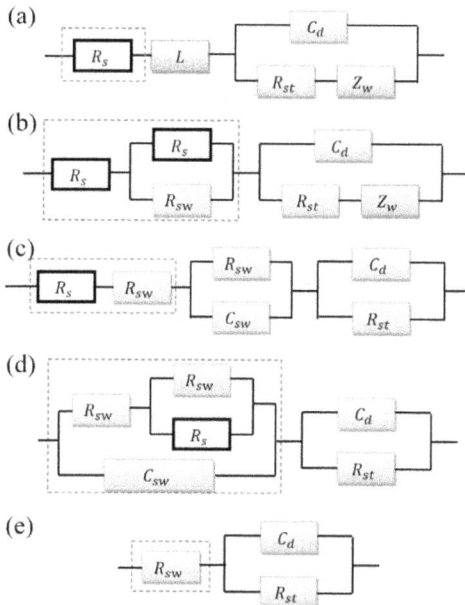

Figure 6.6 Equivalent circuits of SSWs-engineered multifunctional UHPC. (a) W0; (b) W081005; (c) W081010; (d) W081015/W201005; (e) W201010/W201015. R_s-Pore solution resistance; R_{st}-Charge transfer resistance; L-Inductance; C_d-C-S-H gel surface capacitance; Z_w-Diffusion impedance; R_{sw}-SSWs resistance; C_{sw}-SSWs-matrix interface capacitance

Figure 6.5(b) shows that the EIS of UHPC reinforced with 0.5 vol% SSWs at a diameter of 8 μm is a classic Randles curve. The intersection point of the curve and the x-axis at the high frequency is marked as O, and it is determined by the UHPC matrix and the SSWs together. Therefore, the corresponding equivalent circuit is expressed as the parallel connection of pore solution resistance (R_s) and SSWs resistance (R_{sw}), as shown in Figure 6.6(b). This is based on the idea that a large part of the matrix may be bypassed by current flow due to SSWs incorporation. The OA section in Figure 6.5(b) is controlled by the charge transfer resistance (R_{st}) and the C-S-H gel surface capacitance (C_d) together. The AB section in Figure 6.5(b) represents the process of charge diffusion, which is irrelevant to SSWs. This can be attributed to an electric double layer forming at the SSWs' surface and insulating SSWs under the excitation of low-frequency AC voltage [14]. The linear part of the EIS controlled by charge diffusion gradually disappears with increasing SSWs content and diameter. This means that SSWs have gradually become a leading element in the conductive pathway.

It can be seen from Figure 6.5(c) that the EIS of UHPC reinforced with1.0 vol% SSWs at a diameter of 8 μm appears as a dual tangent arc in the Nyquist plot, which represents two time constants. The corresponding equivalent circuit is expressed as two series–parallel connections of resistance–capacitance cells, as shown in Figure 6.6(c). One happens at the SSWs–matrix interface, and the other happens at the C-S-H gel surface. This indicates that SSWs behave as insulators at low frequency and conductors at high frequency [15]. The resistance represented by the point O at high frequency is made up of (R_s) and (R_{sw}) together. The lower frequency is required to collect the resistance produced by the C-S-H gel electric double layer capacitance and diffusion impedance after water curing for 28 days.

Figure 6.5(d) and (e) show that the EIS of UHPC reinforced with1.5 vol% SSWs at a diameter of 8 μm and UHPC reinforced with 0.5 vol% SSWs at a diameter of 20 μm has a similar topological structure, and it is composed of two semicircles. The SSWs and the pore solution behave as conductors in a series–parallel way at high frequency. The corresponding equivalent circuit consists of (R_{sw}), (R_s), and C_{sw}, as shown in Figure 6.6(d), where C_{sw} represents the double layer capacitance at the SSWs–matrix interface. The double layer capacitance of the C-S-H gel surface and the charge transfer behave as conductors at low frequency, and their resistance can be represented by (C_d) and (R_{st}), respectively. Figure 6.6(d) shows that the equivalent circuit has been dominated by overlapped SSWs. This result illustrates that the percolation network has formed when the content of SSWs with a diameter of 8 μm and 20 μm is 1.5 vol% and 0.5 vol%, respectively, in accordance with the test results for electrical resistivity.

Because of low impedance, the EIS test process of UHPC reinforced with 1.0 vol%/1.5 vol% SSWs at a diameter of 20 μm is vulnerable to external and internal conduction connections. Therefore, the inductance appears in the fourth quadrant at high frequency, indicated by OA sections in Figure 6.5(f) and (g). At low frequency, only a flattened semicircle (AB section) appears in the first quadrant. The depression angle of the semicircle can be attributed to a constant phase element, which is sensitive to the pore diameters of composites, especially in conductive fiber reinforced materials [16]. The equivalent circuits of UHPC reinforced with 1.0 vol%/1.5 vol% SSWs at a diameter of 20 μm are shown in Figure 6.6(e). It can be seen from Figure 6.6(e) that the conductive pathway is composed of SSWs entirely at high frequency. Meanwhile, the pathway involves a parallel connection of charge transfer resistance and double layer capacitance of the C-S-H gel surface at low frequency.

Based on this analysis, the conductive pathway of SSWs-engineered multifunctional UHPC can be divided into three categories. (1) When the SSWs content is below the percolation threshold, the electrical transport behavior is dominated by the UHPC matrix. The SSWs are scattered in UHPC and cannot contact each other, so the electrical conduction of the composite is limited. UHPC reinforced with 0.5 vol%/1.0 vol% SSWs at a diameter of 8 μm is of this kind. (2) The SSWs and cement matrix constitute a conductive pathway together with the increase of SSWs content. Meanwhile, the double layer capacitance at the SSWs–matrix interface and the C-S-H gel surface also plays a certain role in electrical conduction. The conductive pathways of UHPC reinforced with 1.5 vol% SSWs at a diameter of 8 μm and UHPC reinforced with 0.5 vol% SSWs at a diameter of 20 μm fall into this category. (3) When the conductive network of SSWs has been formed in composites, the electrical transport behavior is dominated by overlapped SSWs. UHPC reinforced with 1.0 vol%/1.5 vol% SSWs at a diameter of 20 μm belong in this category. The three categories show that the conductive mechanism of SSWs-engineered UHPC mainly depends on the connection of SSWs, which is different from that of carbon fiber reinforced cement-based materials [17].

6.2.2.2 Intrinsic conductivity

The intrinsic conductivity of SSWs-engineered multifunctional UHPC is equal to the composite conductivity (σ) divided by the matrix conductivity (σ_m), as shown in Eq. (6.1). The Fixman equation relating the intrinsic conductivity to the aspect ratio (AR), or length divided by diameter, of SSWs is shown as Eq. (6.2) [15]:

$$\frac{\sigma}{\sigma_m} = \frac{R_m}{R} = 1 + [\sigma]_\Delta \Phi \tag{6.1}$$

$$[\sigma]_\infty = \frac{1}{3}\left(\frac{2(AR)^2}{\left[3\ln\{4AR\} - 7\right]} + 4 \right) \tag{6.2}$$

where AR is the aspect ratio of SSWs, Φ is the volume content of SSWs, R_m is the composite resistivity, and R is the matrix resistivity.

Table 6.3 Calculated and experimental intrinsic conductivity of SSWs-engineered multifunctional UHPC

Group	W081005	W081010	W081015	W201005	W201010	W201015
Experimental values (water curing)	17.5	61.0	218.3	343.4	2777.1	3701.0
Standard deviation	0.94	3.65	30.18	68.00	76.06	542.51
Experimental values (natural curing)	19.0	23.8	105.4	212.3	2708.0	4656.9
Standard deviation	1.92	1.01	6.55	86.00	266.21	397.13
Calculated values	281.8	562.5	843.3	53.7	106.5	159.3

At the same SSWs content and aspect ratio, the deviation of the experimental results from the expected results calculated by Eq. (6.2) can be used to quantify the dispersion of SSWs. The calculated and experimental results of intrinsic conductivity are listed in Table 6.3. Table 6.3 shows that the intrinsic conductivity of SSWs-engineered multifunctional UHPC does meet Eq. (6.2). The experimental intrinsic conductivity of UHPC reinforced with SSWs at a diameter of 8 μm is much lower than the calculated results, especially for specimens after natural curing for 28 days. However, the experimental result of UHPC reinforced with SSWs at a diameter of 20 μm is higher than the calculated values, and it is also considerably greater than the experimental result for UHPC reinforced with SSWs at a diameter of 8 μm. This indicates that the SSWs at a diameter of 8 μm cannot uniformly disperse in UHPC, and they are easily clustered into a bundle. On the contrary, the SSWs at a diameter of 20 μm have better dispersity, and the effective utilization ratio is improved correspondingly. Therefore, the SSWs at a diameter of 20 μm are more suitable for improving the conductivity of UHPC.

6.3 SELF-SENSING PROPERTIES OF SSWS-ENGINEERED MULTIFUNCTIONAL UHPC

6.3.1 Under compressive load

6.3.1.1 Under cyclic compressive load

The relationships between cyclic compressive stress/strain and the fractional change in electrical resistivity of SSWs-engineered multifunctional UHPC after natural curing for 28 days are established, as shown in Figures 6.7–6.12.

The results indicate that SSWs-engineered multifunctional UHPC has piezo-resistive response under cyclic compressive load within the elastic regime, and the characteristic of the piezoresistive response varies with SSWs content and diameter, although the maximum value of the fractional change in electrical resistivity can only achieve 1%. It can be seen from Figures 6.7 and 6.8 that the fractional change in electrical resistivity of UHPC reinforced with 0.5 vol%/1.0 vol% SSWs at a diameter of 8 μm decreases upon loading and increases upon unloading, but the stability and reversibility of the piezore-sistivity are poor. At the beginning of each cycle, the electrical resistivity is higher than the initial value never subjected to external force. This is because some internal defects (e.g. native cracks, holes, gel peristalsis, etc.) are gradu-ally reduced in cycles of compressive loading [18]. As shown in Figure 6.9, the fractional change in electrical resistivity of UHPC reinforced with 1.5 vol% SSWs at a diameter of 8 μm is only 0.3% upon loading and unload-ing. However, the change is stable and reversible. The electrical resistivity at the beginning of each circle is lower than the initial value never subjected to external force, which is different from UHPC reinforced with 0.5 vol%/1.0

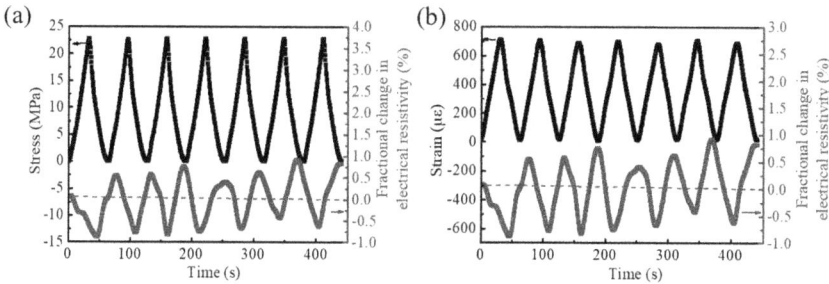

Figure 6.7 Variation of the fractional change in electrical resistivity with stress and strain for W081005 under seven times of cyclic compressive load. (a) Time-Stress-FCR; (b) Time-Strain-FCR

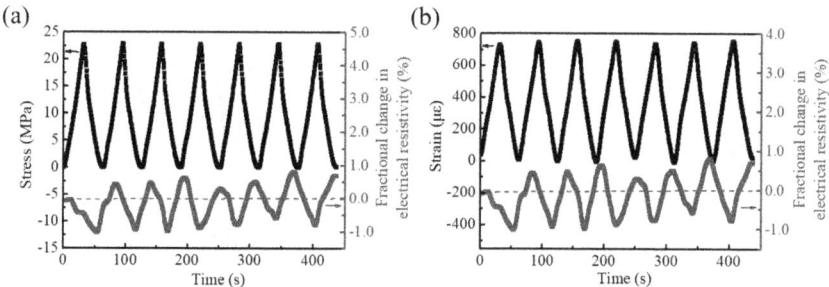

Figure 6.8 Variation of the fractional change in electrical resistivity with stress and strain for W081010 under seven times of cyclic compressive load. (a) Time-Stress-FCR; (b) Time-Strain-FCR

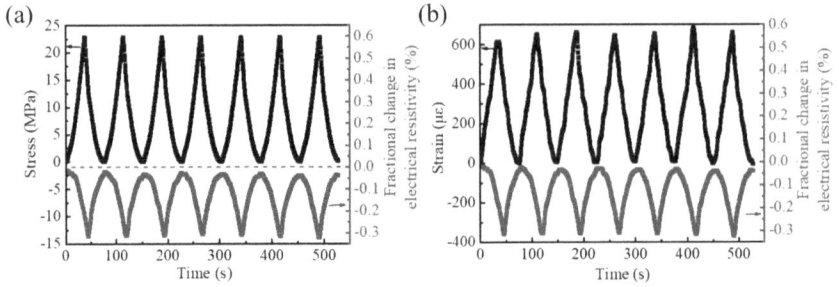

Figure 6.9 Variation of the fractional change in electrical resistivity with stress and strain for W081015 under seven times of cyclic compressive load. (a) Time-Stress-FCR; (b) Time-Strain-FCR

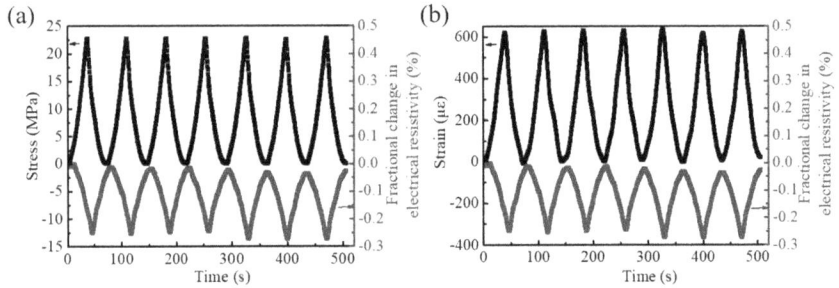

Figure 6.10 Variation of the fractional change in electrical resistivity with stress and strain for W201005 under seven times of cyclic compressive load. (a) Time-Stress-FCR; (b) Time-Strain-FCR

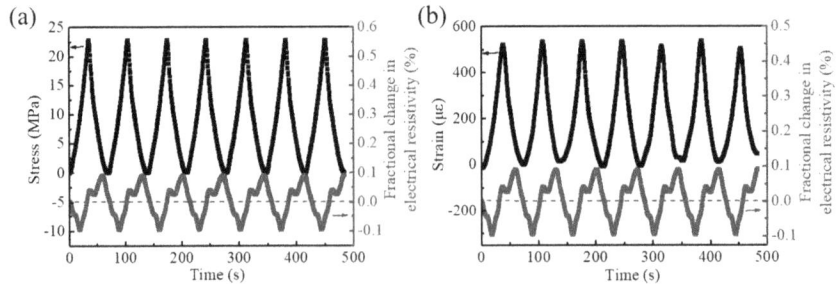

Figure 6.11 Variation of the fractional change in electrical resistivity with stress and strain for W201010 under seven times of cyclic compressive load. (a) Time-Stress-FCR; (b) Time-Strain-FCR

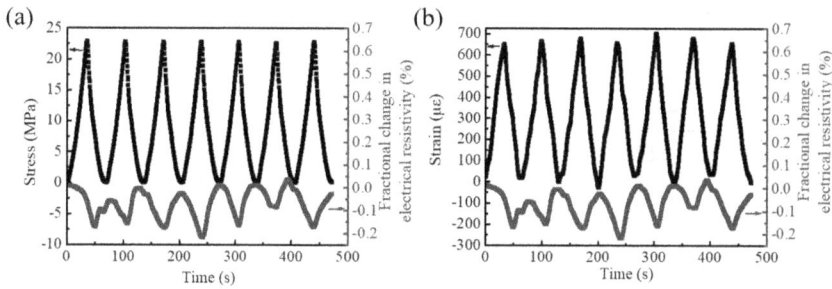

Figure 6.12 Variation of the fractional change in electrical resistivity with stress and strain for W201015 under seven times of cyclic compressive load. (a) Time-Stress-FCR; (b) Time-Strain-FCR

vol% SSWs at a diameter of 8 μm. This results from the increasing overlapping of SSWs due to gel viscous peristalsis under loading. Figure 6.10 shows that the characteristic of the fractional change in electrical resistivity of UHPC reinforced with 0.5 vol% SSWs at a diameter of 20 μm is similar to that of UHPC reinforced with 1.5 vol% SSWs at a diameter of 8 μm. The experimental results also show that the UHPC reinforced with 0.5 vol% SSWs at a diameter of 20 μm has stable piezoresistive response, although the fractional change in electrical resistivity is only 0.3% under cyclic loading. It can be seen from Figure 6.11 that there is good repeatability of the fractional change in electrical resistivity for UHPC reinforced with 1.0 vol% SSWs at a diameter of 20 μm, but the fractional change in electrical resistivity under cyclic loading is less than 0.1%. Figure 6.12 shows that there is no definite corresponding relationship between the fractional change in electrical resistivity and compressive stress/strain for UHPC reinforced with 1.5 vol% SSWs at a diameter of 20 μm.

The fractional change in electrical resistivity of SSWs-engineered multifunctional UHPC decreases with increasing SSWs content and diameter under cyclic compressive load. The stability and reversibility of the piezoresistive response become better with the increase of SSWs content at a diameter of 8 μm. However, there is a reverse change trend with the increase of SSWs content at a diameter of 20 μm. This is because the contact status between SSWs plays a critical role in the piezoresistive behavior of SSWs-engineered multifunctional UHPC. The stable piezoresistive response occurs only at an appropriate SSWs content, when the SSWs are close to each other but cannot contact each other, such as UHPC reinforced with 1.5 vol% SSWs at a diameter of 8 μm and UHPC reinforced with 0.5 vol%/1.0 vol% SSWs at a diameter of 20 μm. When the wire conductive network has been completely formed, the conductivity of the composites is mainly dominated by ohmic contact. Therefore, this kind of composite has no piezoresistivity, such as UHPC reinforced with 1.5 vol% SSWs at a diameter of 20 μm.

Table 6.4 Strain sensitivity of SSWs-engineered multifunctional UHPC under cyclic compressive load

Group	Strain sensitivity	Standard deviation	Group	Strain sensitivity	Standard deviation
W081005	22.52	4.57	W201005	3.91	0.15
W081010	13.82	1.28	W201010	3.71	0.17
W081015	4.61	0.29	W201015	2.66	0.70

The strain sensitivity of SSWs-engineered UHPC under cyclic compressive load is shown in Table 6.4. It is interesting to note that the strain sensitivity of UHPC reinforced with SSWs at a diameter of 20 μm is significantly lower than that of UHPC reinforced with SSWs at a diameter of 8 μm at the same volume content, while the conductivity of the former is better than that of the latter. This indicates that the dispersion of SSWs at a diameter of 20 μm is better than that of SSWs at a diameter of 8 μm. Table 6.4 also shows that the strain sensitivity of UHPC reinforced with SWs at a diameter of 8 μm is considerably larger than that of a commercial metal strain gauge, which has a gauge factor of 2.0–3.0. However, the research of Yu et al. [19] shows that the fractional change in electrical resistance of cement paste containing 0.1 wt% carbon nano-tubes (CNTs) can reach 11.4%, and the research of Azhari et al. [20] shows that the gauge factor of cement paste (with silica fume) containing 15 vol% carbon fiber or 15 vol% carbon fiber and 1 vol% multi-wall CNTs is approximately 445. But, the matrix in these references does not contain aggregate and deforms more obviously under cyclic compressive load compared with UHPC. The experimental results also suggest that the conductive mechanism of SSWs-engineered UHPC is different from that of carbon fiber reinforced cement-based materials [17]. There is no hole conduction in SSWs-engineered multifunctional UHPC, and the electron transport is relatively difficult due to the compact structure of UHPC. As a result, only the adjacent SSWs can make the composites have stable piezoresistivity.

The cyclic compressive stress and fractional change in electrical resistivity of SSWs-engineered multifunctional UHPC have relatively poor quadratic correlation. This phenomenon can be attributed to the dense structure of UHPC and the high bond strength between the SSWs and the UHPC matrix. Therefore, the contact resistance between the SSWs and the UHPC matrix is less affected by the elastic load.

6.3.1.2 Under monotonic compressive load

The fractional change in electrical resistivity of SSWs-engineered multifunctional UHPC after natural curing for 28 days under monotonic compressive load is shown in Figure 6.13. It can be seen from Figure 6.13 that the fractional change in electrical resistivity increases first slowly and then rapidly

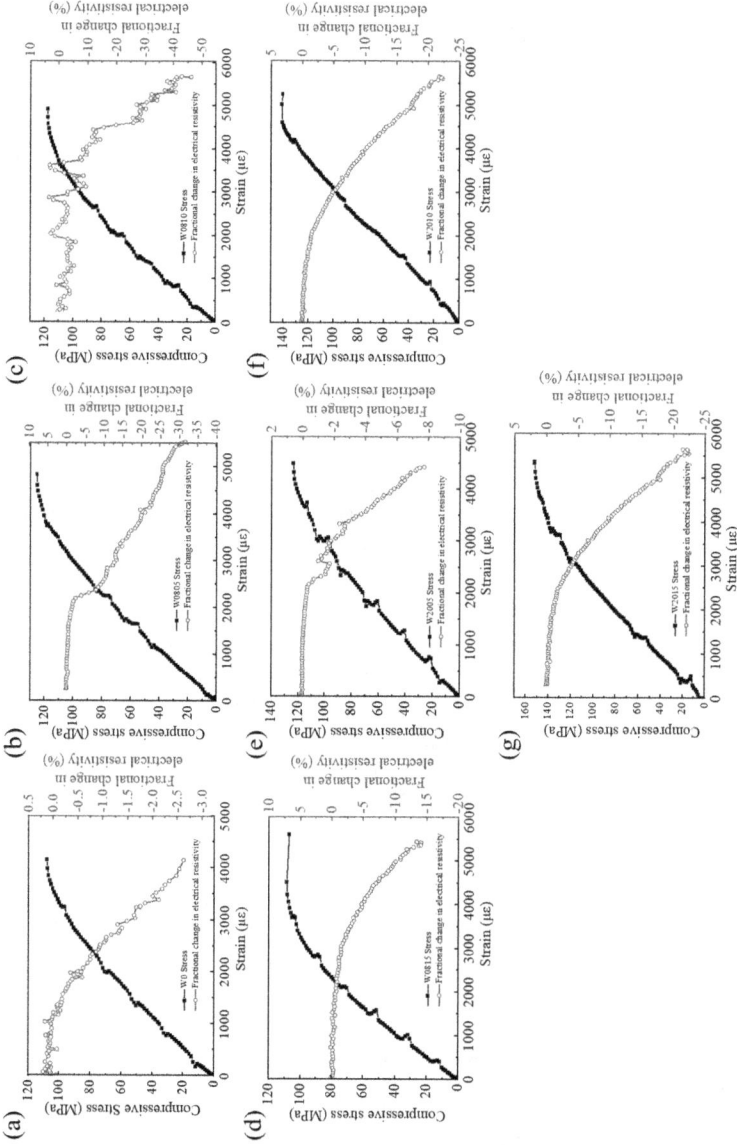

Figure 6.13 Relationship between monotonic compressive strain, stress, and fractional change in electrical resistivity of SSWs-engineered multi-functional UHPC. (a) W0; (b) W081005; (c) W081015; (d) W081010; (e) W201005; (f) W201010; (g) W201015

with the increase of compressive stress/strain, and this can be attributed to the variation of conductive network and specimen deformation under load.

Figure 6.13(a) shows that the fractional change in electrical resistivity of UHPC without SSWs under monotonic compressive load can only reach 2.6%. This is because the UHPC matrix has a low volume fraction of defects, and the dynamics of the defects have only minor effects on the electrical resistivity of UHPC [21]. It can be seen from Figure 6.13(b)–(g) that the fractional change in electrical resistivity corresponding to peak failure stress first increases and then decreases with increasing SSWs content. The resistivity fractional changes of UHPC reinforced with 0.5 vol%/1.0 vol% SSWs at a diameter of 8 μm under peak failure stress are 31.8% and 46.2%, respectively, while the changes of UHPC reinforced with 0.5 vol%/1.0 vol% SSWs at a diameter of 20 μm are only 7.7% and 22.1%, respectively. However, the monotonic compressive stress/strain of UHPC reinforced with 0.5 vol%/1.0 vol% SSWs at a diameter of 20 μm is greater than that of UHPC reinforced with 0.5 vol%/1.0 vol% SSWs at a diameter of 8 μm. This phenomenon illustrates that the effect of strain and damage on resistivity change is less than that of inner SSWs dispersion, especially for a low content of SSWs-engineered multifunctional UHPC. The specimen deformation is crucial to resistivity change until the SSWs have formed a conductive network. This can be used to explain the experimental results that the fractional change in electrical resistivity can reach up to 21.4% for UHPC reinforced with 1.5 vol% SSWs at a diameter of 20 μm, while it is only 14.1% for UHPC reinforced with 1.5 vol% SSWs at a diameter of 8 μm.

The strain sensitivity of SSWs-engineered multifunctional UHPC under monotonic compressive load is shown in Table 6.5. It can be seen from Table 6.5 that the strain sensitivity has the same change trend as the fractional change in electrical resistivity, which is also closely related to the volume content and aspect ratio of SSWs. The strain sensitivity first increases and then decreases with increasing SSWs volume content. UHPC reinforced with 0.5 vol%/1.0 vol% SSWs at a diameter of 8 μm has higher strain sensitivity than UHPC reinforced with 0.5 vol%/1.0 vol% SSWs at a diameter of 20 μm, while the strain sensitivity of UHPC reinforced with 1.5 vol% SSWs at a diameter of 8 μm is lower than that of UHPC reinforced with 1.5 vol% SSWs at a diameter of 20 μm. The former is based on the change of

Table 6.5 Strain sensitivity of SSWs-engineered multifunctional UHPC under monotonic compressive load

Group	Strain sensitivity	Standard deviation	Group	Strain sensitivity	Standard deviation
W081005	62.8	4.14	W201005	17.9	0.92
W081010	94.9	1.20	W201010	42.7	0.78
W081015	29.7	6.29	W201015	42.0	2.90

the SSWs' conductive network under load, and the latter can be attributed to deformation of specimens. The clustering of SSWs at a diameter of 8 μm limits the reinforcing effect of the fibers, while SSWs at a diameter of 20 μm can provide a good lateral restriction effect. The strain sensitivities are considerably larger than the gauge factor of a commercial metal strain gauge. However, the test results are much lower than the results of carbon fiber reinforced cement-based material, in which the fractional change in electrical resistance can reach up to 50% under 110-kN monotonic compressive load (the diameter of the pressure surface is 50.8 mm) [20]. The gap is derived from the differences in the matrix and the conductive mechanisms of carbon fiber and SSWs.

The effect of monotonic compressive stress on the resistivity of SSWs-engineered multifunctional UHPC is also closely related to SSWs content and diameter, and it has the same change rule as strain sensitivity. Generally, fibers with a smaller diameter can contact each other in the matrix more easily at the same content and length, and the corresponding fractional change in electrical resistivity under load is small. It is worth noting that the experimental results of SSWs-engineered multifunctional UHPC are contrary to this general knowledge. This also indicates that the dispersion of SSWs at a diameter of 20 μm in UHPC is better than at a diameter of 8 μm. Therefore, SSWs at a diameter of 20 μm are more suitable to be used as a conductive filler for UHPC.

6.3.2 Under flexural load

The electrical resistance of SSWs-engineered multifunctional UHPC after water and natural curing for 28 days is measured in real time using the two-electrode DC method under flexural load. The relationships between the flexural-tensile stress/strain and fractional change in electrical resistivity are shown in Figure 6.14 and Figure 6.15.

Figure 6.14(a) and Figure 6.15(a) show that the fractional change in electrical resistivity of UHPC without SSWs fluctuates between –2% and 2%, and the resistivity shows an instant rise when the flexural-tensile stress reaches its peak. This indicates that UHPC without SSWs has no piezoresistivity, and its electrical behavior is free from curing conditions. It can be seen from Figure 6.14(b)–(d) that the electrical resistivity change rate of UHPC reinforced with SSWs at a diameter of 8 μm after water curing for 28 days decreases with increasing SSWs content, and the fractional change values under peak failure stress are 1.0%, 1.24%, and 0.54%, respectively. Figure 6.14(e)–(g) show that the electrical resistivity change rate of UHPC reinforced with SSWs at a diameter of 20 μm increases with increasing SSWs content, and the fractional change values under peak failure stress are –0.14%, 0.68%, and 1.43%, respectively. Under flexural load, the bottom part of specimens is tensioned, and the SSWs in this area move away from

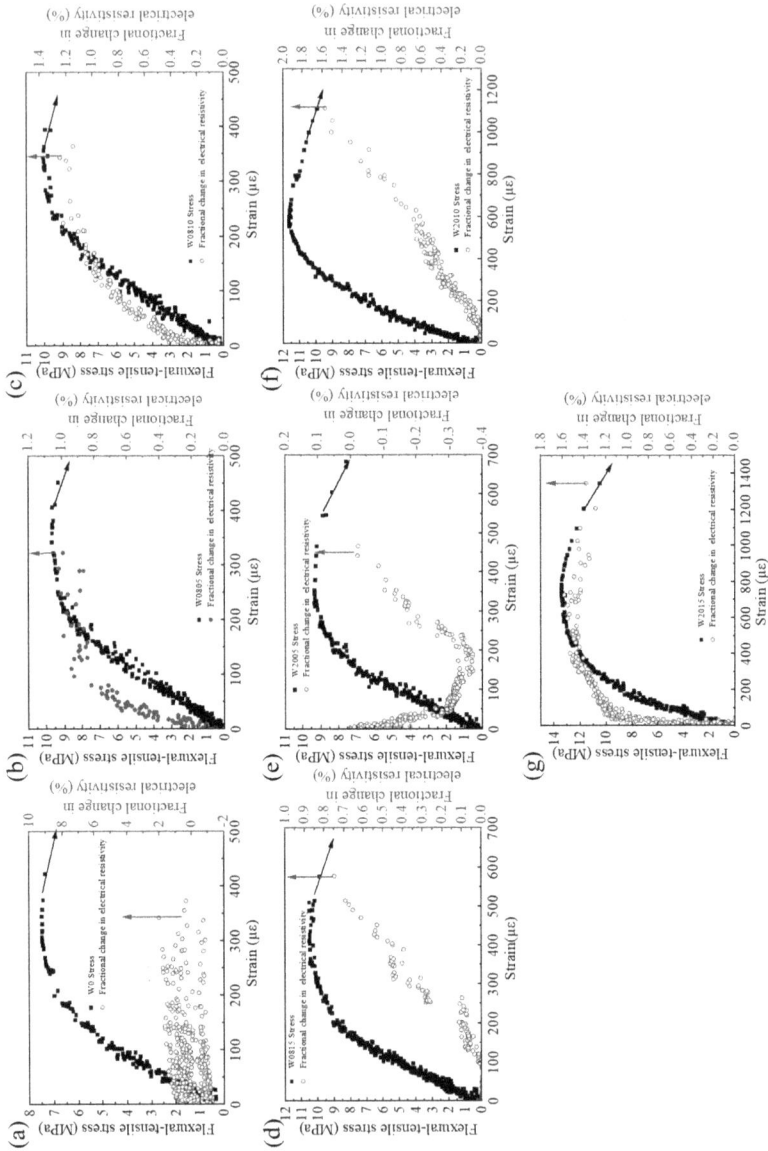

Figure 6.14 Relationships between flexural-tensile strain, stress, and fractional change in electrical resistivity of SSWs-engineered multifunctional UHPC (water curing). (a) W0; (b) W081005; (c) W081010; (d) W081015; (e) W201005; (f) W201010; (g) W201015

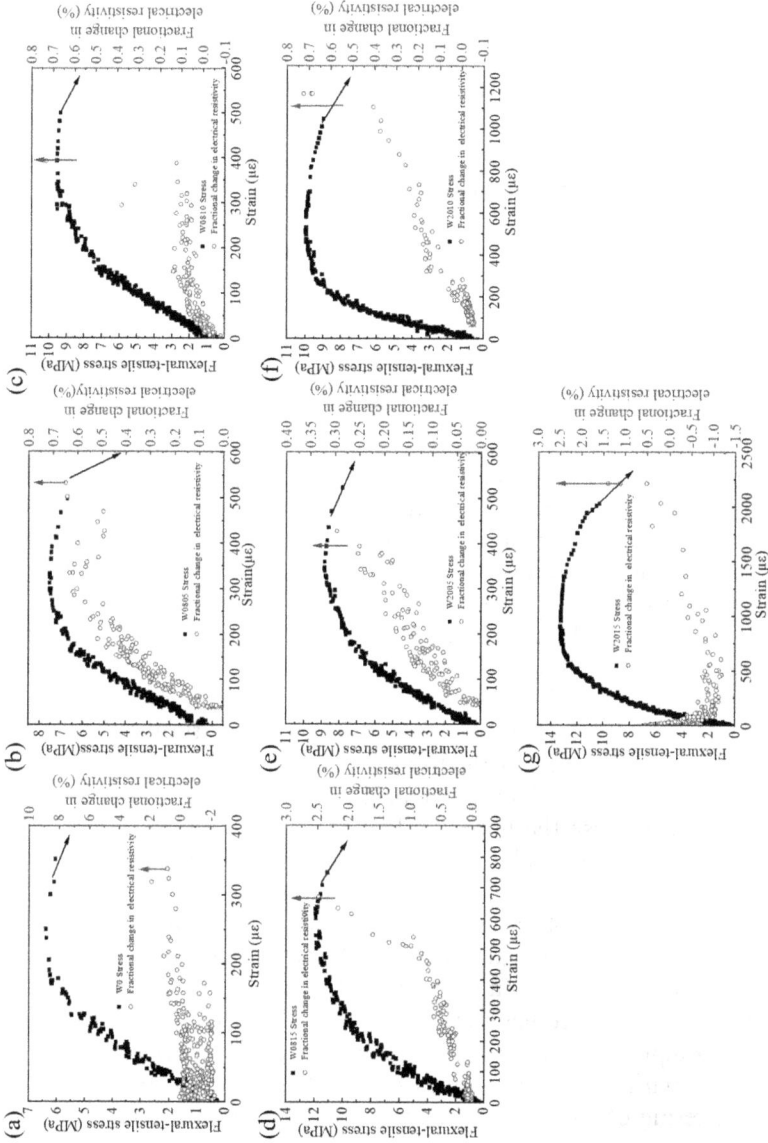

Figure 6.15 Relationships between flexural-tensile strain, stress, and fractional change in electrical resistivity of SSWs-engineered multifunctional UHPC (natural curing). (a) W0; (b) W081005; (c) W081010; (d) W081015; (e) W201005; (f) W201010; (g) W201015

each other or are even pulled out. This phenomenon leads to the increase of electrical resistivity. The top part of specimens is compressed, and the SSWs tend to become close to each other or even contact each other, which leads to a decrease of electrical resistivity. Therefore, the electrical resistivity of SSWs-engineered multifunctional UHPC under flexural load presents an offset phenomenon [22]. The offset phenomenon is particularly evident for UHPC reinforced with 1.5 vol% SSWs at a diameter of 8 μm and UHPC reinforced with 0.5 vol% SSWs at a diameter of 20 μm due to the adjacent SSWs. When the SSWs are sporadically distributed in the UHPC matrix or have formed an extensive conductive network, the increasing distance between the SSWs at the bottom part of specimens dominates the fractional change in electrical resistivity.

It can be seen from Figure 6.15(b)–(d) that the electrical resistivity change rate of UHPC reinforced with SSWs at a diameter of 8 μm after natural curing for 28 days has the same trend as that of composites after water curing for 28 days. However, the fractional change values of UHPC reinforced with SSWs at a diameter of 8 μm increase with increasing SSWs content. This can be attributed to the pull-out of SSWs at the bottom part of specimens under flexural load, based on microcracks generated by natural curing and the SSWs' dispersion status. Figure 6.15(e)–(g) show that the electrical resistivity change rate of UHPC reinforced with SSWs at a diameter of 20 μm decreases with increasing SSWs content, and the change values under peak failure stress are lower than 0.2%. UHPC reinforced with 1.5 vol% SSWs at a diameter of 20 μm even has a negative fractional change in electrical resistivity. This is completely different from that of composites after water curing for 28 days. It is because the SSWs have formed a completely conductive network in specimens of UHPC reinforced with SSWs at a diameter of 20 μm, and the microcracks generated by natural curing make the insertion of SSWs easier at the top part of these specimens. Therefore, the offset effect is obvious. This phenomenon is the same as in the research of Azhari [23], because the upper half of the specimen's cross section is compressed and the lower half undergoes tension when the specimen is subjected to flexure.

The strain sensitivity of SSWs-engineered multifunctional UHPC under flexure after water and natural curing for 28 days is shown in Table 6.6. Table 6.6 shows that the strain sensitivity of specimens after water curing for 28 days under flexure has the same trend as that of specimens under monotonic compressive load. This is because the pull-out effect is increasingly obvious with the content increase of SSWs at a diameter of 20 μm, and the clustering of SSWs at a diameter of 8 μm at high content hampers the presentation of fiber bridging. However, the strain sensitivity of specimens after natural curing for 28 days has no evident change-rule because of the existence of shrinkage cracks. UHPC reinforced with SSWs at a diameter of 8 μm has higher strain sensitivity than UHPC reinforced with SSWs at a diameter of 20 μm, and the strain sensitivity of SSWs-engineered

Table 6.6 Strain sensitivity of SSWs-engineered multifunctional UHPC under flexural load

Group (water curing)	Strain sensitivity	Standard deviation	Group (natural curing)	Strain sensitivity	Standard deviation
W081005	27.8	5.04	W081005	21.4	0.39
W081010	36.9	2.30	W081010	21.5	3.06
W081015	12.6	2.63	W081015	43.6	8.03
W201005	3.73	0.26	W201005	5.0	0.71
W201010	13.0	1.27	W201010	3.5	0.36
W201015	19.5	0.64	W201015	5.5	0.39

multifunctional UHPC under flexure is always larger than the gauge factor of a commercial metal strain gauge.

Based on this analysis, the fractional change in electrical resistivity of SSWs-engineered multifunctional UHPC under flexural load is small and is easily affected by curing conditions. However, the incorporation of SSWs makes the fractional change in electrical resistivity of SSWs-engineered multifunctional UHPC correspond to the combined effect of compression and tension under flexure. Therefore, strain, stress (or external force), cracking, and damage of SSWs-engineered multifunctional UHPC can be monitored through measuring the electrical resistivity of the composites.

6.3.3 Under fracture load

6.3.3.1 Under three-point bending fracture load

As shown in Figure 6.16, three-point bending fracture load was applied to specimens of SSWs-engineered multifunctional UHPC with sizes of 40 mm × 40 mm × 160 mm and crack length/depth ratio of 0.25/0.5.

At the crack length/depth ratio of 0.25, the self-sensing properties of SSWs-engineered multifunctional UHPC obtained by the three-point bending fracture load test are plotted in Figure 6.17. The variation of fractional change in electrical resistivity (FCR) with load is the same as that of cracking mouth opening displacement (CMOD) with load, which can be divided into three stages: the stationary stage, the rapid rising stage, and the sharp rising stage.

It can be seen from Figure 6.17(a) that the CMOD and FCR of UHPC without SSWs are in stationary stage when the load is smaller than 0.43 kN. Then, the UHPC matrix enters the crack initiation stage. The corresponding CMOD and FCR are in the stage of rapid rising. The CMOD and FCR are 0.06 mm and 0.12%, respectively, at the peak load of 1.36 kN. After that, the CMOD and FCR enter the sharp rising stage as the load decreases. The CMOD and FCR are 0.38 mm and 1.04%, respectively, as the load

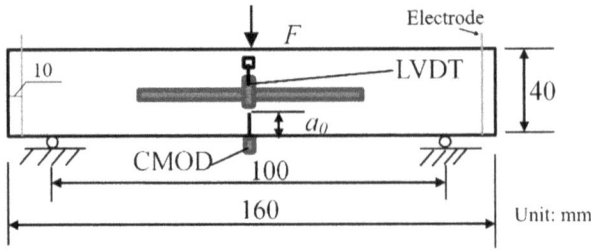

Figure 6.16 Three-point bending fracture test setup

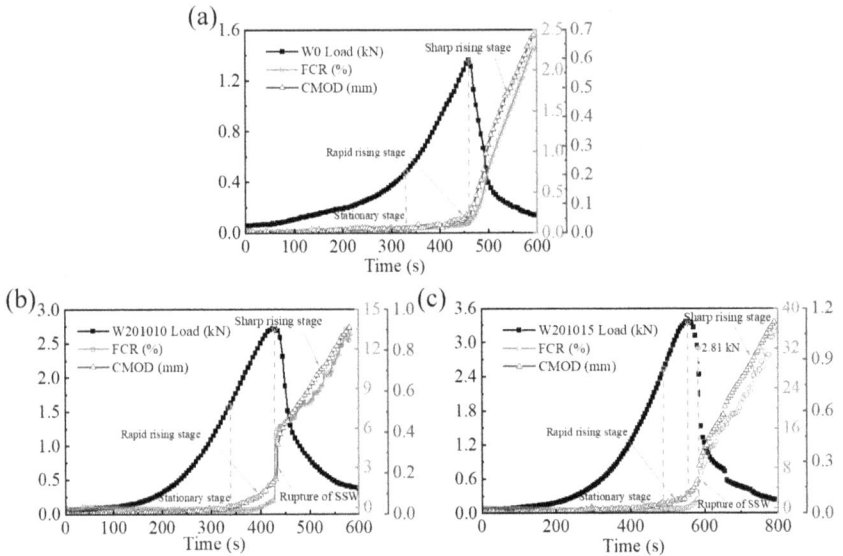

Figure 6.17 a_0/h = 0.25, three-point bending fracture sensibility of SSWs-engineered multifunctional UHPC. (a) W0; (b) W201010; (c) W201015

drops to 0.3 kN. This phenomenon indicates that the FCR of UHPC without SSWs is mainly determined by CMOD. Figure 6.17(b) shows that the CMOD and FCR of UHPC reinforced with 1 vol% SSWs are in the stationary stage when the load is smaller than 1.19 kN. And then, the CMOD and FCR enter the rapid rising stage with the cracking of the UHPC matrix. The CMOD and FCR are 0.4 mm and 6.2%, respectively, at the peak load of 2.7 kN. Meanwhile, the CMOD and FCR have been in the stage of sharp rising because of the convergence of cracks and the rupture of SSWs. When the load drops to 0.44 kN, the CMOD and FCR are 0.89 mm and 13.6%, respectively.

As displayed in Figure 6.17(c), the CMOD and FCR of UHPC reinforced with 1.5 vol% SSWs begin to rise rapidly when the load is 2.48 kN. The FCR enters the rapid rising stage while the CMOD enters the sharp rising stage after peak load of 3.4 kN. The corresponding values of CMOD and FCR are 0.10 mm and 0.14%, respectively. Due to the rupture of SSWs, the FCR of the composite enters the sharp rising stage as the load drops to 2.81 kN. The CMOD and FCR are 0.98 mm and 28.7%, respectively, as the load drops to 0.3 kN. It can be concluded that the FCR of SSWs-engineered UHPC can reflect the CMOD development and the residual bearing capacity.

At the crack length/depth ratio of 0.5, the self-sensing properties of SSWs-engineered multifunctional UHPC obtained by the three-point bending fracturing test are demonstrated in Figure 6.18. As can be seen from Figure 6.18, the variation of CMOD and FCR of the composites still includes three stages: the stationary stage, the rapid rising stage, and the sharp rising stage. It is known from Figure 6.18(a) that the CMOD and FCR of UHPC without SSWs develop from the stationary stage into the rapid rising stage when the load is larger than 0.4 kN. The values of CMOD and FCR are 0.09 mm and 0.02%, respectively, at peak load of 0.8 kN. And then, the CMOD and FCR enter the sharp rising stage with the unloading process because of the unstable propagation and convergence of cracks. Even if the load drops to 0.19 kN, the corresponding FCR of UHPC without SSWs is only 0.19%.

As shown in Figure 6.18(b), the rapid growth of CMOD and FCR of UHPC reinforced with 1.0 vol% SSWs occurs at the load of 0.6 kN. The values of CMOD and FCR are 0.13 mm and 0.25%, respectively, at the peak load of

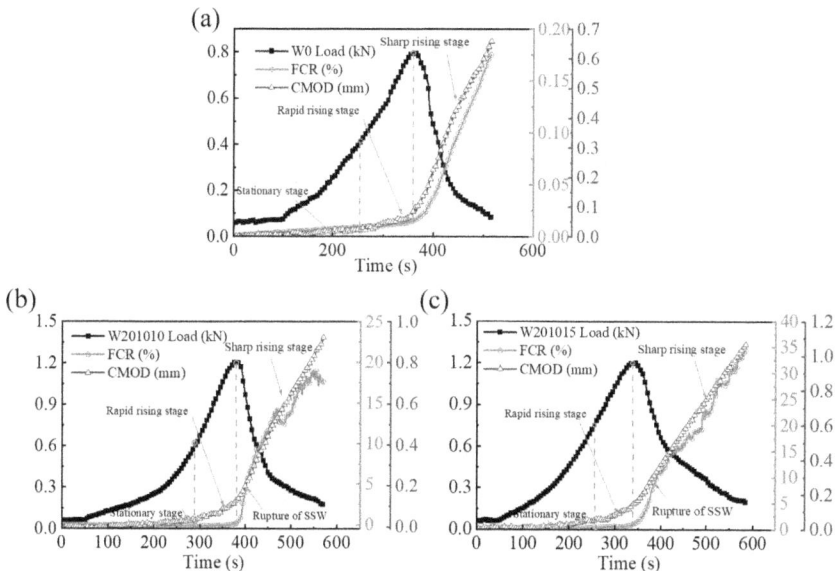

Figure 6.18 a_0/h = 0.5, three-point bending fracture sensibility of SSWs-engineered multifunctional UHPC. (a) W0; (b) W201010; (c) W201015

1.2 kN. As the load decreases, the CMOD and FCR enter the sharp rising stage, and the cracks extend rapidly with the rupture of SSWs. When the load drops to 0.19 kN, the CMOD and FCR are 0.92 mm and 17.6%, respectively. Figure 6.18(c) demonstrates that the CMOD and FCR of UHPC reinforced with 1.5 vol% SSWs enter the rapid rising stage at the load of 0.76 kN. When the load reaches the peak value of 1.2 kN, the CMOD and FCR are 0.15 mm and 1.2%, respectively. Then, the CMOD and FCR develop into the sharp rising stage with the process of unloading. When the load drops to 0.19 kN, the CMOD and FCR are 1.06 mm and 34.4%, respectively.

It is summarized that the change rules of CMOD and FCR versus load for SSWs-engineered multifunctional UHPC are the same. The development of CMOD can be diagnosed and monitored by the change of electrical resistivity. The FCR is mainly dominated by the opening width of cracks at the stage of rapid rising, and then, the FCR enters the sharp rising stage as SSWs are ruptured. The ultimate FCR of SSWs-engineered multifunctional UHPC far exceeds that of UHPC without SSWs due to the disconnection of the conductive pathway.

6.3.3.2 Under four-point shearing fracture load

As shown in Figure 6.19, a four-point shearing fracture load was applied to specimens of SSWs-engineered multifunctional UHPC with sizes of 40 mm × 40 mm × 160 mm and crack length/depth ratio of 0.25/0.5.

It can be seen from Figure 6.20(a) that at the crack length/depth ratio of 0.25, the CMOD of UHPC without SSWs shows a rapidly rising trend at the load of 5.8 kN. The specimens of UHPC without SSWs have been in the cracking state. The FCR of UHPC without SSWs grows with loading time due to polarization. The CMOD and FCR increase sharply after the load reaches the peak value of 6.95 kN. Then, the load drops instantaneously, and the specimens fall into failure state. UHPC without SSWs does not have self-sensing ability to the four-point shearing load.

Figure 6.20(b) shows that the FCR of UHPC reinforced with 1 vol% SSWs fluctuates up and down after the load reaches 1.05 kN. Then, it moves into

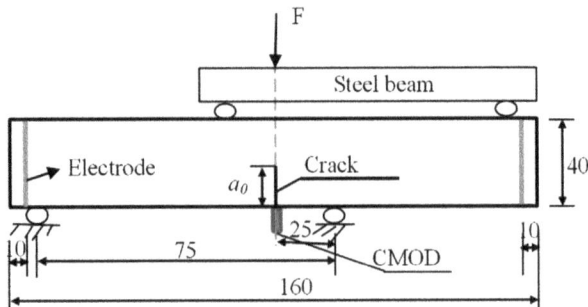

Figure 6.19 Four-point shearing fracture test setup

the rapid declining stage due to crack initiation in the composite. After that, the FCR goes into the sharp declining stage and the CMOD enters the rapid rising stage because the specimens have begun to show macrocracks. The FCR reaches the minimum value when the load is 18.3 kN. The conductive pathway is destroyed, and the FCR begins to increase with the propagation of cracks. The corresponding CMOD and FCR are 0.011 mm and 6.58%, respectively, at the peak load of 19.6 kN. Figure 6.20(c) demonstrates that the FCR of UHPC reinforced with 1.5 vol% SSWs reaches the minimum value and the CMOD begins to increase rapidly when the load is 20.5 kN. The specimens begin to crack from the precast position. This can be attributed to the occurrence of macrocracks, and the corresponding FCR is in the rapid rising stage. The FCR and CMOD enter the sharp rising stage when the load is 25.8 kN due to the rupture of SSWs.

At the crack length/depth ratio of 0.5, the FCR of UHPC without SSWs increases with time because of polarization, as shown in Figure 6.21(a). The FCR begin to decrease at the load of 2.2 kN. The FCR reaches the minimum value of 0.44% when the peak load is 7.2 kN. Then, the load drops instantaneously, and the specimens are in failure state. Figure 6.21(b) shows that when the load is 9.4 kN, the FCR of UHPC reinforced with 1 vol% SSWs is in the rapid declining stage. The CMOD moves into the rapid rising stage because of the emergence of initial cracks. The minimum value of FCR is obtained as the load reaches 12.9 kN, and then, the FCR turns to increase due to the appearance of macrocracks. The FCR and CMOD enter

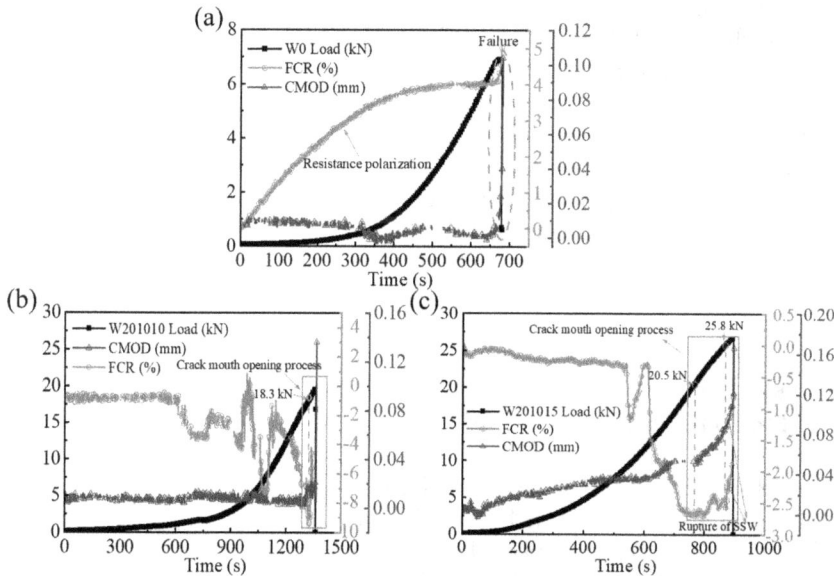

Figure 6.20 $a_0/h = 0.25$, four-point shearing fracture sensibility of SSWs-engineered multifunctional UHPC. (a) W0; (b) W201010; (c) W201015

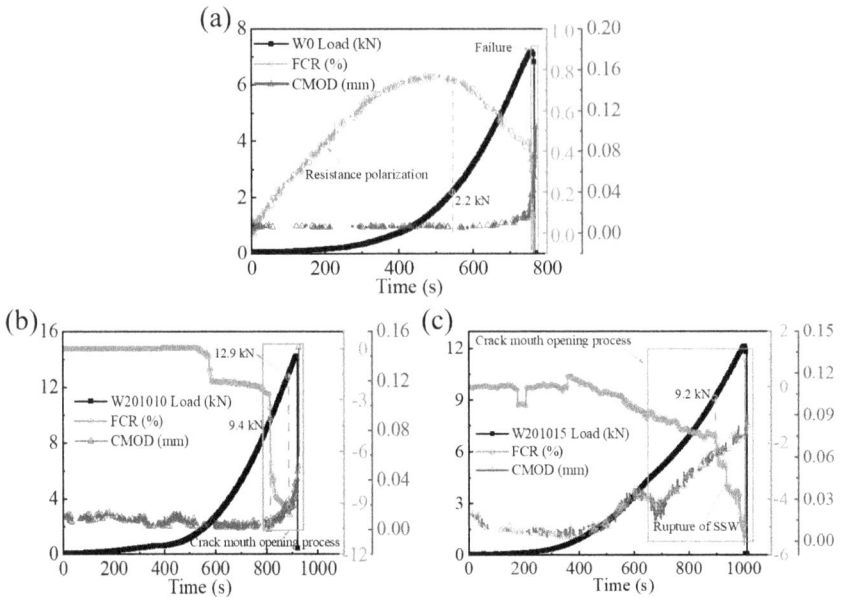

Figure 6.21 a_0/h = 0.5, four-point shearing fracture sensibility of SSWs-engineered multifunctional UHPC (a) W0; (b) W201010; (c) W201015

the sharp rising stage when the load decreases. Figure 6.21(c) shows that the FCR of UHPC reinforced with 1.5 vol% SSWs is in the rapid declining stage at the load of 9.2 kN. With the rupture of SSWs, the FCR of composite reaches the minimum value of –5.4%. Then, the FCR and CMOD enter the sharp rising stage after the peak load of 12.2 kN.

It can be concluded that the variation of FCR can be used for sensing the generation and propagation of cracks in SSWs-engineered multifunctional UHPC. The FCR begins to decrease when the initial crack occurs, and it enters the rapid development stage after the appearance of macrocracks. After that, the SSWs are gradually ruptured, and the FCR is in the sharp rising stage.

6.4 SELF-SENSING MECHANISMS OF SSWS-ENGINEERED MULTIFUNCTIONAL UHPC

The self-sensing mechanisms of SSWs-engineered UHPC under different load forms can be explained through analyzing the development process of deformation cracks. Especially, under monotonic failure load, the corresponding mechanisms can be revealed through investigating the initiation and development of cracks at the upper end of the precast crack, as shown

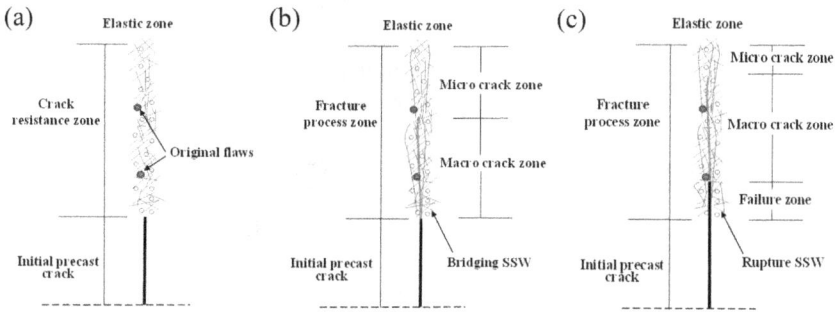

(a) Elastic zone / Crack resistance zone / Original flaws / Initial precast crack

(b) Elastic zone / Micro crack zone / Fracture process zone / Macro crack zone / Initial precast crack / Bridging SSW

(c) Elastic zone / Micro crack zone / Fracture process zone / Macro crack zone / Failure zone / Initial precast crack / Rupture SSW

Figure 6.22 Self-sensing mechanisms of SSWs-engineered multifunctional UHPC under monotonic failure load. (a) Crack initiation; (b) Stable crack propagation; (c) Unstable crack propagation

in Figure 6.22. The development of cracks includes three different stages: crack initiation, stable crack propagation, and unstable crack propagation.

As demonstrated in Figure 6.22(a), the SSWs and the UHPC matrix work together to form a crack resistance zone in SSWs-engineered multifunctional UHPC. Due to the high aspect ratio and large specific surface area of SSWs, lots of hydration products gather on the surface of SSWs, and the original flaws of UHPC are reduced, leading to refined grains, homogeneous structures, and excellent interfacial transition zone performance of the composites. At this stage, the FCR of the composites is in the stationary stage and shows no obvious change. Figure 6.22(b) demonstrates that the stress concentration leads to the generation of microcracks, and the saturation of microcracks results in the formation of macrocracks. The SSWs can relieve the stress concentration on the tip of the microcracks and prevent the generation and propagation of macrocracks. Then, the SSWs bridge adjacent macrocracks, and the tensile stress is shared by the SSWs and the UHPC matrix together. During this process, the FCR of the composites is in the rapid development stage, which is mainly affected by the opening of crack mouths and the increase of mid-span deflection. The emergence of initial cracks can be detected by variation in the inflection point of FCR due to the change of conductive pathway under loading [24, 25]. Figure 6.22(c) illustrates that the SSWs are gradually ruptured with the propagation of macrocracks. The FCR of the composites is in the sharp rising stage due to the rupture of SSWs. The failure process of SSWs-engineered multifunctional UHPC can be detected by the change of FCR.

6.5 SUMMARY

In this chapter, the electrical resistivity measured using different test methods and the self-sensing properties under cyclic compressive, monotonic compressive, flexural, three-point bending fracture, and four-point shearing fracture load of SSWs-engineered multifunctional UHPC are introduced. Meanwhile, the electrically conductive and self-sensing mechanisms are elucidated. The main conclusions are as follows:

1) The effect of SSWs content and diameter on the electrical resistivity of SSWs-engineered multifunctional UHPC is not subject to the test method. The electrical resistivity of SSWs-engineered multifunctional UHPC decreases with increasing SSWs content and diameter, and only 0.5 vol% SSWs at a diameter of 20 μm can make the reduction of electrical resistivity reach up to four orders of magnitude. Percolation has occurred when the content of SSWs at a diameter of 8 μm and 20 μm is 1.5 vol% and 0.5 vol%, respectively.

2) The stability and reversibility of the piezoresistive response of SSWs-engineered multifunctional UHPC under cyclic compressive load increase with increasing SSWs content at a diameter of 8 μm, while there is almost no piezoresistivity when the content of SSWs at a diameter of 20 μm is 1.5 vol%. The strain sensitivity of SSWs-engineered multifunctional UHPC can reach 22.5 and 94.9 under cyclic and monotonic compressive load, respectively. The strain sensitivity of SSWs-engineered multifunctional UHPC under flexure can reach up to 43.6, and it is sensitive to natural curing. The generation and propagation of cracks in the composites can be monitored by the variation of FCR under three-point bending and four-point shearing fracture load.

3) SSWs at a diameter of 20 μm are more suitable to be used as a conductive filler for UHPC compared with SSWs at a diameter of 8 μm. The conductive and self-sensing mechanisms of SSWs-engineered multifunctional UHPC are closely related to the formation of an overlapping network of SSWs.

REFERENCES

1. S. Dong, D. Zhou, Z.X. Yu, B. Han. Super-fine stainless wires enabled multifunctional and smart reactive powder concrete, *Smart Materials and Structures*. 28(12) (2019) 125009.
2. S. Dong, W. Zhang, D. Wang, X. Wang, B. Han. Modifying self-sensing cement-based composites through multiscale composition, *Measurement Science and Technology*. 32(7) (2021) 074002.
3. S. Ding, Y. Xiang, Y. Ni, V. Thakur, X. Wang, B. Han, J. Ou. In-situ synthesizing carbon nanotubes on cement to develop self-sensing cementitious composites for smart high-speed rail infrastructures, *Nano Today*. 43 (2022) 101438.

4. B. Han, X. Yu, J. Ou. *Self-Sensing Concrete in Smart Structures*, Elsevier, 2014.

5. S. Ding, X. Wang, L. Qiu, Y. Ni, X. Dong, Y. Cui, A. Ashour, B. Han, J. Ou. Self-sensing cementitious composites with hierarchical carbon fiber-carbon nanotube composite fillers for crack development monitoring of a maglev girder, *Small*. 19(9) (2023) 2206258.

6. D. Wang, S. Dong, X. Wang, N. Maimaitituersun, S. Shao, W. Yang, B. Han. Sensing performances of hybrid steel wires and fibers reinforced ultra-high performance concrete for in-situ monitoring of infrastructures, *Journal of Building Engineering*. 58 (2022) 105022.

7. S. Ding, S. Dong, X. Wang, S. Ding, B. Han, J. Ou. Self-heating ultra-high performance concrete with stainless steel wires for active deicing and snow-melting of transportation infrastructures, *Cement and Concrete Composites*. 138 (2023) 105005.

8. S. Dong, B. Han, J. Ou, Z. Li, L. Han, X. Yu. Electrically conductive behaviors and mechanisms of short-cut super-fine stainless wire reinforced reactive powder concrete, *Cement and Concrete Composites*. 72 (2016) 48–65.

9. S. Dong, X. Dong, A. Ashour, B. Han, J. Ou. Fracture and self-sensing characteristics of super-fine stainless wire reinforced reactive powder concrete, *Cement and Concrete Composites*. 105 (2020) 103427.

10. J. Ou, B. Han. Piezoresistive cement-based strain sensors and self-sensing concrete components, *Journal of Intelligent Material Systems and Structures*. 20(3) (2009) 329–336.

11. M. D'Alessandro, F. Rallini, A.L. Ubertini, J.M. Materazzi, J.M. Kenny. Investigations on scalable fabrication procedures for self-sensing carbon nanotube cement-matrix composites for SHM applications, *Cement and Concrete Composites*. 65 (2016) 200–213.

12. S. Huang, J. Chang, L. Lu, F. Liu, Z. Ye, X. Cheng. Preparation and polarization of 0–3 cement based piezoelectric composites, *Materials Research Bulletin*. 41(2) (2006) 291–297.

13. C. Cao, J. Zhang. *An Introduction to Electrochemical Impendence Spectroscopy*, Science Press, 2002.

14. J.M. Torrents, T.O. Mason, A. Peled, S.P. Shah, E.J. Garboczi. Analysis of the impedance spectra of short conductive fiber-reinforced composites, *Journal of Materials Science*. 36(16) (2001) 4003–4012.

15. L.Y. Woo, S. Wansom, N. Ozyurt, B. Mu, S.P. Shah, T.O. Mason. Characterizing fiber dispersion in cement composites using AC-impedance spectroscopy, *Cement and Concrete Composites*. 27(6) (2005) 627–636.

16. B.T. Tamtsia, J.J. Beaudoin, J. Marchand. A coupled AC impedance-creep and shrinkage investigation of hardened cement paste, *Materials and Structures*. 36(3) (2003) 147–155.

17. S. Wang, L. Lu, X. Cheng. Effect of preparation processes on the piezoresistivity effect of CFSC, *Advanced Materials Letters*. 2(2) (2011) 136–141.

18. B. Han, S. Ding, X. Yu. Intrinsic self-sensing concrete and structures: A review, *Measurement*. 59 (2015) 110–128.

19. B. Han, X. Yu, J. Ou. Nanotechnology in civil infrastructure: A paradigm shift // *Multifunctional and Smart Carbon Nanotube Reinforced Cement-Based Materials*, Springer, 276 (2011) 1–47.

20. F. Azhari, N. Banthia. Cement-based sensors with carbon fibers and carbon nanotubes for piezoresistive sensing, *Cement and Concrete Composites.* 34(7) (2012) 866–873.
21. P. Stynoski, P. Mondal, C. Marsh. Effects of silica additives on fracture properties of carbon nanotube and carbon fiber reinforced Portland cement mortar, *Cement and Concrete Composites.* 55 (2015) 232–240.
22. S. Zhu, D.D.L. Chung. Theory of piezoresistivity for strain sensing in carbon fiber reinforced cement under flexure, *Journal of Materials Science.* 42(15) (2007) 6222–6233.
23. F. Azhari. *Cement-Based Sensors for Structural Health Monitoring,* University of British Columbia, 2008.
24. E. García-Macías, A. D'Alessandro, R. Castro-Triguero, D. Pérez-Mira, F. Ubertini. Micromechanics modeling of the electrical conductivity of carbon nanotube cement-matrix composites, *Composites Part B: Engineering.* 108 (2017) 451–469.
25. B. Han, Y. Wang, S. Dong, L. Zhang, S. Ding, X. Yu, J. Ou. Smart concrete and structures: A review, *Journal of Intelligent Material Systems and Structures.* 26(11) (2015) 1–43.

Chapter 7

Thermal and electrothermal properties of stainless steel wires-engineered multifunctional ultra-high performance concrete

7.1 INTRODUCTION

As the basic physical performance indexes, the thermal conductivity and specific heat of concrete directly affect its application range and the selection of structural measures. Improving the thermal conductivity of concrete can reduce the probability of cracks caused by the temperature difference stress of hydration heat when it is used in large-volume structures [1, 2]. Meanwhile, concrete with high thermal conductivity can be used for energy conversion and storage. Electrical energy is converted into heat by applying a voltage to high–thermal conductivity concrete, thus endowing concrete with a self-heating property and derived self-deicing/snow-melting properties. The key to realizing these functions is to endow concrete with stable and high thermal conductivity by adding fillers that can form a heat transfer network [2, 3]. Stainless steel wires (SSWs) are preferred fillers due to their high thermal conductivity, microdiameter, good dispersion, and high corrosion resistance. Combined with the dense microstructure of ultra-high performance concrete (UHPC), SSWs-engineered multifunctional UHPC is expected to obtain stable and good heat transfer performance, and has the potential to be applied to large-volume structures and infrastructure requiring snow or ice melting.

Therefore, the effects of SSWs on the thermal conductivity and specific heat of UHPC, and the performance to control thermal cracking caused by hydration heat of UHPC specimens, pavement slab, and pier cap, are demonstrated in this chapter. Meanwhile, the technical feasibility of using SSWs-engineered multifunctional UHPC to develop self-heating and active deicing/snow-melting slab is presented. The diameter of SSWs used in this chapter is 20 μm and the length is 10 mm. The main contents of this chapter are shown in Figure 7.1.

7.2 THERMAL PROPERTIES OF SSWS-ENGINEERED MULTIFUNCTIONAL UHPC

The specimen size used for the thermal conductivity and specific heat test was 90 mm × 90 mm × 10 mm. The heating voltage was set as 18 V, and

DOI: 10.1201/9781003276357-7

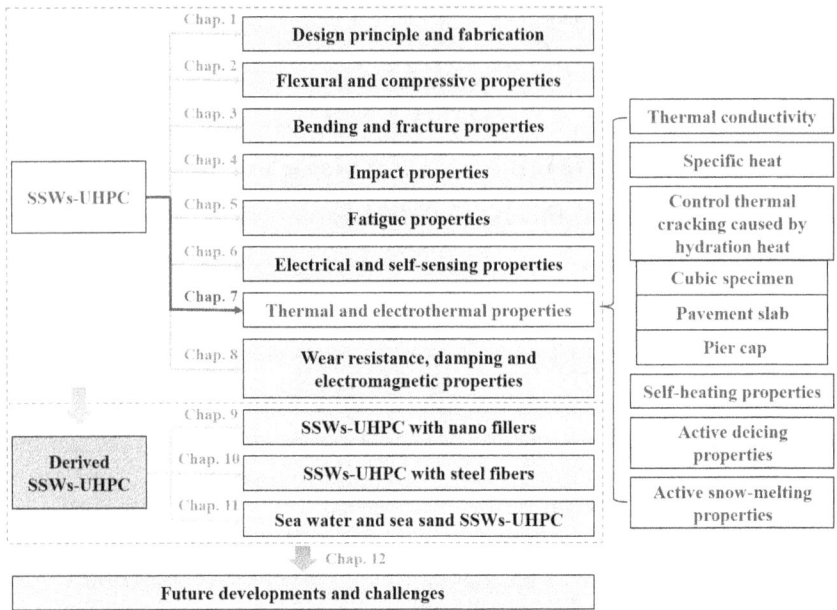

Figure 7.1 Main contents of Chapter 7

the resistance of the heater was 110Ω. The temperature difference thermo-electric potential at quasi-steady state was used to calculate the temperature difference, and the temperature rise thermoelectric potential was used to obtain the temperature rise rate. The thermal conductivity and specific heat of SSWs-engineered multifunctional UHPC were calculated in accordance with Sukontasukkul et al. [4].

Test results show that the density of UHPC reinforced with 0 vol%, 1 vol%, and 1.5 vol% SSWs is 2.14 g/cm³, 2.29 g/cm³, and 2.34 g/cm³, respectively. The thermal conductivity and specific heat of SSWs-engineered multifunctional UHPC are presented in Figure 7.2. It can be seen from Figure 7.2 that the thermal conductivity of UHPC without SSWs is only 1.07 W/(m·K), which increases to 1.86 W/(m·K) because of the addition of 1.5 vol% SSWs. The thermal conductivity of UHPC reinforced with 1 vol% and 1.5 vol% SSWs is 20.6% and 73.8% higher, respectively, than that of UHPC without SSWs. Meanwhile, the thermal conductivity of SSWs-engineered multifunctional UHPC increases by 44.2% due to the increase of SSWs content. Due to the excellent heat conduction performance of SSWs, the widely distributed SSWs can form a thermal conductive pathway in UHPC, leading to improvement of the thermal conductivity of UHPC. Figure 7.2 also shows that the specific heat of UHPC is reduced due to the addition of SSWs. The specific heat of UHPC without SSWs is 1523.0 J/(kg·K), which decreases to 1499.8 J/(kg·K) and 1439.3 J/(kg·K), respectively, due to the

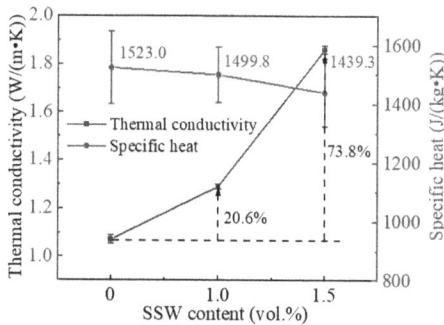

Figure 7.2 Thermal conductivity and specific heat of SSWs-engineered multifunctional UHPC

addition of 1 vol% and 1.5 vol% SSWs, indicating 1.5% and 5.5% decrements. This means that the specific heat of UHPC is less affected by SSWs.

The increased thermal conductivity performance of SSWs-engineered multifunctional UHPC can be attributed to the following aspects. (1) Due to the characteristics of stainless steel, the thermal conductivity of SSWs reaches 16.3 W/(m·K). (2) The thermal conductivity pathway in SSWs-engineered multifunctional UHPC can be composed of the UHPC matrix and SSWs, and can also be composed of overlapped SSWs, which depends on the content of SSWs and their dispersion state in the UHPC. The proportion of the thermal conductivity pathway occupied by SSWs increases with increasing SSWs content; therefore, a large content of SSWs endows UHPC with high thermal conductivity and small specific heat. (3) Due to the high aspect ratio and large specific surface area of SSWs, the structural compactness of the UHPC matrix is enhanced, and there is no weak interface structure between the SSWs and the UHPC matrix. This limits the improvement effect of thermal conductivity for SSWs-engineered UHPC.

7.3 PERFORMANCE TO CONTROL THERMAL CRACKING CAUSED BY HYDRATION HEAT OF SSWS-ENGINEERED MULTIFUNCTIONAL UHPC

7.3.1 Hydration heat–induced temperature and stress fields of cubic specimens

In order to monitor the temperature rise in SSWs-engineered multifunctional UHPC specimens caused by heat of hydration, type T copper-constantan thermocouples with a precision of 0.1 °C and temperature ranging from 60 °C to 220 °C were attached on the cross wooden frame and then embedded in different locations of specimens with sizes of 70.7 mm × 70.7 mm × 70.7 mm. The arrangements of the four thermocouples are shown in Figure 7.3(a).

Meanwhile, a thermocouple was located out of the UHPC specimens to obtain the ambient temperature. After casting, a data logger was used to record the thermal data at 1-minute intervals until 72 hours. In order to obtain the temperature and stress fields of specimens fabricated with SSWs-engineered multifunctional UHPC, a finite element (FE) model was established on the basis of heat transfer theory, as shown in the following Eq. (7.1) [5]:P

$$\lambda \left(\frac{\partial^2 T}{\partial x^2} + \frac{\partial^2 T}{\partial y^2} + \frac{\partial^2 T}{\partial z^2} \right) + Q_h = \rho \cdot C_p \cdot \frac{\partial T}{\partial t} \tag{7.1}$$

where λ , ρ , and C_p are the thermal conductivity, density, and specific heat of concrete, respectively, and Q_h is the heat production rate per unit volume. Assuming that the heat production rate of UHPC was not affected by SSWs, the adiabatic temperature rise of composites was determined according to An et al. [6], as shown in Figure 7.3(b). Meanwhile, Dirichlet and Neumann boundary conditions were used as boundary conditions in the calculation process of the temperature field [7].

The FE model for SSWs-engineered multifunctional UHPC specimens was established on the basis of experiments to verify the accuracy of numerical simulation, as demonstrated in Figure 7.3(c). The mold was acrylonitrile butadiene styrene (ABS) plastic with a density of 1100 kg/m³, a thermal conductivity of 0.25 W/(m·K), and a specific heat capacity of 1470 J/(kg·K). The surface of the formwork was in contact with air, which satisfied the Neumann boundary condition. The wind speed (v_a) was taken as 0.3 m/s, and the surface heat release coefficient (β) was 25.85 W/(m²·K) (this value can be obtained according to Li et al. [7] and the equation $\beta = 21.8 + 13.53 v_a$). In addition, based on experimental results, the molding temperature of UHPC mixtures without and with 0.5 vol% SSWs were 18 °C and 19 °C, respectively, and the ambient temperature was the actual measured value. The density of UHPC without and with 0.5 vol% SSWs is 2.14 g/cm³ and 2.22 g/cm³, respectively, and the conductivity coefficient

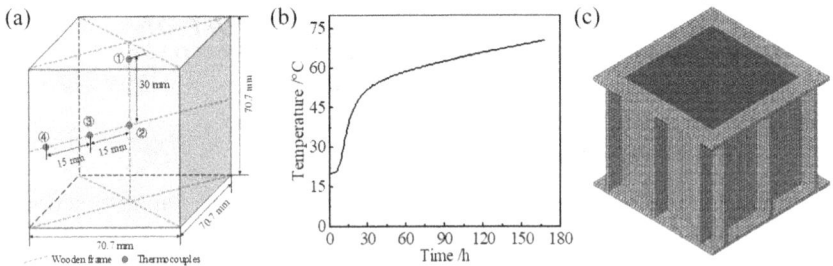

Figure 7.3 Hydration heat test and simulation of SSWs-engineered multifunctional UHPC. (a) Arrangements of thermocouples (①-Upper surface center position; ②-Core position); (b) Adiabatic temperature rise curve of composites; (c) FE model of specimen

and specific heat of composites are 1.07 W/(m·K)/1.18 W/(m·K) and 1523 J/(kg·K)/1512 J/(kg·K), respectively. Meanwhile, the elastic modulus at the curing ages of 3 days and 28 days and the tensile strength at the curing age of 3 days for these composites are 22.8 GPa/24.8 GPa, 31.3 GPa/34.0 GPa, and 2.79 MPa/4.00 MPa, respectively [8–12]. It is worth noting that the homogeneous and elastic model was used, and the cracking criterion was not considered in the FE analysis. When the calculated thermal stress is greater than the tensile strength of composites, the composites will crack. In contrast, temperature cracks do not occur when the calculated thermal stress is less than the tensile strength of composites. Meanwhile, the influence of SSWs is reflected by the change of performance of composites in the modelling process, including the change of thermal conductivity, elastic modulus, and tensile strength.

The temperature–time history measurements and FE simulation results at positions of points 1 and 2 in specimens of UHPC without SSWs and 0.5 vol% SSWs-engineered multifunctional UHPC are shown in Figure 7.4. It can be seen that the measured temperature–time history relationships at different positions are consistent with the FE simulation results, indicating

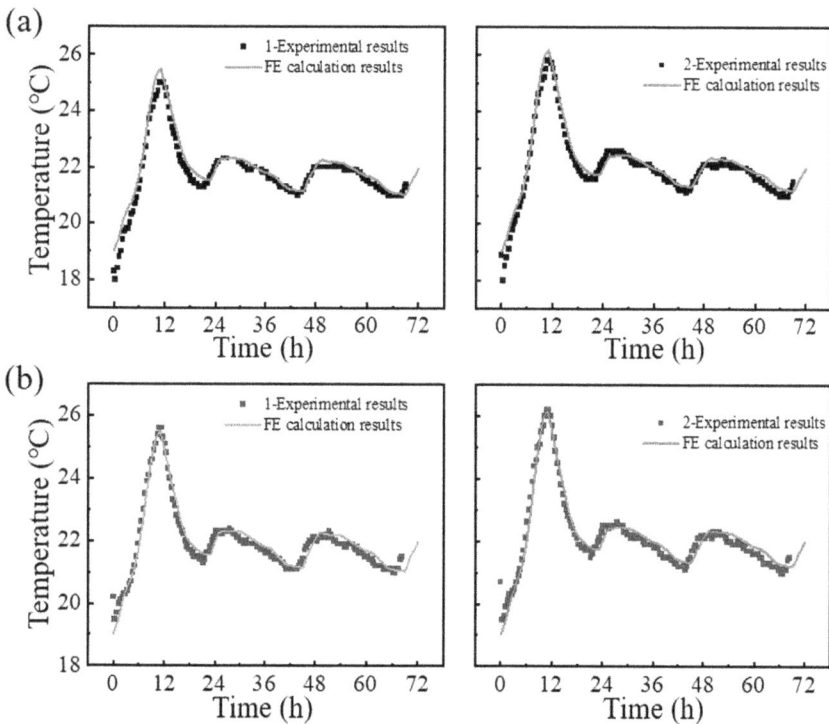

Figure 7.4 Temperature history measurements and FE simulation results at different positions of specimens. (a) UHPC without SSWs; (b) 0.5 vol% SSWs-engineered multifunctional UHPC

that the FE model can accurately describe the temperature history of composites caused by heat of hydration. The core temperature of composites is significantly higher than the upper surface temperature because the heat of hydration at the surface is dissipated more slowly than it is generated. The temperature peak for 0.5 vol% SSWs-engineered multifunctional UHPC appears at the time of 11 hours, earlier than that for UHPC without SSWs, indicating that the incorporation of SSWs is conducive to early cement hydration and the formation of heat of hydration, which is consistent with the conclusion of Cecini et al. [13].

The temperatures obtained by experiments and simulation of SSWs-engineered multifunctional UHPC caused by hydration heat are shown in Table 7.1. The initial casting temperature of SSWs-engineered multifunctional UHPC is higher than that of UHPC without SSWs, resulting from the increased frictional resistance during stirring. Incorporating SSWs is able to reduce the temperature rise at the different positions of UHPC specimens:

Table 7.1 Experimental and simulation temperatures of SSWs-engineered multifunctional UHPC specimens

		Peak temperature (°)	*Temperature rise ΔT (°)*	*Temperature gradient$\Delta T'$ (°)*	*Time from casting to peak temperature (hours)*
		UHPC without SSWs			
Position	*Value*				
Point 1 (Upper	E	25.0	6.7	−0.7	11.6
surface center)	S	25.1	7.1	−0.8	11.0
Point 2 (Core)	E	25.7	6.8	–	11.6
	S	25.9	7.9	–	11.0
Point 3	E	25.7	7.0	0	11.6
	S	25.7	7.7	−0.2	11.0
Point 4	E	25.6	7.2	−0.1	11.6
	S	25.5	7.5	−0.4	11.0
		0.5 vol% SSWs-engineered multifunctional UHPC			
Point 1 (Upper	E	25.6	5.4	−0.6	11.3
surface center)	S	25.5	6.5	−0.7	11.0
Point 2 (Core)	E	26.2	5.5	–	11.3
	S	26.2	7.2	–	11.0
Point 3	E	26.3	6.8	0.1	11.3
	S	26.1	7.1	−0.1	11.0
Point 4	E	26.1	6.0	−0.1	11.3
	S	25.9	6.9	−0.3	11.0

Abbreviations: E represents experimental results; S represents FE simulation results; Temperature gradient is the peak temperature difference between point 2 and the other three points 1, 3, and 4.

1.3 °C and 1.3 °C decrements can be achieved for the temperature rise values of the core and upper surface center positions, respectively. The experimental values of the temperature gradient between the core and the upper surface center positions are 0.7 °C and 0.6 °C for UHPC without and with 0.5 vol% SSWs, respectively. The decrease of the temperature gradient is conducive to reducing thermal stress and then inhibiting the initiation of thermal cracks. Table 7.2 also shows that the FE model is accurate and can be used to analyze the thermal and stress fields inside SSWs-engineered multifunctional UHPC structures.

Considering the symmetry of the FE model, the simulation results for the temperature fields of half-size specimens at the time of 11 hours are demonstrated in Figure 7.5(a) and (b). It can be seen that the temperature on the core positions is higher than that on the upper surface positions, leading to the formation of a temperature gradient. The high-temperature area is decreased, and the low-temperature area has a growing tendency, due to the addition of SSWs. This indicates improvement in the uniformity of temperature distribution, and this phenomenon can be attributed to the increase of thermal conductivity of UHPC caused by the addition of SSWs.

The thermal stress fields in a half-size FE model of specimens at the time of 11 hours are demonstrated in Figure 7.5(c) and (d). There is a similar thermal stress distribution contour for specimens prepared by UHPC without and with 0.5 vol% SSWs. The upper surface of specimens is under pressure, and excessive local tensile stress is likely to occur at the edges of specimens. The tensile thermal stress at the core of specimens is lower than that at the side due to mold constraint. In theory, the thermal stress can be calculated by Eq. (7.2), as follows:

$$\sigma_t = \beta \frac{E(t)\alpha\Delta T}{1-\mu} \tag{7.2}$$

where β is the relaxation coefficient, $E(t)$ is the elastic modulus of composites, α is the composite coefficient of thermal expansion, ΔT is the temperature rise, and μ is the Poisson's ratio of UHPC. This indicates that the thermal stress is influenced by elastic modulus, Poisson's ratio, and temperature gradient simultaneously. This explains why the temperature gradient is reduced but the thermal stress is increased when 0.5 vol% SSWs are added into UHPC.

Figure 7.5(c) shows that 11 hours' setting time and thermal stress 1.40 MPa are the maximum time and maximum tensile thermal stress attained by UHPC without SSWs at the corner position of specimens. In the case of 0.5 vol% SSWs-engineered multifunctional UHPC, the maximum tensile stress of 1.51 MPa is seen when the setting time is 11 hours. It can be concluded that the maximum thermal tensile stress of 0.5 vol% SSWs-engineered multifunctional UHPC is higher than that of UHPC without SSWs, and this happens because of the high elastic modulus of composites.

Table 7.2 Properties of different layers in the FE model for pavement slab and pier cap [17]

FE model	Structural layer	Geometry size (m^3)	Density (kg/m^3)	Thermal conductivity ($W/(m \cdot K)$)	Specific heat ($J/(kg \cdot K)$)	Elastic modulus (MPa)	Poisson's ratio	Coefficient of thermal expansion (/K)
SSWs-engineered UHPC pavement slab	Cement stabilized macadam base	$6.5 \times 7 \times 0.2$	2200	1.56	911.7	1400	0.2	6.7×10^{-6}
	Low content cement stabilized macadam base	$6.5 \times 7 \times 0.2$	2100	1.43	900	1200	0.2	5×10^{-6}
	Graded crushed stone	$6.5 \times 7 \times 0.3$	1800	0.86	921.1	500	0.35	1×10^{-6}
	Soil base	$6.5 \times 7 \times 1$	1800	0.9	1040	60	0.35	2×10^{-5}
SSWs-engineered UHPC pier cap	Graded crushed stone	$5 \times 4 \times 1$	1800	0.86	921.1	500	0.35	1×10^{-6}

(a)   (b)

(c)   (d)

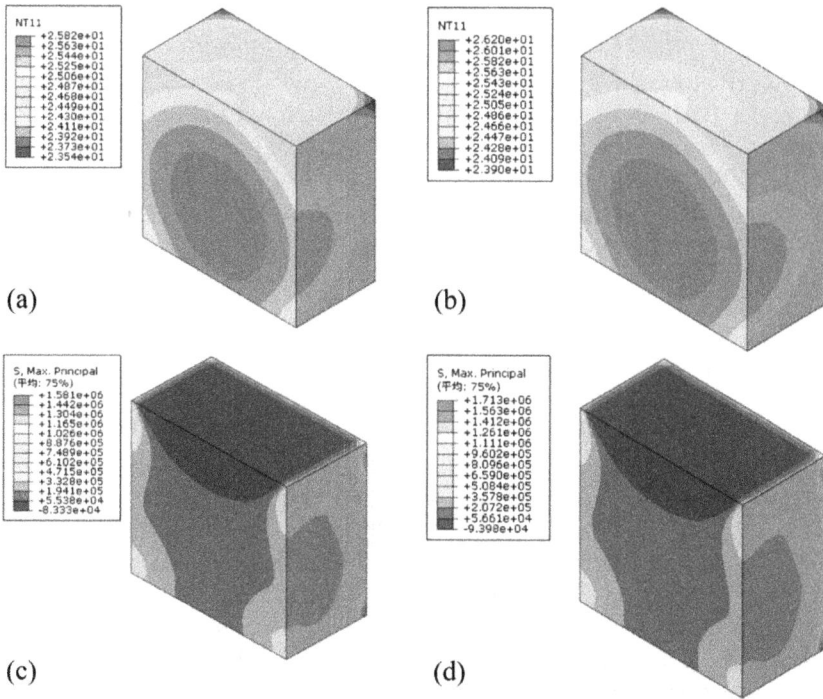

Figure 7.5 Temperature fields and thermal stress fields inside SSWs-engineered multifunctional UHPC specimens. (a) Temperature fields of UHPC without SSWs;
(b) Temperature fields of 0.5 vol% SSWs-engineered multifunctional UHPC;
(c) Thermal stress fields of UHPC without SSWs; (d) Thermal stress fields of
0.5 vol% SSWs-engineered multifunctional UHPC

However, whether the thermally induced stresses were greater than the
tensile strength of composites at a given age plays an important role in
the occurrence of thermal cracking. Similarly to the development rule
of elastic modulus shown in Han and Kim [11], the tensile strengths of
UHPC without and with 0.5 vol% SSWs at the curing age of 11 hours are
0.67 MPa and 0.96 MPa, respectively. This means that thermal cracking
only occurs on the corners of specimens of SSWs-engineered multifunctional UHPC, while there is a tendency to large area cracking on the
lateral surface of UHPC without SSWs. It can therefore be deduced that
incorporating SSWs may be an effective way to reduce the hydration thermal cracking risk of UHPC. In addition, the high thermal conductivity
of composites is conducive to improving the temperature uniformity and
reducing the temperature gradient, which may effectively inhibit thermal
cracking of structures with large cross-sectional dimensions.

7.3.2 Hydration heat–induced temperature and stress fields of pavement slab

The thermal cracking risk of 0.5 vol%/1.5 vol% SSWs-engineered multifunctional UHPC pavement slab was investigated in the following FE analysis. The density, thermal conductivity coefficient, and specific heat of 1.5 vol% SSWs-engineered multifunctional UHPC are 2.34 g/cm^3, 1.86 W/(m·K), and 1439 J/(kg·K), respectively. The elastic modulus at the curing ages of 3/28 days and the tensile strength at the curing ages of 3 days for 1.5 vol% SSWs-engineered multifunctional UHPC are 22.7 GPa/31.1 GPa and 5.60 MPa, respectively.

As shown in Table 7.2, the FE model for pavement slabs with a size of 4.5 m × 5 m × 0.4 m consists of five layers: pavement layer (fabricated with SSWs-engineered multifunctional UHPC), cement stabilized macadam base layer, low content cement stabilized macadam base layer, graded crushed stone layer, and soil base layer. In order to eliminate the boundary effect, the horizontal size of lower layers was larger than the concrete pavement [14]. Tie constraints were used between the layers, i.e. there was no heat and force transfer loss between layers. The sides and bottom of structures were fixed in the process of thermal stress calculation. The surface heat dissipation coefficient was 67.4 W/(m^2·K) assuming the wind speed near the upper surface of the pavement layer was 3 m/s [7]. The ambient temperature, the temperature around the layers below the pavement, and the temperature of the bottom of the soil base were assumed to be constant at 20 °C.

The temperature fields and temperature–time history relationships obtained by FE analyses for pavement slabs fabricated with SSWs-engineered UHPC are displayed in Figure 7.6. It can be found that the temperatures at different positions of the structures all first increase to the maximum values and then decrease as the times increase. The maximum temperature at the core of pavement slab caused by heat of hydration is reduced by 1.2 °C and 4.6 °C, respectively, due to the addition of 0.5 vol% and 1.5 vol% SSWs. Meanwhile, the area of the high-temperature region is reduced, and the corresponding maximum temperature has a downward trend. The upper surface center temperature of UHPC with 1.5 vol% SSWs is increased by 2.2 °C compared with that of UHPC without SSWs, mostly depending on high thermal conductivity of composites. In addition, the time that the core positions take to reach the maximum temperature is significantly shortened. The times since casting for core positions to reach peak temperature for UHPC without SSWs, UHPC with 0.5 vol% SSWs, and 1.5 vol% SSWs are 27 hours, 27 hours, and 25 hours, respectively, indicating that SSWs can speed up the hydration process. Meanwhile, the time since casting for the upper surface center position to reach the maximum temperature is increased by 1 hour, benefiting from improvement of the heat conduction ability of the composites. The maximum values of the temperature gradient between the core and the upper surface center positions in the composites are 25.4 °C, 24.0 °C, and 18.5 °C, respectively, reduced by 6.9 °C due to

(a)

(b)

(c)

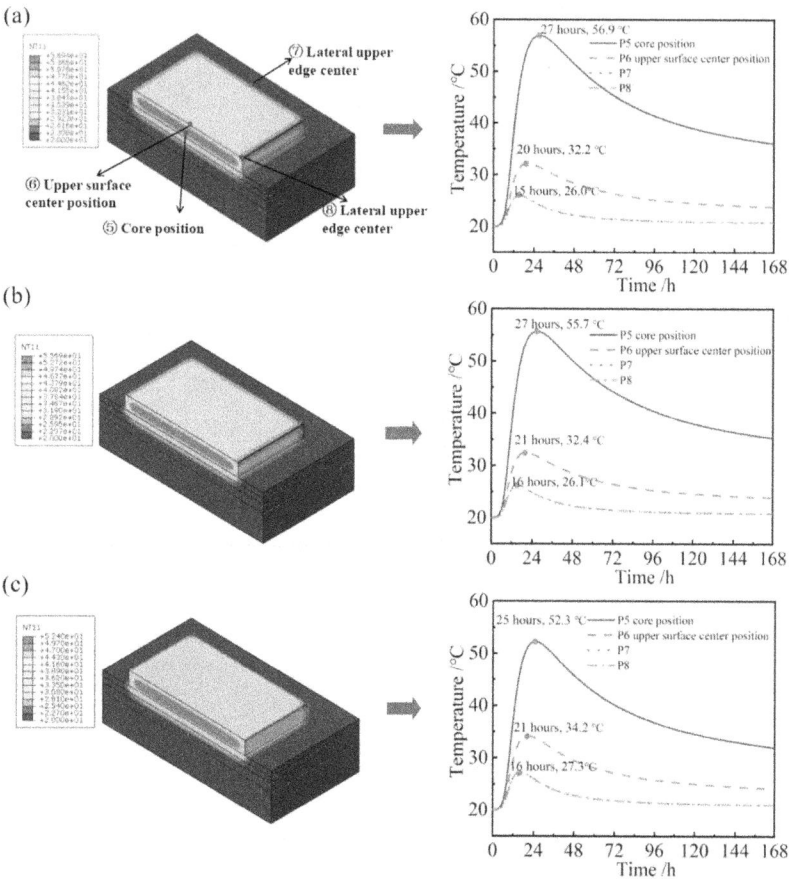

Figure 7.6 Temperature fields and temperature–time history relationships inside pavement slabs fabricated from SSWs-engineered multifunctional UHPC. (a) UHPC without SSWs, at the time of 27 hours; (b) 0.5 vol% SSWs-engineered multifunctional UHPC, at the time of 27 hours; (c) 1.5 vol% SSWs-engineered multifunctional UHPC, at the time of 25 hours

the addition of 1.5 vol% SSWs. This indicates that the temperature distribution is more uniform in pavement slabs fabricated with SSWs-engineered multifunctional UHPC. Due to the effect of external conditions, the temperature around the four edges of the upper surface is lower than in other positions.

The thermal stress fields and thermal stress–time history relationships inside pavement slabs fabricated with different composites are shown in Figure 7.7. It can be seen that the upper surface and the core positions of pavement slabs are in the state of tension and compression, respectively. Meanwhile, the thermal tensile stress on the edges is higher than that on the surface. However, the thermal stress on the four corners is lower than that

(a)

(b)

(c)

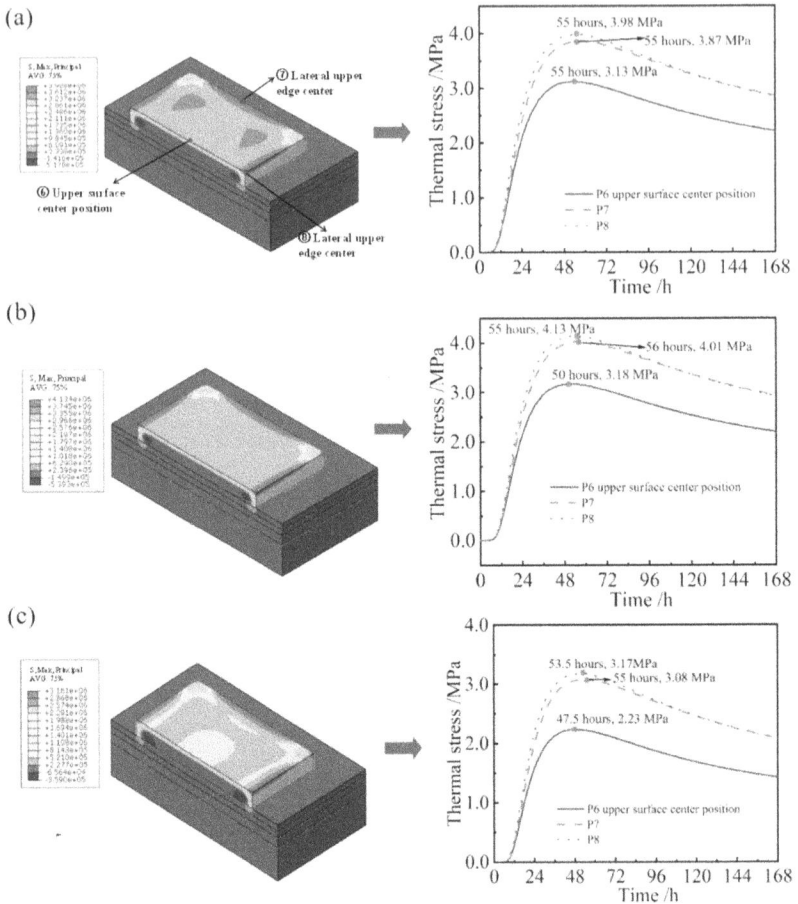

Figure 7.7 Thermal stress fields and thermal stress–time history relationships inside pavement slabs fabricated from SSWs-engineered multifunctional UHPC. (a) UHPC without SSWs, at the time of 51.5 hours; (b) 0.5 vol% SSWs-engineered multifunctional UHPC, at the time of 49.5 hours; (c) 1.5 vol% SSWs-engineered multifunctional UHPC, at the time of 40 hours

on the surface. Incorporating SSWs has no influence on these characteristics of thermal stress distribution. With the increase of SSWs content, the tension stresses on the different positions of the structures present a decreasing trend. At the time of 55 hours, the maximum thermal tensile stress appearing at the upper surface center position of UHPC pavement slab is 3.13 MPa. For pavement slabs fabricated from UHPC with 0.5 vol% and 1.5 vol% SSWs, the maximum thermal stresses with values of 3.18 MPa and 2.23 MPa at the same positions occur at 50 hours and 47.5 hours after casting, respectively, which are later than the time corresponding to the peak temperature. The upper surface center maximum thermal tensile stress is reduced by 28.8% due to the addition of 1.5 vol% SSWs. The maximum thermal stresses at the lateral upper edge center positions are 3.99 MPa, 4.13 MPa, and 3.17 MPa, and the corresponding times since casting are 55 hours, 55 hours, and 53.5 hours when 0 vol%. 0.5 vol%, and 1.5 vol% SSWs are incorporated, respectively. Based on the references of Incropera and DeWitt [5], the tensile stresses for the composites at the same curing times are 2.08 MPa, 2.99 MPa, and 4.13 MPa. This illustrates that the pavement slab fabricated with UHPC without SSWs incurs early cracking due to heat of hydration; meanwhile, the probability of thermal cracking is reduced due to the incorporation of 0.5 vol% SSWs. Furthermore, the addition of 1.5 vol% SSWs can eliminate the thermal cracking risk inside pavement slab.

7.3.3 Hydration heat–induced temperature and stress fields of pier cap

When SSWs-engineered multifunctional UHPC was used for pier caps with a size of 3 m × 2 m × 1 m, graded crushed stone was adopted for the foundation with a size of 5 m × 4 m × 1 m to eliminate the boundary effect, as demonstrated in Table 7.2. Tie constraints were also used between the pier cap and the foundation, i.e. the heat and force were transferred without loss. Meanwhile, the sides and bottom of the structures were also fixed in the process of thermal stress calculation. In addition, the wind speed near the upper surface and the surface heat dissipation coefficient were the same as those used for pavement slab simulation. The ambient temperature, the temperature around the foundation, and the temperature of the bottom of the foundation were also assumed to be constant at 20 °C.

The FE simulation temperature fields and temperature–time history relationships of pier cap fabricated with SSWs-engineered multifunctional UHPC are shown in Figure 7.8. As the core temperature reaches the maximum value, the temperature of the upper surface center position increases with increasing SSWs content, representing reduction of the thermal gradient. Meanwhile, the addition of SSWs causes the area of the high-temperature zone to decrease, resulting from the high heat conductivity of SSWs-engineered multifunctional UHPC. Figure 7.8 shows that the temperature first increases rapidly to a maximum value and then decreases as the time increases. Due to the incorporation of SSWs, the time for the maximum temperature at the core of structures appears earlier. At the time of 51.5 hours, the maximum temperature at the core of the pier cap fabricated with UHPC without SSWs can reach 69.5 °C. The temperature gradient between the core and the upper surface center of the structure is as high as 40.4 °C. At the time of 49.5 hours, the maximum temperature at the core of the pier cap fabricated with 0.5 vol% SSWs-engineered multifunctional UHPC rises to 68.0 °C, which is 1.5 °C lower than that for the UHPC structure. Meanwhile, the temperature gradient is reduced by 1.7 °C due

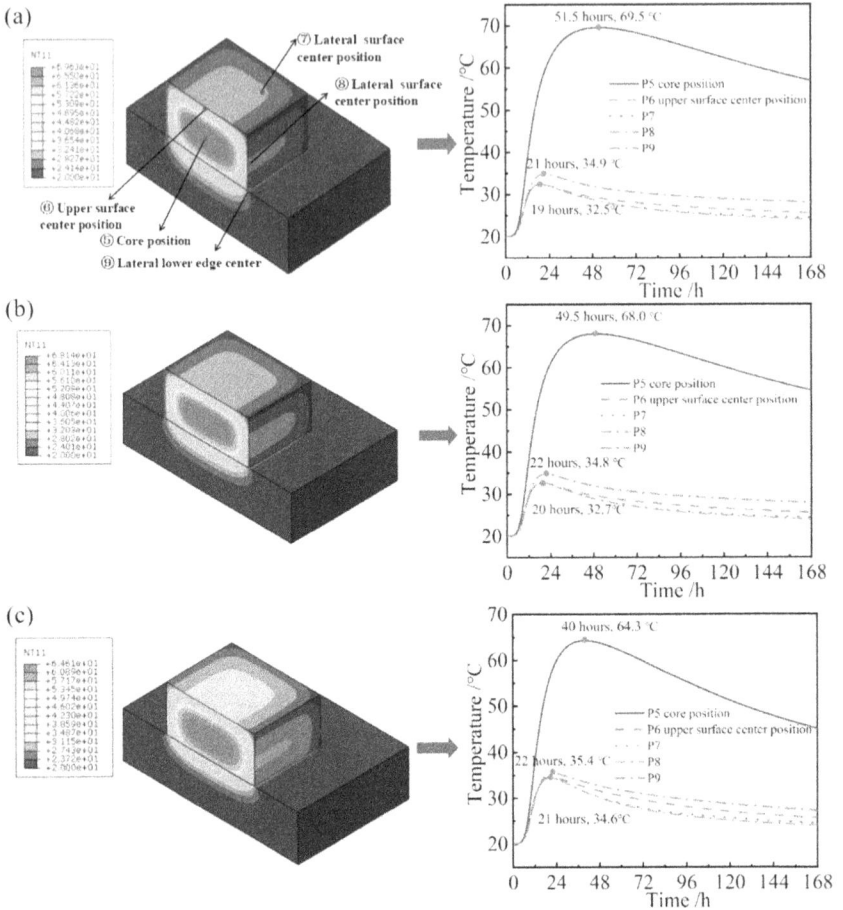

Figure 7.8 Temperature fields and temperature–time history relationships inside pier caps fabricated from SSWs-engineered multifunctional UHPC/ (a) UHPC without SSWs, at the time of 51.5 hours; (b) 0.5 vol% SSWs-engineered multifunctional UHPC, at the time of 49.5 hours; (c) 1.5 vol% SSWs-engineered multifunctional UHPC, at the time of 40 hours

to the addition of 0.5 vol% SSWs. At the time of 40 hours, the high-temperature region is remarkably concentrated in a small region of the pier cap fabricated from 1.5 vol% SSWs-engineered multifunctional UHPC, and the high-temperature region is moving to the lower part of the pier cap. Meanwhile, the maximum temperature at the core of the pier cap and the temperature gradient between the core and the upper surface have dropped to 64.3 °C and 29.7 °C, respectively. This can be attributed to the thermal conductivity of 1.5 vol% SSWs-engineered multifunctional UHPC increasing by 73.8% compared with that of UHPC without SSWs. Therefore, the heat of hydration can be easily transferred to the outside of the structure in the same external environment.

When the thermal tensile stress on the upper surface center reaches the maximum value, the thermal stress fields inside pier caps are as demonstrated in Figure 7.9. Similar to that for pavement slab, the upper surface and the core position of the pier cap are in the state of tension and compression, respectively. Meanwhile, the thermal tensile stress of the upper surface presents a decreasing trend with the increasing SSWs content. Furthermore,

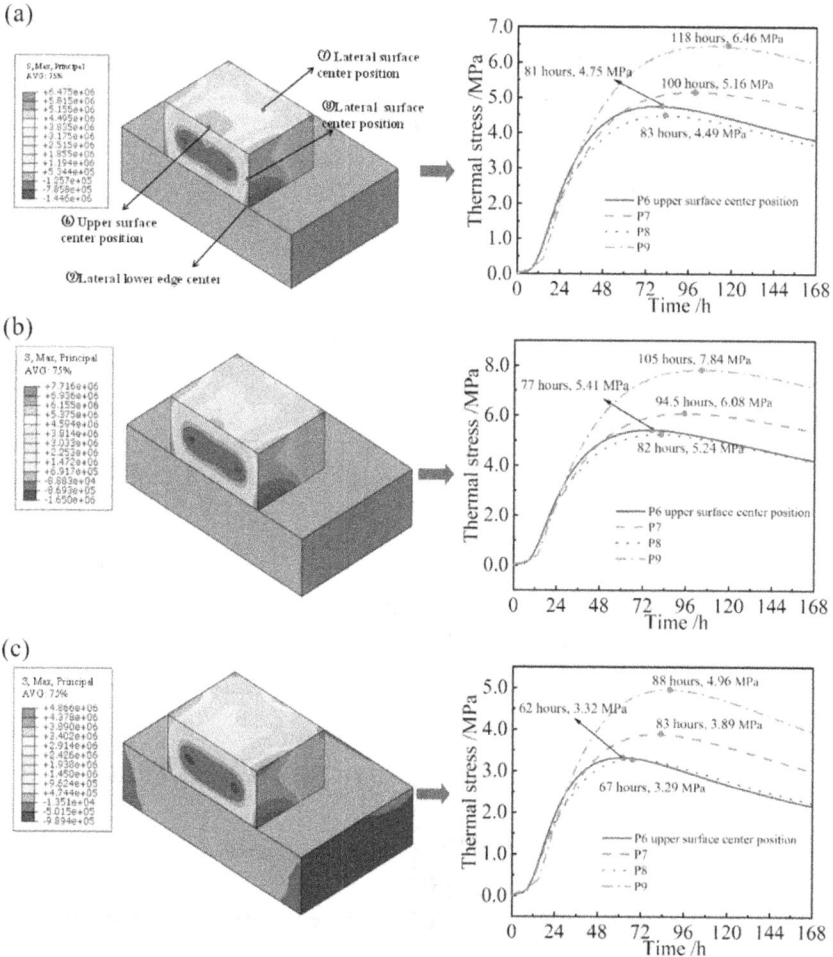

Figure 7.9 Thermal stress fields and thermal stress–time history relationships inside pier caps fabricated from SSWs-engineered multifunctional UHPC. (a) UHPC without SSWs, at the time of 81 hours; (b) 0.5 vol% SSWs-engineered multifunctional UHPC, at the time of 77 hours; (c) 1.5 vol% SSWs-engineered multifunctional UHPC, at the time of 62 hours

the maximum tensile stress on the lateral surface of the pier cap has a higher value than that on the upper surface.

The maximum tensile stress of 4.75 MPa is achieved on the upper surface center position of the pier cap fabricated with UHPC without SSWs when the curing time is 81 hours. Meanwhile, at the curing times of 100 hours and 83 hours, 5.16 MPa and 4.42 MPa maximum tensile stresses, respectively, occur on the lateral surface center positions of the structures. These values are all larger than the tensile strength of UHPC without SSWs at the curing age of 3 days, indicating that thermal cracking occurs on the surface of the pier cap fabricated with UHPC without SSWs due to heat of hydration. For the pier cap made of 0.5 vol% SSWs-engineered multifunctional UHPC, the maximum tensile stress is 5.41 MPa on the upper surface center position at the curing age of 77 hours and 6.08 MPa and 5.24 MPa on the lateral surface center positions, larger than the tensile strength of 0.5 vol% SSWs-engineered multifunctional UHPC at the curing age of 3 days. This means that pier cap fabricated from 0.5 vol% SSWs-engineered multifunctional UHPC also has the tendency to crack. However, the difference between the maximum tensile thermal stress and the tensile strength for 0.5 vol% SSWs-engineered multifunctional UHPC is significantly smaller than that for UHPC without SSWs, indicating that thermal cracks can be reduced due to the addition of 0.5 vol% SSWs. At 62 hours, 83 hours, and 67 hours, tensile stresses of 3.32 MPa, 3.89 MPa, and 3.29 MPa, respectively, appear on the upper and lateral surface center positions of the pier cap fabricated with 1.5 vol% SSWs-engineered multifunctional UHPC. Meanwhile, the tensile stress can reach 4.96 MPa at the lateral lower edge center position at the curing age of 88 hours. These values indicate that the maximum tensile stress caused by heat of hydration is less than the tensile strength of 1.5 vol% SSWs-engineered multifunctional UHPC. Therefore, it can be concluded that pier cap fabricated with 1.5 vol% SSWs-engineered multifunctional UHPC does not crack because of heat of hydration.

7.4 ELECTROTHERMAL PROPERTIES OF SSWS-ENGINEERED MULTIFUNCTIONAL UHPC

7.4.1 Self-heating properties

Specimens of 1.0 vol%/1.5 vol% SSWs-engineered multifunctional UHPC with sizes of 400 mm × 100 mm × 15 mm were used for the self-heating properties test. As shown in Figure 7.10(a), stainless electrode meshes (with sizes of 95 mm × 13 mm) were embedded into the slab, paralleling the cross section of slabs with a distance of 10 mm from each edge. According to the hydration characteristics of concrete and heat conduction theory, if the temperatures on the upper surface (UE)and lower surface (LE) reached 80 °C, the middle layer would be over-heated. Hence, temperatures in the middle

(a)

(b)

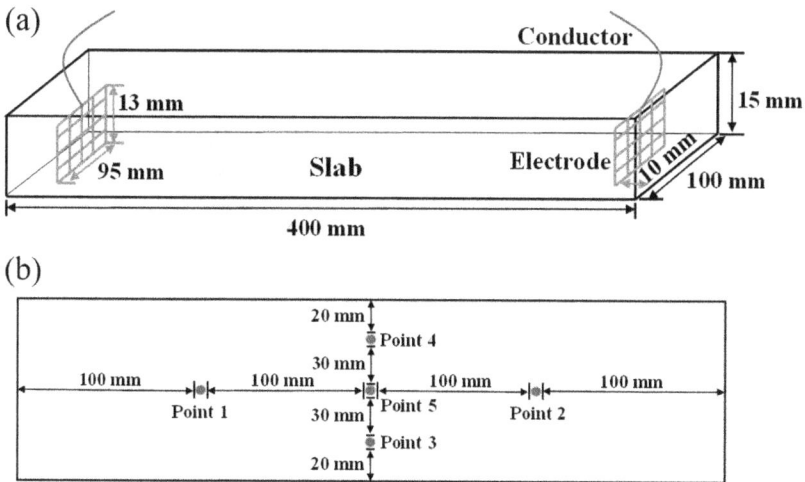

Figure 7.10 Overall plan view and sectional view of measuring point layouts on self-heating SSWs-engineered multifunctional UHPC slab. (a) Overall plan view; (b) Sectional view

layer have to be measured as the control temperature. Meanwhile, the temperatures on the LE and UE are also collected. Five representative temperature measuring points were selected to reflect the temperature distribution in each plane. Five thermocouple wires (series K produced by Omega, with the precision of 0.1 °C) were embedded in the middle layer of the slab according to the arrangement in Figure 7.10(b). This embedded method of arrangement can effectively avoid inaccurate measurement results due to heat transfer between the surface of the slab and the surrounding air.

The temperature collecting equipment employed was a thermometer produced by Omega Co. Ltd., and the electrical resistance involved in this study was measured by a Keithley 2100 Digital Multimeter using the method of two-electrode direct current (DC). The electrical resistivity can be calculated by the equation $\rho = RS/L$ (where ρ and R are the resistivity and resistance of the slab, respectively, S is the cross-sectional area (taken as 1500 mm^2), and L is the distance between two electrodes (taken as 380 mm).

As shown in Figure 7.11(a), the slab was erected vertically on two concrete prismoid supports, and 10 temperature measuring points were arranged on the UE and LE, which was the same as the middle layer. The measuring points on the LE, UE, and middle layer are named LP, UP, and MP, respectively. A total of 16 channels of temperature data were acquired, including 15 channels of measuring points and 1 channel of ambient temperature. In order to avoid reduction in mechanical properties due to excessively high temperature, the upper limit of the self-heating temperature was 80 °C [15]. The experiment was accomplished if the slab could be continuously heated to

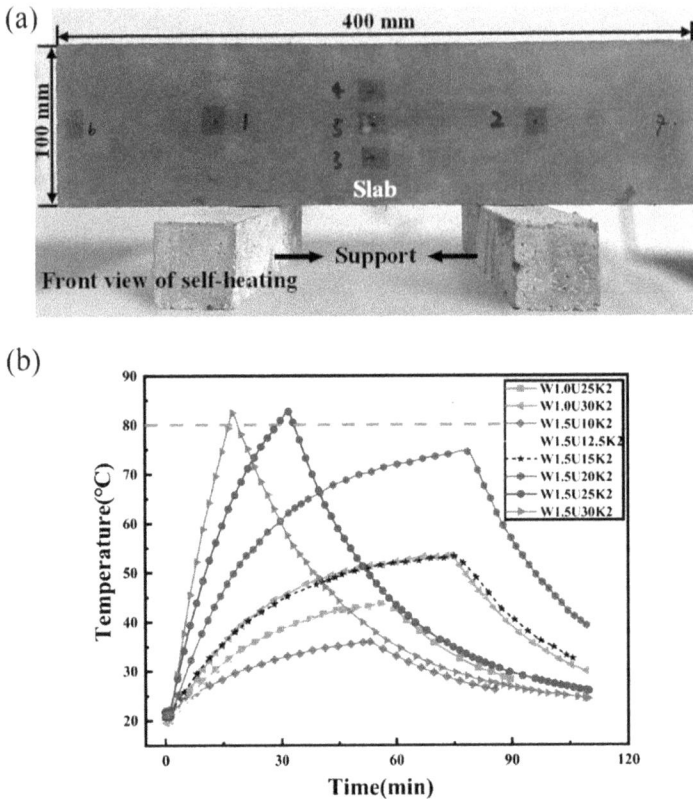

Figure 7.11 Schematic diagram of self-heating slab and temperature time histories of LP5 (K2 out of K1–K5). (a) Schematic diagram of self-heating slab; (b) Temperature time histories of LP5 (K2 out of K1–K5)

80 °C, and then, the second to fifth (K2–K5) cyclic experiments of self-heating experiments were carried out after complete cooling. Otherwise, the test was terminated, as the slab cannot be heated by more than 1 °C in 10 min, and the subsequent cyclic experiments were not conducted. This is because in practical applications, it is necessary for the self-heating UHPC with SSWs to remove ice and snow from the pavement in a timely way, so combinations showing high self-heating efficiency can handle different complicated environmental conditions. Therefore, the subsequent cyclic experiments were only carried out on the highly efficient combinations. A series of pre-experiments have been conducted, and the results showed that the 1.0 vol% SSWs-engineered multifunctional UHPC slab could not display a fast self-heating rate when the loading voltage was under 25 V. In order to ascertain the influence of SSWs content on the self-heating property of UHPC at the same self-heating power, the loading voltages for the 1.0 vol% SSWs-engineered

multifunctional UHPC slab were set as 25 V and 30 V (marked as U25 and U30), and those of the 1.5 vol% SSWs-engineered multifunctional UHPC slab were set as 10 V, 12.5 V, 15 V, 20 V, 25 V, and 30 V, respectively, according to basically equal power values. All the tests were conducted in order of voltage, from the smallest to the largest. The resistance of the 1.5 vol%/1.0 vol% SSWs-engineered multifunctional UHPC slabs was taken to be 4.82Ω and 17.84Ω, respectively. Figure 7.11(b) depicts the temperature–time histories (K2 out of K1–K5) of measuring point 5 on LE, i.e. LP5. It is worthwhile to note that W1.0U25K2 means that the self-heating properties of 1.0 vol% SSWs-engineered multifunctional UHPC slab were tested under the voltage of 25 V, and the data was recorded from the second cyclic heating experiment. Other symbols use the same naming principles.

The thermograms were taken by a thermal imager (type Fotric 287-L20) with a measuring accuracy of ± 2 °C/$\pm 2\%$. The infrared resolution of the thermal imager was 512×384 with the noise equivalent temperature difference of 0.03 °C at 30 °C. Self-heating experiments were conducted with the ambient temperature ranging from 19.4 °C to 22.0 °C. Figure 7.12 shows the thermograms of the self-heating process conducted on the 1.5 vol% SSWs-engineered multifunctional UHPC slab with the loading voltage of

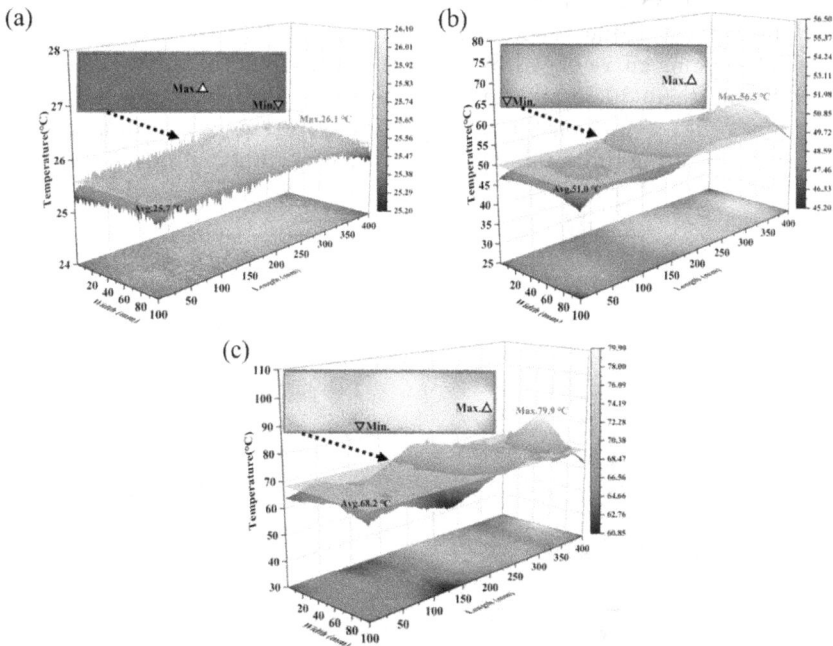

Figure 7.12 Thermogram of 1.5 vol% SSWs-engineered UHPC slab with the loading voltage of 30 V. (a) Average temperature of slab is 25.6 °C; (b) Average temperature of slab is 51.0 °C; (c) Average temperature of slab is 68.2 °C

30 V. Before self-heating, the average temperature of this slab is 25.6 °C. Then, as revealed in Figure 7.12(b), the average temperature of the slab reaches 51.0 °C in 15.74 min, and the maximum temperature is 56.5 °C. Similarly, when the slab is heated to 68.2 °C in 16.02 min, the maximum temperature is up to 79.9 °C.

During the experiments, the self-heating rate decreases as the temperature rises, so it is not always constant. The average self-heating rate of the SSWs-engineered multifunctional UHPC slabs can be calculated by using Eq. (7.3):

$$V_s = \frac{VT_1 - VT_0}{t_s} \qquad (7.3)$$

where V_s is the average self-heating rate of the slab, ΔT_1 is the temperature peak of the slab, ΔT_0 is the initial temperature of the slab, and t_s is the duration of powering.

Figure 7.13 shows the average self-heating rate and energy consumption of SSWs-engineered multifunctional UHPC slabs under different voltages, taking point 5 in the lower layer as an example. It can be found that the temperature peaks increase with the growth of SSWs content when the voltages are equal. When the contents of SSWs are the same, the average self-heating rate grows with the increase of voltage. Therefore, it can be concluded that the content of SSWs determines the upper limit of the temperature peak, and the average self-heating rate is highly dependent on the loading voltage. The self-heating curve of W1.5U30K2 is steeper than that of W1.5U25K2, and it takes them 30.55 min and 16.18 min, respectively, to reach the temperature peaks (both over 80 °C). In addition, the energy consumption (the product of heating time and power) of these two combinations is 166.52 kJ

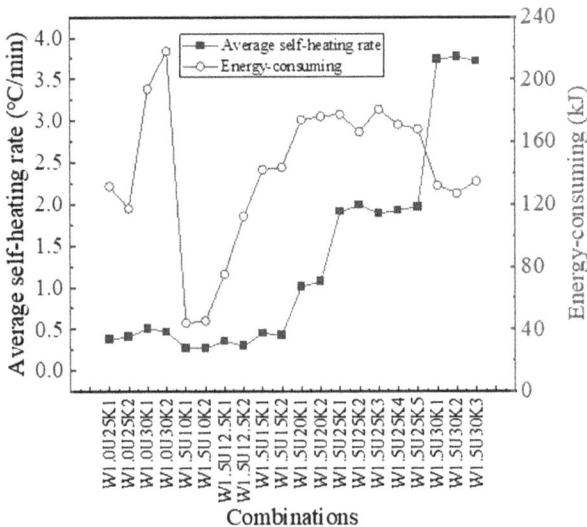

Figure 7.13 Average self-heating rate and energy consumption of different combinations

and 127.00 kJ, respectively. Consequently, the combination of W1.5U30K2 results in a time saving of 47.04% and an energy conservation of 23.73% compared with W1.5U25K2. Hence, it can be figured out that properly increasing the loading voltage can effectively save energy and improve the self-heating efficiency. It can be observed that the self-heating curves of W1.0U30K2 and W1.5U15K2 are highly coincident with each other. The phenomenon that the self-heating rate slows down with increasing temperature can be observed on all self-heating curves for the slabs. As the temperature difference increases, the thermal convection effect between the slab and the surrounding air is enhanced, which boosts the heat loss.

According to a previous study by Shishegaran et al. [16], conductive concrete with steel fibers can only realize a self-heating rate of 0.13 °C/min under the power density of 172.85 W/m². Mohammed et al. [17] found that the temperature of carbon fiber reinforced concrete slabs (sized 30 cm × 25 cm × 8 cm) increased by only about 7.5 °C and 16.5 °C within 150 min when the power density was 200 W/m² and 400 W/m², respectively. The mix proportioning of these studies is not exactly the same as that in this chapter, but as the electrically conductive property is dominated by conductive fillers, and self-heating performance is determined by electrical conductivity and self-heating efficiency of the composite, it can be indicated that the SSWs-engineered $U_{1.5}$UHPC displays more rapid self-heating performance (increased by 61.0 °C in 16.18 min) than carbon fiber reinforced conductive concrete. When the loading voltage is 25 V, the average self-heating rate of 1.5 vol% SSWs-engineered multifunctional UHPC slab is around 1.94 °C/min, and it increases to 3.75 °C/min when 30 V of voltage is loaded. It is noteworthy that 1.5 vol% SSWs-engineered multifunctional UHPC slab can be heated to over 80 °C under the voltage of 25 V and 30 V, while the other groups cannot meet this condition. Hence, the following active deicing and snow-melting experiments will be conducted on 1.5 vol% SSWs-engineered multifunctional UHPC slab with a loading voltage of 25 V or 30 V.

Figures 7.14–7.16 display the average self-heating rate during five cyclic experiments of 1.5 vol% SSWs-engineered multifunctional UHPC slab. It is demonstrated that 1.5 vol% SSWs-engineered multifunctional UHPC slab can maintain a stable average self-heating rate after five cyclic experiments when different voltages are loaded, and this phenomenon is generated by the steady self-heating capability of SSWs-engineered multifunctional UHPC. The dispersion degree of experimental data can be estimated by using the coefficient of variation (CV), which can be calculated by Eq. (7.4):

$$CV = \frac{\sigma}{T_m} \times 100\% \qquad (7.4)$$

where σ and T_m are the standard deviation and the mean value of temperature, respectively. Consequently, the results calculated for loading voltages of 25 V and 30 V are $CV_{U25} = 1.91\%$ and $CV_{U30} = 1.19\%$, respectively. Generally, the data can be considered homogeneous when the coefficient of

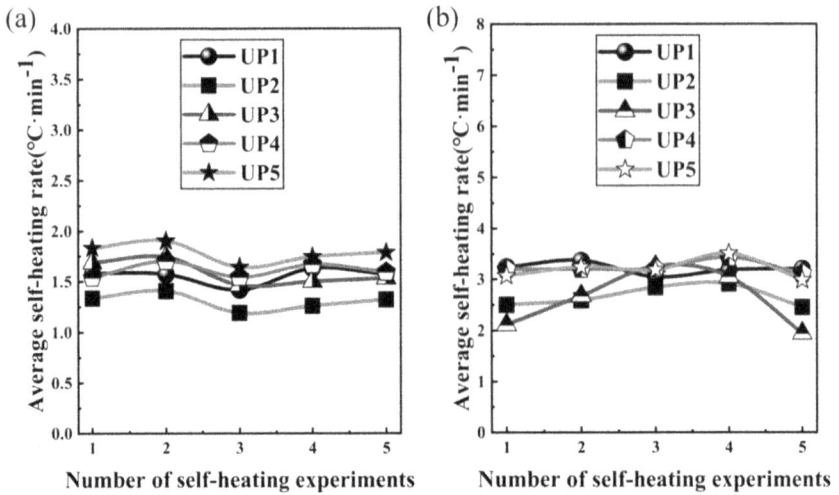

Figure 7.14 Average self-heating rate during five cycles of heating on upper surface of 1.5 vol% SSWs-engineered multifunctional UHPC slab. (a) 25V; (b) 30V

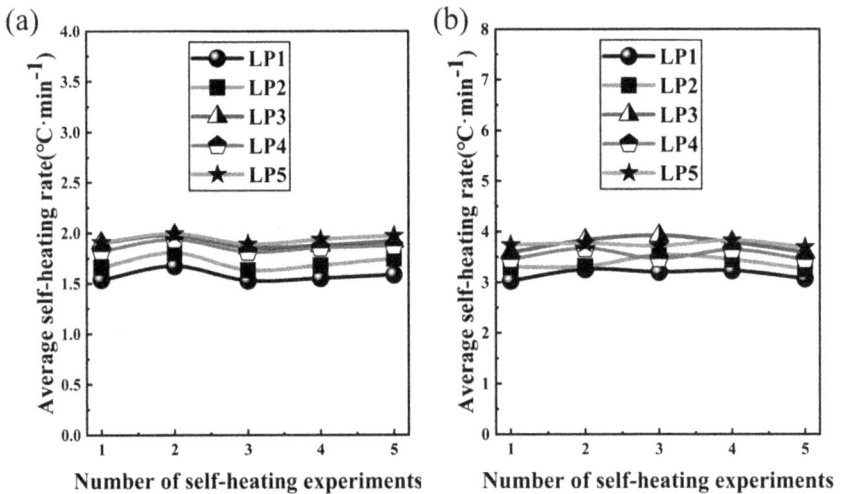

Figure 7.15 Average self-heating rate during five cycles of heating on lower surface of 1.5 vol% SSWs-engineered multifunctional UHPC slab. (a) 25V; (b) 30V

variation is no more than 10%, so it can be proved that the SSWs-engineered multifunctional UHPC slab possesses good experimental repeatability during the cyclic self-heating experiments [18].

Figure 7.17 shows the resistivity of SSWs-engineered multifunctional UHPC slabs under different voltages. It can be seen that the average

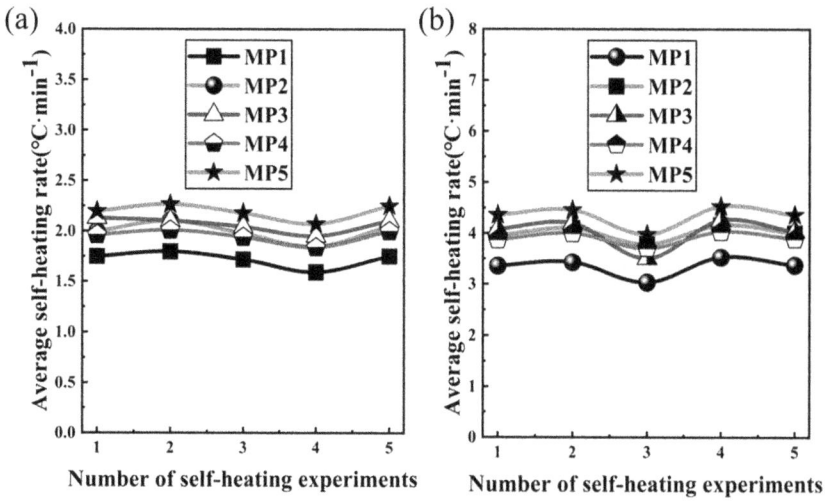

Figure 7.16 Average self-heating rate during five cycles of heating on middle surface of 1.5 vol% SSWs-engineered UHPC slab. (a) 25V; (b) 30V

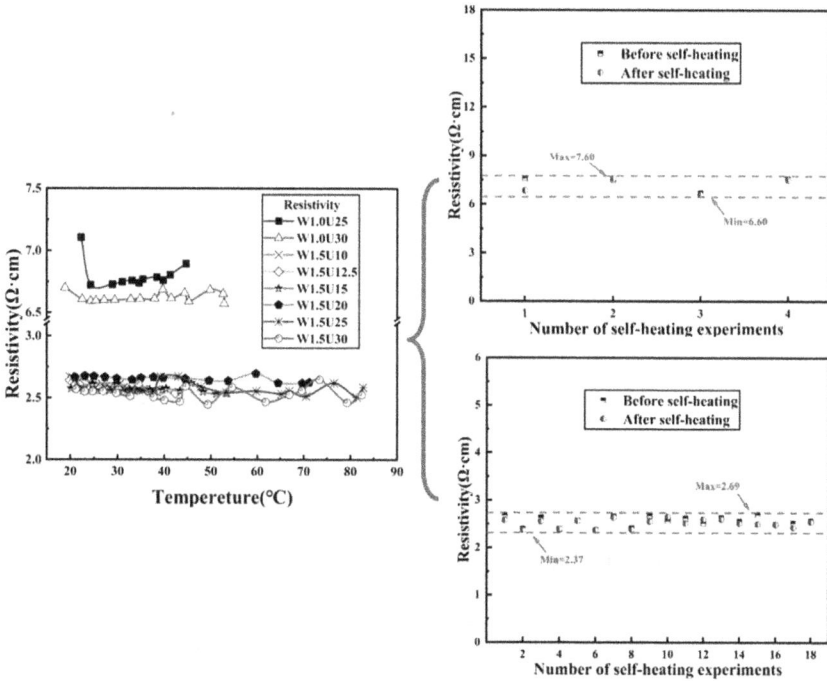

Figure 7.17 Resistivity curves of 1.0 vol%/1.5 vol% SSWs-engineered multifunctional UHPC slabs during and after heating

resistivity drops from 6.67Ω•cm to 2.58Ω•cm as the content of SSWs increases from 1.0 vol% to 1.5 vol%. This indicates that the higher volume content of SSWs greatly reduces the resistance and resistivity of UHPC. It can be noted that the resistivity of the slabs remains stable under different voltages and temperatures, reflecting the thermal stability of SSWs-engineered multifunctional UHPC. The resistivity slightly fluctuates and the range is fairly small after cyclic self-heating experiments, displaying the stable self-heating capability of the slabs. This can be attributed to SSWs bridging each other to form a complete electrical and thermal conductive network inside the UHPC, and the arrangement of SSWs being closer with the increase of volume content. According to the report of Hong et al. [19], the resistivities of graphite slurry infiltrated steel fiber concrete specimens are 26.3Ω•cm, 25.3Ω•cm, and 18.2Ω•cm, respectively, with the content of steel fiber at 4 vol%, 6 vol%, and 8 vol%. This indicates that the reinforcing capability of SSWs is better than that of graphite slurry infiltrated steel fiber, since the UHPC with SSWs can achieve better electrical conductivity at a lower volume content [20, 21].

7.4.2 Active deicing properties

The deicing experiments (marked D) were carried out in the freezer with temperatures ranging from –30 °C to 0 °C. The temperatures selected to simulate common cold weather and extremely frozen weather were –15 °C (labelled O1) and –25 °C (labelled O2), respectively. Uniform ice layers with thicknesses of 5 mm (denoted C5) and 9 mm (denoted C9) on LE were prepared by sprinkling water successively, and the loading voltages selected were 25 V and 30 V. Since 1.5 vol% SSWs-engineered multifunctional UHPC (W1.5) slab displays higher self-heating efficiency than 1.0 vol% SSWs-engineered multifunctional UHPC (W1.0) slab, the active deicing and snow-melting experiments were carried out on 1.5 vol% SSWs-engineered multifunctional UHPC slab. There were 16 groups of deicing experiments performed using the traversal combination of voltage, ambient temperature, and ice thickness on 1.5 vol% SSWs-engineered multifunctional UHPC slab. In the arrangement of measuring points of active deicing and snow-melting experiments, the measuring points set in the middle layer were cancelled, and another two new points, 6 and 7, were added along the longitudinal direction of UE and LE, as shown in Figure 7.18(a). According to Boyd [22] and Morita and Tago [23], one side of the slab was raised by 10% by two wooden spacers to simulate the drainage slope in practical applications, as shown in Figure 7.18(b). During the tests, the slab, along with two supports and two spacers, was placed on a low temperature–resistant electronic balance. The numerical reading of current size and total mass was acquired every 5 minutes, so the mass loss equaled the weight of melted ice layer. Three key time points were recorded: time point A (TA) when the ice layer began to be melted, time point B (TB) when

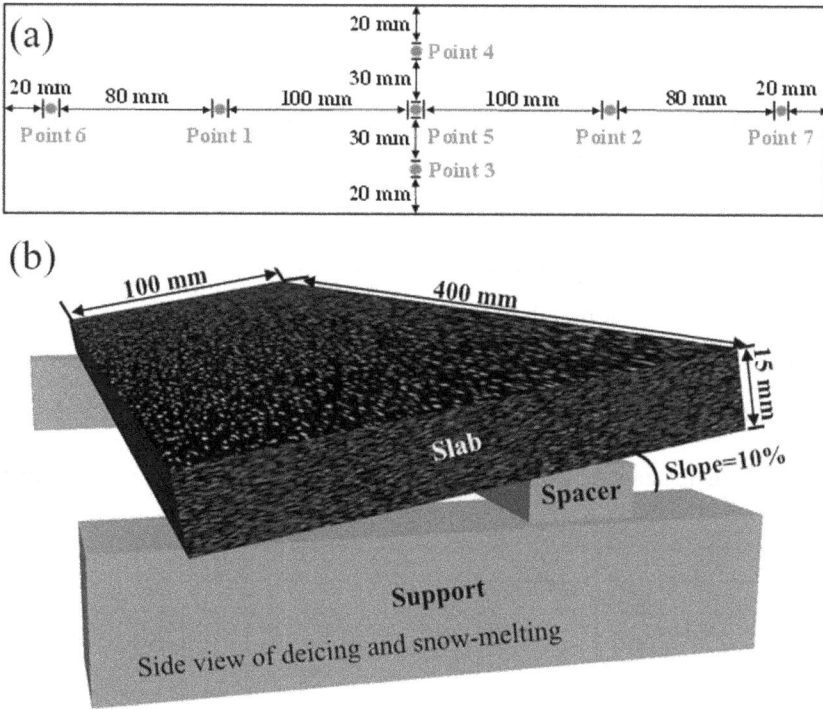

Figure 7.18 Sectional view of measuring point layouts and side view of setups for active deicing and snow-melting experiments. (a) Sectional view; (b) Side view

the ice layer was completely melted, and time point C (TC) when the LE was totally dried. Since the deicing curves are fairly similar, the groups of DW1.5U25O1C5K2 and DW1.5U30O2C9K2 cover the two distinct and representative conditions (i.e. applying lower voltage in common cold weather and higher voltage in extremely frigid weather, respectively).

The 1.5 vol% SSWs-engineered multifunctional UHPC slab was heated under a voltage of 25 V and 30 V and at a temperature of –15 °C and –25 °C to melt ice with a thickness of 5 mm and 9 mm, respectively; the corresponding experiments were marked DW1.5U25O1C5K2 and DW1.5U30O2C9K2. It can be seen from Figure 7.18 that the test course can be divided into three phases: phase (I) is the self-heating stage, phase (II) is the state-transforming stage, and phase (III) is the drying stage, which is consistent with the curves of snow-melting pavement with embedded pipelines [24]. During phase (I), the slab is heated up, and the thermal energy is continuously transferred to the ice layer through heat conduction. In phase (II), the temperature of the ice layer continues to rise, and the ice starts to melt when it reaches 0 °C; hence, the platform segment of the curve appears. After the ice layer is mostly melted, the status of the slab enters phase (III). Due to the uninterrupted input of thermal energy, the remaining water is evaporated by continuously rising temperature, and finally, the slab is totally dried. Therefore, the three key time points A, B, and C are used to evaluate the deicing efficiency. Remarkably, the temperature curves do not rise steadily all the time, as shown in Figure 7.19. For instance, in the deicing process of DW1.5U30O2C9K2, the temperature experiences ups and downs as time goes by. It can also be seen that the interval between time nodes B and C is fairly short. As shown in Figure 7.20, a cavity is formed by the voids beneath the ice layer, thus blocking the heat conduction upwards.

This results in a tiny residual ice layer when the whole slab is basically dried out. Furthermore, when the ice layer is broken and reattaches to LE, a large fluctuation of temperature emerges. Additionally, the temperature on UE decreases sharply when the melted water flows over it. The ice layer always begins to be melted along the center of the long side, then spreads from the middle to the surrounding area, and finally, leaves some residual ice near measuring points 6 and 7. This phenomenon is also consistent

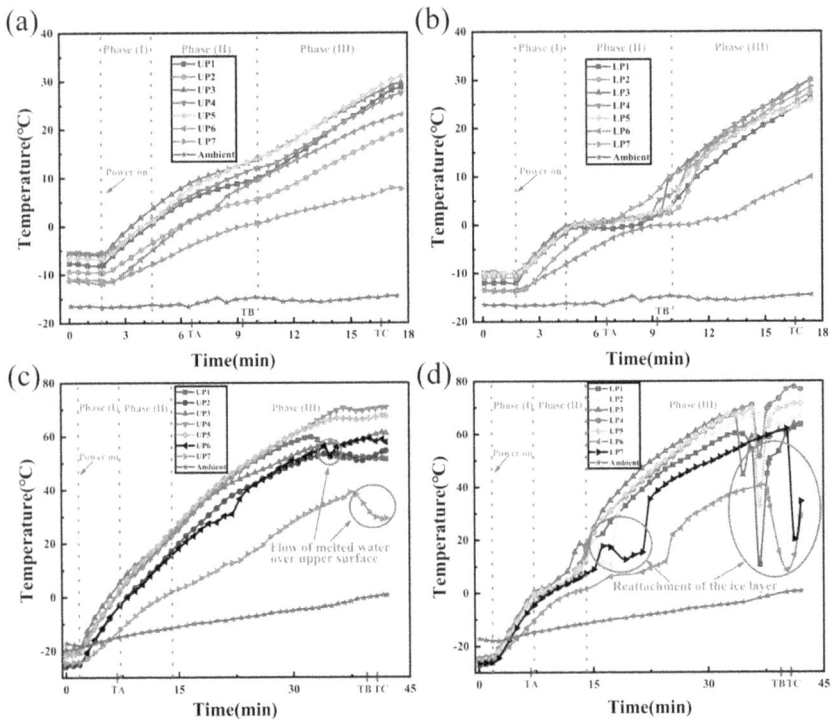

Figure 7.19 Temperature–time histories of DW1.5U25O1C5K2 and DW1.5U30O2C 9K2. (a) Upper surface of DW1.5U25O1C5K2; (b) Lower surface of DW1.5U25O1C5K2; (c) Upper surface of DW1.5U30O2C9K2; (d) Lower surface of DW1.5U30O2C9K2

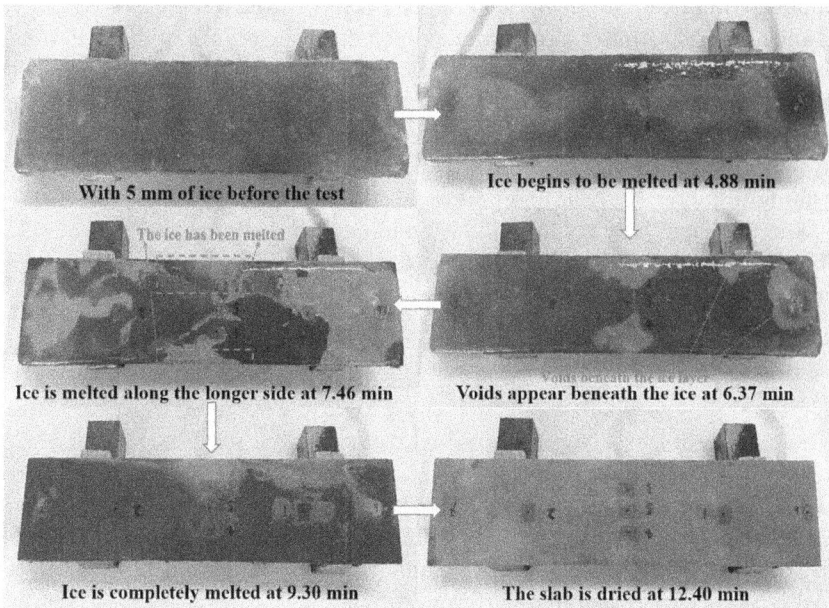

Figure 7.20 Photographic recording of DWI.5U30O1C5K2

with the regularity of temperature distribution. The values of temperature decrease from the center to the edges, and the temperature gradient along the longer side changes more obviously, since the aspect ratio of the slab is 4. Figure 7.20 depicts the photographic recording of DW1.5U30O1C5K2. Before the test, the slab is covered by a 5-mm layer of ice. It begins to melt after 4.88 min of heating, the ice layer is completely melted after 9.30 min of heating, and finally, the slab is dried out after 12.40 min of heating. It is vividly reflected in Figure 7.19 that the time–temperature relationship on LE (i.e., the iced surface) displays an obvious three-segment curve. Temperature increases linearly in phase (I), which is slightly different from the curves of self-heating experiments. As the internal pores of the slabs are filled with water, the temperature grows faster than that of self-heating experiments. During the platform segment of Phase (II), the temperature of the iced surface remains at around 0 °C until the bottom of the ice layer is detached due to the existence of the cavity formed by voids. The temperature in the cavity fluctuates, and the ice layer is melted from bottom to top. In phase (III), the temperature continues to rise near-linearly, and the slope of the curve improves with the increase of loading voltage. On UE of the slab (i.e. the non-iced surface), the temperature curves are not significantly segmented. The curves of the measuring points on UE increase roughly linearly, and the slopes of the curves are also positively correlated with the applied voltage. It can be noticed that the temperatures of measuring points 6 and 7 increase more slowly than those of other points in all situations, and the amplitude is relatively small, but the overall temperature uniformity is within an acceptable range. Although the deicing rate is accelerated under higher voltage, the speed of the temperature rise is much too fast, resulting in uneven heating of the ice layer. The cavity formed by the void enlarges, leading to increased heat loss and reduced energy efficiency.

The percentage of residual ice is employed to demonstrate the relationship between ice-melting degree and time consumption, and it can be calculated according to Eq. (7.5):

$$\omega_i = \frac{m_{i1}}{m_{i0}} \tag{7.5}$$

where ω_i is the percentage of residual ice, m_{i1} is the weight of residual ice, and m_{i0} is the initial weight of the ice layer.

The ice-melting rate is used to measure the ice-melting efficiency of the self-heating slab, and the unit is ml/(m²•min), which represents the volume of solid ice melted into water per unit time per unit area [25]. The ice-melting rate can be calculated using Eq. (7.6):

$$\Theta_i = \frac{V_w}{S_0 \times t} \tag{7.6}$$

where Θ_i is the ice melting rate, V_w is the volume of ice melting into water, S_0 is the area of the ice receiving surface (taken as 400 cm²), and t is the duration of powering.

According to the law of conservation of energy, input and output of energy are always equal. So in the process of deicing, the energy efficiency Θ_i can be calculated using Eq. (7.7) [26]:

$$\eta_i = \frac{m_c C_{pc} \Delta T_c + m_i C_{pi} \Delta T_i + \Delta_{sol} H m_i}{P \Delta t} \tag{7.7}$$

where P is the loading power, Δt is the duration of powering, m_c and m_i are the weights of the slab and ice layer, respectively (m_c value 1565.4 g for the W1.5 slab), C_{pc} and C_{pi} are the specific heat capacities of the slab and the ice layer (taken as 1.44 J/(g•K) and 2.10 J/(g•K), respectively), ΔT_c and ΔT_i are the temperature rises of the slab and the ice layer, and $\Delta_{sol} H$ is the fusion heat of ice (taken as 333.33 J/g). It can be concluded from Eq. (7.7) that the overall heat energy (Q_s) generated by powering is equal to the sum of four components: the heat energy required for heating the slab (Q_c) and the ice layer (Q_i), the latent heat of fusion for ice layer (Q_f), and the heat loss of the system (Q_l).

Figure 7.21(a) displays the percentage of residual ice (8 out of 16 cyclic experiments) during the deicing experiments. It can be noted from the

(a)

(b)

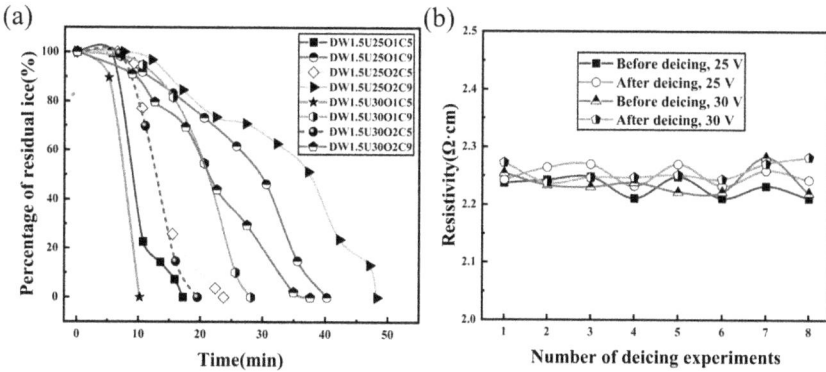

Figure 7.21 Deicing properties. (a) Percentage of residual ice; (b) Resistivity of the W1.5 slab after 16 cyclic deicing experiments

photograph that the thickness of the ice layer shows more influence on the deicing process than the loading voltage. With increasing ice thickness, the time is prolonged, and the slope of the curves has decreased. The fastest self-heating combination is DW1.5U30O1C5, and the slowest is DW1.5U25O2C9. In order to prove that the deicing performance of the slab is hardly affected by its electrical properties, the resistivities before and after each deicing experiment were recorded. Figure 7.21(b) depicts the resistivity of the 1.5 vol% SSWs-engineered multifunctional UHPC slab after 16 cyclic deicing experiments (8 under the loading voltage of 25 V and another 8 under the loading voltage of 30 V). It can be noted that the range of resistivities is less than 0.07Ω •cm, showing that the resistance stability of the self-heating slab is high after 16 active deicing experiments.

The power density of the deicing experiments can be calculated using Eq. (7.8):

$$P_d = \frac{P_i}{S_{slab}} \tag{7.8}$$

where P_d is the power density, P_i is the loading power of deicing experiments, and S_{slab} is the surface area of the slab (taken as 0.095 m^2).

Furthermore, the results of the ergodic combination of the deicing experiments are shown in Table 7.3.

It can be seen from Table 7.3 that the average power densities of 1.5 vol% SSWs-engineered multifunctional UHPC slab are 1099.09 W/m^2 and 1568.04 W/m^2 under the loading voltages of 25 V and 30 V, respectively. The average ice-melting rate of 16 cyclic deicing experiments is 125.37 ml/(m^2•min), with 102.31 ml/(m^2•min) under the loading voltage of 25 V and 148.43 ml/(m^2•min) under the loading voltage of 30 V. The ice melting rate is positively correlated with the ice thickness and

Table 7.3 Data record of the deicing properties of 1.5 vol% SSWs-engineered multifunctional UHPC

Experiment Title	Ice-melting rate (ml/ (m²•min))	TA (min)	TB (min)	TC (min)	Power density (W/m²)	Energy efficiency (%)
DW1.5U25O1C5K1	80.78	5.75	17.70	20.45	1101.18	99.22
DW1.5U25O1C5K2	58.71	6.58	9.21	16.61	1092.84	97.18
DW1.5U25O1C9K1	122.69	3.93	37.48	39.21	1090.80	86.27
DW1.5U25O1C9K2	147.81	8.65	38.19	39.65	1108.05	80.78
DW1.5U25O2C5K1	72.62	9.02	21.24	23.35	1091.97	99.96
DW1.5U25O2C5K2	77.12	10.68	23.61	25.26	1110.83	95.24
DW1.5U25O2C9K1	135.13	11.39	46.77	48.70	1091.97	78.28
DW1.5U25O2C9K2	123.58	10.35	36.56	37.70	1105.13	86.37
DW1.5U30O1C5K1	128.59	4.95	20.70	22.13	1570.33	90.11
DW1.5U30O1C5K2	105.50	4.88	9.30	12.40	1578.73	97.85
DW1.5U30O1C9K1	181.12	8.90	29.07	31.77	1572.69	82.55
DW1.5U30O1C9K2	192.53	7.02	25.47	27.65	1570.79	78.35
DW1.5U30O2C5K1	103.27	6.90	17.59	18.79	1572.01	97.84
DW1.5U30O2C5K2	90.37	5.69	16.50	17.12	1566.97	84.74
DW1.5U30O2C9K1	247.09	7.37	28.98	31.97	1551.73	89.74
DW1.5U30O2C9K2	138.95	6.75	39.48	40.80	1561.07	75.03

ambient temperature at the same voltage, which is consistent with the research results of Huang's study [25]. When the thickness of the ice layer is 5 mm and 10 mm, peak values of the ice melting rate of the thermal fluid circulation snow and ice melting system (TFCSIMS) are only about 42 ml/(m²•min) and 65 ml/(m²•min), respectively [25], indicating that the SSWs-engineered multifunctional UHPC features better deicing performance than the TFCSIMS.

The average energy efficiency of 16 cyclic deicing experiments is 88.72%, and it has been significantly boosted compared with the previous study (50–66%) [27]. The average energy efficiency is 90.41% under the loading voltage of 25 V, and it slightly decreases to 87.03% under the loading voltage of 30 V. It can be concluded that with the increase in power density, the value of the ice melting rate has improved while that of energy efficiency has decreased due to uneven temperature rise.

7.4.3 Active snow-melting properties

The snow-melting experiments (marked M) were conducted outdoors in winter with the ambient temperature ranging from −6.7 °C to −0.5 °C and the wind speed ranging from 0 m/s to 5.36 m/s. In the deicing experiments, the ice melting rates are very similar under the loading voltages of 25 V and 30 V, while the lower voltage is more energy-efficient. Consequently, the loading voltage of 25 V was selected for the snow-melting experiments. Snow layers of 2 cm and 5 cm were selected to simulate light to medium snow weather and heavy snow weather, respectively, and the stacking status can be divided into two states: natural accumulation and artificial compaction (denoted SN and SA, respectively). The layout of temperature measuring points, drainage slope design, and mass loss collection method of the snow-melting experiments were the same as those of the deicing experiments. During the testing period, the wind speed was collected synchronously, and three key time points, TA, TB, and TC, were also recorded.

According to the nomenclature of this chapter, DW1.5U30O2C9 means the deicing experiment on the 1.5 vol% SSWs-engineered multifunctional UHPC slab with an ice thickness of 9 mm and an ambient temperature of –25 °C under the loading voltage of 30 V. Meanwhile, MW1.5U25SA5 stands for the snow-melting experiment with an artificial compacted snow layer of 5 cm under the voltage of 25 V. Measuring point layouts are the same as for deicing experiments, as shown in Figure 7.18.

Figure 7.22 shows the temperature–time histories and wind speed curves of MW1.5U25SN2K2 and MW1.5U25SA5K2. SN2 and SA5 stand for the snow layers that are easiest and hardest to melt, respectively, so these two combinations are enough to evaluate the snow-melting performance of the developed composite. It can be clearly observed in Figure 7.22 that the temperatures of UE and LE fluctuate greatly due to the influence of the wind, and the amplitude and frequency of fluctuation on LE are higher than on UE. The interval between TB and TC is also very close because the snow layer can be detached from the bottom during the melting process, just like the ice layer. As shown in Figure 7.23, graph (c) features a partial close-up view of the void beneath the snow layer during the experimental course, and it can be clearly seen that there is a distinct gap between the snow layer and LE of the slab. As a result, the snow-melting rate is slowed down, and the surface is basically dried before the snow layer is completely melted. It can be figured out from Figure 7.23 that the fluctuations on LE principally result from the reattachment of the snow layer, while those on UE are primarily ascribed to the flow of melted water. The outdoor wind heightens heat loss and decreases the snow-melting rate, and it also makes the temperature curves unstable. Snow is relatively fluffy, and the contact area with the air is much larger than for ice, so it is much easier to melt. The high porosity of snow contributes to lower energy efficiency compared with that of the deicing experiments, but the overall snow-melting rate has been improved.

Figure 7.23 also photographically shows the snow-melting process of MW1.5U25SA2K1. Before the experiment, the slab is covered with 2 cm

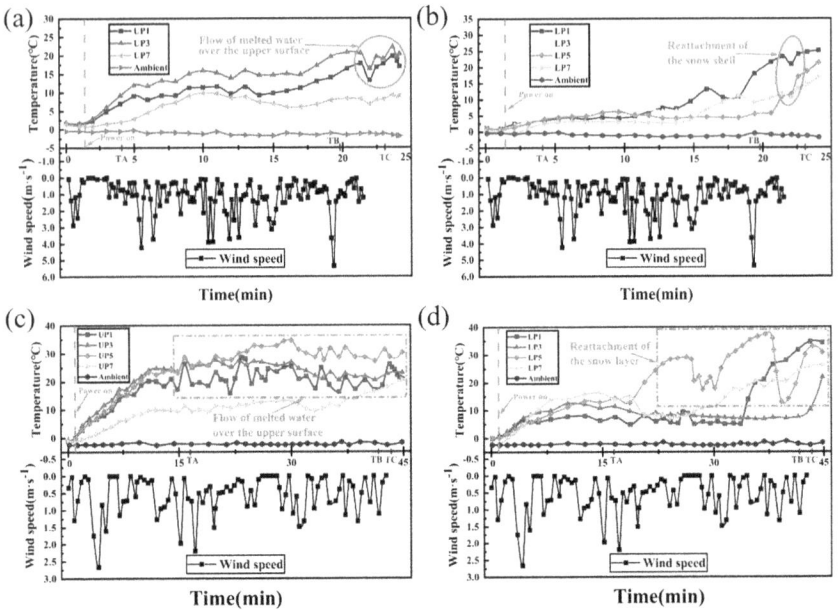

Figure 7.22 Temperature and wind speed curves of MWI.5U25SN2K2 and MWI.5U25SA5K2. (a) Upper surface of MWI.5U25SN2K2; (b) Lower surface of MWI.5U25SN2K2; (c) Upper surface of MWI.5U25SA5K2; (d) Lower surface of MWI.5U25SA5K2

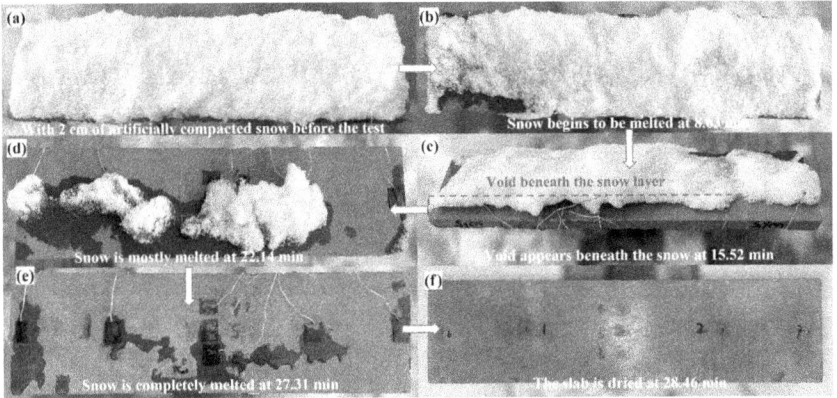

Figure 7.23 Photographic recording of MWI.5U25SA2KI

of artificially compacted snow layer. This snow layer began to melt from the corner of LE at 8.63 min; it was completely melted at 27.31 min of self-heating, and finally, the slab was dried at 28.46 min.

The percentage of residual snow (four out of eight cyclic snow-melting experiments) is shown in Figure 7.24. It can be detected from Figure 7.24(a)

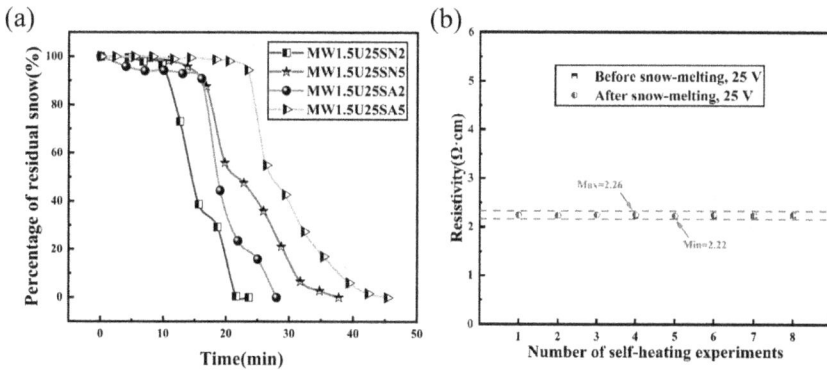

Figure 7.24 Snow-melting capacity and stability of SSWs-engineered multifunctional UHPC. (a) Percentage of residual snow; (b) Resistivity

that the time consumption is extended with the increase in thickness. For each thickness of the snow layer, the natural/artificial compaction state has a remarkable effect on the efficiency of snow-melting. Artificially compacted (AC) snow layers are more difficult to melt than naturally accumulated snow layers, because the void beneath the snow layer is more likely to be formed in AC conditions. In order to guarantee that the snow-melting performance is not affected by the electrical properties of the slab, the resistivity before and after each snow-melting experiment was also recorded. Figure 7.24(b) reveals the resistivity of the 1.5 vol% SSWs-engineered multifunctional UHPC slab after eight cyclic snow-melting experiments. It is noteworthy that the curves of resistivity are almost constant with a range of less than 0.04Ω•cm, indicating that the developed self-heating UHPC shows stable resistivity in both indoor and outdoor circumstances.

The key parameters of the snow-melting experiments are shown in Table 7.4, in which all the calculation formulas are the same as those of the deicing experiments It can be noted from Table 7.4 that the average snow-melting

Table 7.4 Key parameters of snow-melting tests

Experiment title	Snow-melting rate (ml/ (m²·min))	TA (min)	TB (min)	TC (min)	Power density (W/m²)	Energy efficiency (%)
MW1.5U25SN2K1	122.92	4.94	10.99	12.97	1097.23	82.37
MW1.5U25SN2K2	124.62	4.03	19.86	23.18	1109.80	58.05
MW1.5U25SN5K1	157.66	6.34	15.20	17.12	1100.62	86.52
MW1.5U25SN5K2	173.24	10.65	33.65	38.74	1100.62	65.88
MW1.5U25SA2K1	175.26	8.63	27.31	28.46	1098.19	62.62
MW1.5U25SA2K2	142.40	16.65	38.54	41.99	1095.48	60.61
MW1.5U25SA5K1	171.31	17.85	40.55	41.40	1182.23	55.44
MW1.5U25SA5K2	217.37	17.60	42.69	45.45	1096.90	72.34

rate of eight cyclic experiments is 160.60 ml/(m²•min), which is 56.97% higher than that of the deicing experiments (102.31 ml/(m²•min)) under the voltage of 25 V. The energy efficiency of snow-melting obtained in this study for SSWs-engineered multifunctional UHPC is 67.98%, higher than that obtained by Zhu et al. [28] for an electric cable heating system (26.2–46.1%) and Hong et al. [19] for graphite slurry infiltrated steel fiber concrete (less than 20%). The snow-melting rate grows with increasing snow thickness, while the energy efficiency is just augmented slightly. When the snow layer is altered from natural accumulation to artificial compaction, the energy efficiency is greatly reduced.

7.5 SUMMARY

The chapter introduced the effect of SSWs on the thermal conductivity and specific heat of UHPC, and the performance to control thermal cracking caused by hydration heat of UHPC specimens and structural elements. The self-heating as well as active deicing and snow-melting performances of SSWs-engineered multifunctional UHPC slab were presented. The main conclusions can be drawn as follows:

1) SSWs can increase the thermal conductivity of UHPC by 70.3%. Incorporating 0.5 vol% SSWs can decrease the temperature rise on the center position of a UHPC specimen with a size of 70.7 mm × 70.7 mm × 70.7 mm by 1.3 °C due to heat of hydration, and can effectively inhibit the hydration heat–induced thermal cracking of UHPC pavement slab with a size of 4.5 m × 5 m × 0.4 m. 1.5 vol% SSWs can decrease the peak temperature at the core position and the temperature gradient between the core and the upper surface center positions of pavement slab by 4.6 °C and 6.9 °C, respectively, thus decreasing the thermal stress by 0.9 MPa. The peak temperature at the core position and the temperature gradient between the core and the upper surface center of pier cap with a size of 3 m × 2 m × 1 m fabricated from 1.5 vol% SSWs-engineered multifunctional UHPC are reduced by 5.2 °C and 10.7 °C, respectively, and the maximum thermal stress is decreased by 1.43 MPa, thus inhibiting the early thermal cracking caused by heat of hydration.

2) Under the voltage of 30 V, a 1.5 vol% SSWs-engineered multifunctional UHPC slab with a size of 400 mm × 100 mm × 15 mm can be heated from about 21 °C to over 80 °C within 16.18 min in windless conditions with a temperature of about 21 °C. The average self-heating rate and electrical resistivity of SSWs-engineered multifunctional UHPC are barely influenced by temperature and self-heating cycles.

3) The deicing rate of 1.5 vol% SSWs-engineered multifunctional UHPC slab to melt a 9-mm layer of ice in a freezer with a temperature of about –25 °C can reach up to 247.09 ml/(m²•min). The average energy efficiency of 16 cycles of deicing experiments for 1.5 vol% SSWs-engineered multifunctional UHPC slab is 88.72%. Under the voltage of 25 V, the snow-melting rates for 1.5 vol% of SSWs-engineered multifunctional UHPC slab to melt 5 cm of naturally accumulated snow and 2 cm of artificially compacted snow are 157.66 ml/(m²•min) and 175.26 ml/(m²•min), respectively. The average energy efficiency of snow-melting performance for 1.5 vol% of SSWs-engineered multifunctional UHPC slab is 67.98%.

REFERENCES

1. S. Dong, X. Wang, H. Xu, J. Wang, B. Han. Incorporating super-finer stainless wires to control thermal cracking of concrete structures caused by heat of hydration, *Construction and Building Materials*. 271 (2021) 121896.
2. B. Han, L. Zhang, J. Ou. *Smart and Multifunctional Concrete Toward Sustainable Infrastructures*, Springer, 2017.
3. S. Ding, S. Dong, X. Wang, S. Ding, B. Han, J. Ou. Self-heating ultra-high performance concrete with stainless steel wires for active deicing and snow-melting of transportation infrastructures, *Cement and Concrete Composites*. 138 (2023) 105005.
4. P. Sukontasukkul, P. Uthaichotirat, T. Sangpet, K. Sisomphon, M. Newlands, A. Siripanichgorn, P. Chindaprasirt. Thermal properties of lightweight concrete incorporating high contents of phase change materials, *Construction and Building Materials*. 207 (2019) 431–439.
5. F.P. Incropera, D.P. DeWitt. *Fundamentals of Heat and Mass Transfer* (4th edition), Wiley, 1996.
6. G. An, J. Park, S. Cha, J. Kim. Development of a portable device and compensation method for the prediction of the adiabatic temperature rise of concrete, *Construction and Building Materials*. 102 (2016) 640–647.
7. Y. Li, L. Nie, B. Wang. A numerical simulation of the temperature cracking propagation process when pouring mass concrete, *Automation in Construction*. 37 (2014) 203–210.
8. S. Dong, D. Zhou, Z. Li, X. Yu, B. Han. Super-fine stainless wires enabled multifunctional and smart reactive powder concrete, *Smart Materials and Structures*. 28(12) (2019) 125009.
9. B. Han, S. Dong, J. Ou, C. Zhang, Y. Wang, X. Yu, S. Ding. Microstructure related mechanical behaviors of short-cut super-fine stainless wire reinforced reactive powder concrete, *Materials and Design*. 96 (2016) 16–26.
10. S. Dong, B. Han, X. Yu, J. Ou. Constitutive model and reinforcing mechanisms of uniaxial compressive property for reactive powder concrete with super-fine stainless wire, *Composites Part B: Engineering*. 166 (2019) 298–309.

11. S. Han, J. Kim. Effect of temperature and age on the relationship between dynamic and static elastic modulus of concrete, *Cement and Concrete Research.* 34(7) (2004) 1219–1227.

12. S. Dong. *Mechanical and Functional Performances of Super-Fine Stainless Wire Reinforced Reactive Powder Concrete,* Dalian University of Technology, 2018. (In Chinese).

13. D. Cecini, S.A. Austin, S. Cavalaro, A. Palmeri. Accelerated electric curing of steel-fibre reinforced concrete, *Construction and Building Materials.* 189 (2018) 192–204.

14. P. Jin, J. Gao, P. An, Y. Sheng. Finite element analysis of shrinkage strain of cement stabilized crushed stone in early age, *Journal of Nanjing University of Aeronautics and Astronautics.* 50(6) (2018) 866–870. (In Chinese).

15. S. Ahmad, M. Rasul, S.K. Adekunle, S.U. Al-Dulaijan, M. Maslehuddin, S.I. Ali. Mechanical properties of steel fiber-reinforced UHPC mixtures exposed to elevated temperature: Effects of exposure duration and fiber content, *Composites Part B: Engineering.* 168 (2019) 291–301.

16. A. Shishegaran, M.A. Naghsh, H. Taghavizade, M.H. Afsharmovahed, A. Shishegaran, M. Babaei Lavasani. Sustainability evaluation of conductive concrete for pavement deicing: The case study of parkway Bridge, Tehran, Iran, *Arabian Journal for Science and Engineering.* 46(5) (2021) 4543–4562.

17. A.G. Mohammed, G. Ozgur, E. Sevkat. Electrical resistance heating for deicing and snow melting applications: Experimental study, *Cold Regions, Science and Technology.* 160 (2019) 128–138.

18. S. Qing, L. Wang, H. Tao. *Statistics,* Tsinghua University Publishing House, 2010. (In Chinese).

19. L. Hong, Y. Zhao. The electrical properties and snow melting of graphite slurry infiltrated steel fiber concrete, *Journal of Wuhan University of Technology-(Material Science Edition).* 25(4) (2010) 609–612.

20. S. Ding, X. Wang, L. Qiu, Y. Ni, X. Dong, Y. Cui, A. Ashour, B. Han, J. Ou. Self-sensing cementitious composites with hierarchical carbon fiber-carbon nanotube composite fillers for crack development monitoring of a maglev girder, *Small.* 19(9) (2022) 2206258.

21. S. Ding, Y. Xiang, Y. Ni, V.K. Thakur, X. Wang, B. Han, J. Ou. In-situ synthesizing carbon nanotubes on cement to develop self-sensing cementitious composites for smart high-speed rail infrastructures, *Nano Today.* 43 (2022) 101438.

22. T.L. Boyd. New snow melt projects in Klamath Falls, OR, *Geo-heat Center Quarterly Bulletin.* 27 (2003) 77-80 .

23. K. Morita, M. Tago. Operational characteristics of the Gaia snow-melting system in Ninohe, Iwate, Japan-Development of a snow-melting system which utilizes thermal functions of the ground, Proceedings World Geothermal Congress, Tohoku, Japan, 2000.

24. W. Zhao, L. Wang, Y. Zhang, X. Cao, W. Wang, Y. Liu, B. Li. Snow melting on a road unit as affected by thermal fluids in different embedded pipes, *Sustainable Energy Technologies and Assessments.* 46 (2021) 101221.

25. Y. Huang. *Study on Ice-Snow Melting and Heat Absorption of Solar Radiation in Road,* Jilin University, 2010. (In Chinese).

26. L. Gong. *Graphite Slurry Infiltrated Fiber Concrete for Snow Melting and Finite Element Simulation*, Dalian University of Technology, 2009. (In Chinese).

27. A. Sassani, A. Arabzadeh, H. Ceylan, S. Kim, S.M.S. Sadati, K. Gopalakrishnan, P.C. Taylor, H. Abdualla. Carbon fiber-based electrically conductive concrete for salt-free deicing of pavements, *Journal of Cleaner Production*. 203 (2018) 799–809.

28. X. Zhu, Q. Zhang, Z. Du, H. Wu, Y. Sun. Snow-melting pavement design strategy with electric cable heating system balancing snow melting, energy conservation, and mechanical performance, *Resources, Conservation and Recycling*. 177 (2022) 105970.

Chapter 8

Wear resistance, damping, and electromagnetic properties of stainless steel wires-engineered ultra-high performance concrete

8.1 INTRODUCTION

As demonstrated in Chapters 2–7, stainless steel wires (SSWs)-engineered multifunctional ultra-high performance concrete (UHPC) has not only excellent static (including flexural, compressive, bending, and fracture properties) and dynamic mechanical properties (including impact and fatigue properties), but also good electrically and thermally conductive properties. This greatly broadens the application field of UHPC, but also puts forward higher and more comprehensive performance requirements for UHPC. For example, wear resistance is an important index to ensure service performance when SSWs-engineered multifunctional UHPC is used for constructing transportation infrastructures such as pavements, airport runways, and bridge decks. Meanwhile, the self-damping characteristic of SSWs-engineered multifunctional UHPC is valuable for structures because it can mitigate hazards (whether due to accidental loading, wind, ocean waves, or earthquakes), increases the comfort of the people who use the structures, and enhances the reliability and performance of structures [1–3]. Furthermore, the electromagnetic properties provide a good possibility for the application of SSWs-engineered multifunctional UHPC in bunker engineering [4–6].

Therefore, the wear resistance, self-damping, and electromagnetic properties of SSWs-engineered multifunctional UHPC are introduced in this chapter. The mechanisms by which SSWs modify these functional properties are also displayed. The diameter of SSWs used in this chapter is 20 μm and the length is 10 mm. The main contents of this chapter are shown in Figure 8.1.

8.2 WEAR RESISTANCE OF SSWS-ENGINEERED MULTIFUNCTIONAL UHPC

The wear resistance of SSWs-engineered multifunctional UHPC was tested according to the China test standard of T 0567-2005. However, the wear revolutions were increased to 180 after dust brushing because of the

DOI: 10.1201/9781003276357-8

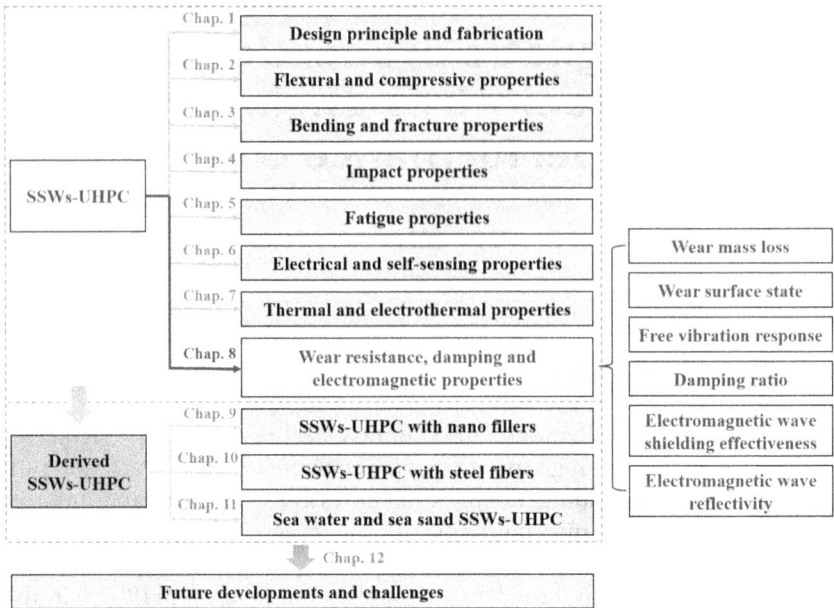

Figure 8.1 Main contents of Chapter 8

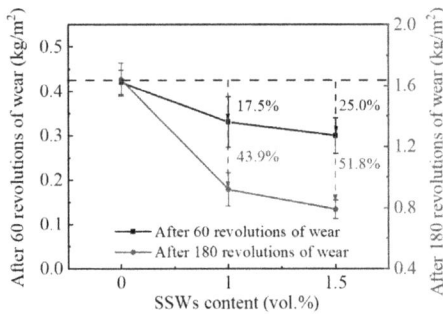

Figure 8.2 Wear mass loss per unit area of SSWs-engineered multifunctional UHPC

excellent mechanical behavior of UHPC. The average value of wear mass loss per unit area of three specimens was taken as the test result. The wear mass loss per unit area of SSWs-engineered multifunctional UHPC after 60 and 180 revolutions of wear are displayed in Figure 8.2. Figure 8.2 shows that the wear mass loss per unit area of UHPC presents a decreasing trend with increasing SSWs content. After 60 revolutions of wear, the wear mass loss per unit area of UHPC without SSWs is 0.40 kg/m², and it decreases to 0.33 kg/m² and 0.30 kg/m², respectively, for UHPC reinforced with 1 vol% and 1.5 vol% SSWs, indicating a 17.5% and 25.0% increase in wear resistance. After 180 revolutions of wear, the wear mass loss per unit area of UHPC reinforced with 0 vol%, 1 vol%, and 1.5 vol% SSWs is 1.64 kg/m^2, 0.92 kg/m^2, and 0.79 kg/m^2, respectively. This means that the addition of 1 vol% and 1.5 vol% SSWs leads to a 43.9% and 51.8% increase, respectively, in the wear resistance of UHPC. The wear mass loss per unit area of UHPC is only reduced by 7.5% and 7.9%, respectively, as the SSWs content increases from 1 vol% to 1.5 vol% after 60 and 180 revolutions of wear. This means that 1 vol% SSWs endows UHPC with enough wear resistance. The increment of UHPC wear resistance caused by 1.5 vol% SSWs is higher than that caused by steel fibers (SFs) for concrete [7, 8].

The surface state of SSWs-engineered multifunctional UHPC after 180 revolutions of wear is presented in Figure 8.3. It is clear from Figure 8.3(a) that lots of small pit slots appear on the surface of UHPC without SSWs after 180 revolutions of wear. This is because the fine aggregate in UHPC is removed under the action of the rotary cutting force of the flower wheel blade. Figsure 8.3(a) and (b) show that the wear degree of the UHPC surface is significantly reduced due to the incorporation of SSWs. The reinforcing network formed by SSWs increases the rotary resistance of the blade, thus reducing the cutting force on the UHPC surface. The hindering effect of the SSWs network on wear is enhanced by the increasing SSWs content, until only scratches can be observed on the surface of UHPC reinforced with 1.5 vol% SSWs after 180 revolutions of wear. Scanning electron microscopic

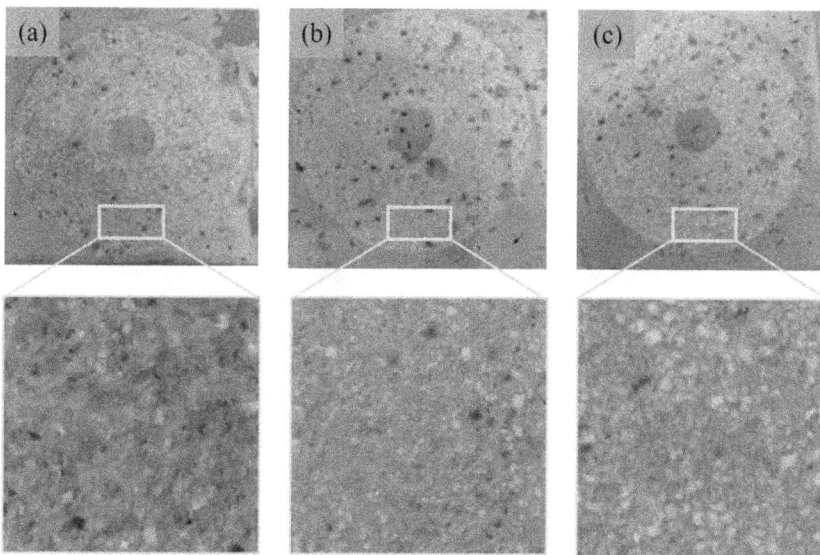

Figure 8.3 Surface state of SSWs-engineered multifunctional UHPC after 180 revolutions of wear. (a) UHPC without SSWs; (b) UHPC with 1.0 vol% SSWs; (c) UHPC with 1.5 vol% SSWs

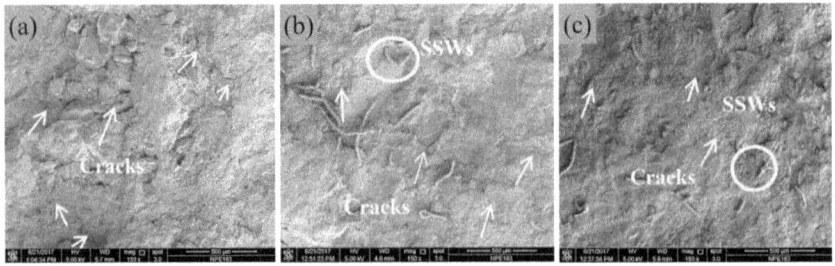

Figure 8.4 SEM images of SSWs-engineered multifunctional UHPC after 180 revolutions of wear. (a) UHPC without SSWs; (b) UHPC with 1.0 vol% SSWs; (c) UHPC with 1.5 vol% SSWs

(SEM) images of SSWs-engineered multifunctional UHPC after 180 revolutions of wear are shown in Figure 8.4. Figure 8.4(a) demonstrates that there are many cracks on the wear surface of UHPC without SSWs. As shown in Figure 8.4(b), cracks can be observed at the root of the SSWs, but the number of cracks is reduced. In addition, SSWs show a warping state under the action of wear force; then, they are ruptured or pulled out. Figure 8.4(c) illustrates that the number of cracks caused by wear decreases with increasing SSWs content.

The wear resistance enhancement of SSWs-engineered multifunctional UHPC can be explained by the following three aspects. (1) The improvement of compressive and flexural strength caused by SSWs can enhance the friction resistance of UHPC to the blade. Then, SSWs-engineered multifunctional UHPC is not easy to crack under the same wear force compared with UHPC without SSWs. (2) The cutting force of the blade is weakened by the SSWs network during the wear process. (3) SSWs can play the role of pulling back on UHPC in the process of high-speed rotation of the blade.

8.3 SELF-DAMPING PROPERTY OF SSWS-ENGINEERED MULTIFUNCTIONAL UHPC

The self-damping behavior of SSWs-engineered multifunctional UHPC was tested and analyzed according to Ruan et al. [9]. Cotton rope was adopted to hang specimens with a size of 260 mm × 20 mm × 20 mm. The free vibration response time–history delay curves of SSWs-engineered multifunctional UHPC are displayed in Figure 8.5. The time-domain damping ratio of composites is calculated on the basis of Figure 8.5 and Ruan et al. [9]. Meanwhile, the effect of various factors on the damping ratio of SSWs-engineered multifunctional UHPC is discussed, including peak starting point, number of peak intervals, different knock times of the same specimen, and SSWs volume fraction. In addition, the free vibration response

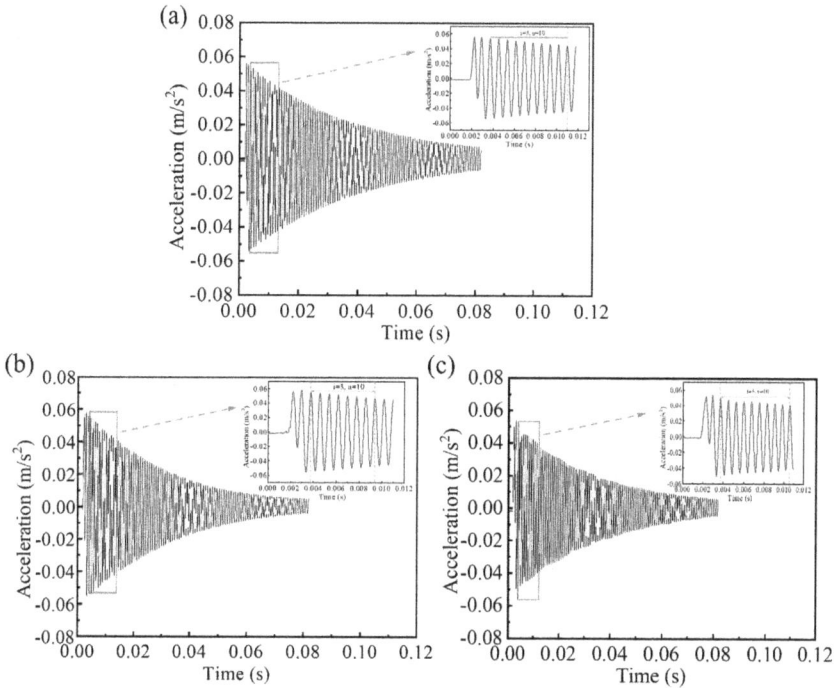

Figure 8.5 Free vibration response time-history delay curve of SSWs-engineered multi-functional UHPC. (a) UHPC without SSWs; (b) UHPC with 1.0 vol% SSWs; (c) UHPC with 1.5 vol% SSWs

time–history delay curves of SSWs-engineered multifunctional UHPC are converted into a frequency-domain signal by Fourier transform (FFT) to obtain a spectrum feature curve. Then, the frequency-domain damping ratio of composites is achieved in accordance with the half-power bandwidth method [9].

In order to evaluate the effect of peak starting point on the time-domain damping ratio, the peak interval number i is 10, and the peak starting points n are selected as 3, 6, 9, 12, 15, and 18. The average value of the damping ratio for three specimens is determined as the final value. The damping ratios of SSWs-engineered multifunctional UHPC calculated by different peak starting points are shown in Figure 8.6(a). Figure 8.6(a) shows that the average value of the damping ratio obtained by peak starting points of 3, 6, 9, 12, 15, and 18 is 0.00348, 0.00410, and 0.00408, respectively, for UHPC reinforced with 0 vol%, 1 vol%, and 1.5 vol% SSWs. The relative error between the damping ratio of different peak starting points and the average value is less than 3.4%, 2.7%, and 4.4%, respectively, for the composites. This means that the damping ratio of composites measured by the time-domain method is independent of peak starting point.

In the case of peak starting point is 3, peak interval numbers i are chosen as 5, 10, 15, 20, and 25 to determine its effect on the time-domain damping ratio. The damping ratio of SSWs-engineered multifunctional UHPC calculated by different peak interval numbers is presented in Figure 8.6(b). It can be seen from Figure 8.6(b) that the damping ratio of SSWs-engineered multifunctional UHPC fluctuates above and below the average value. The relative error between damping ratios obtained by different peak interval numbers and the average value is less than 5.0%, indicating that the influence of peak interval numbers on the damping ratio can be neglected when the time-domain method is used.

Taking the peak starting point and peak interval numbers as 3 and 10, respectively, the damping ratio of SSWs-engineered multifunctional UHPC obtained by the time-domain method and different specimen numbers is shown in Figure 8.7. It can be observed from Figure 8.7 that the damping ratio of each specimen in the same group is approximated. The standard deviation of the damping ratio between the values calculated by different

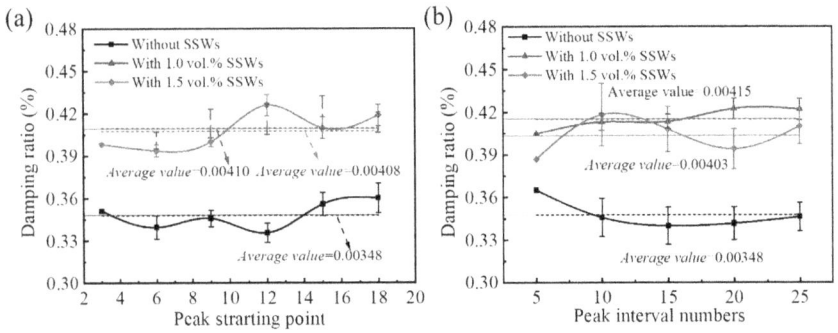

Figure 8.6 Damping ratio of SSWs-engineered multifunctional UHPC. (a) Different peak starting point; (b) Different peak interval numbers

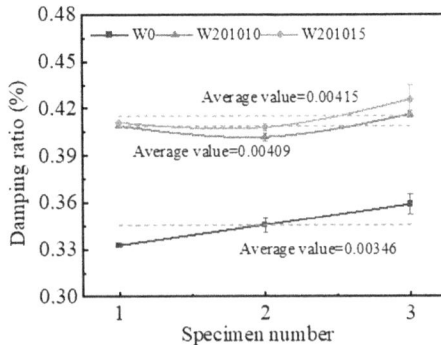

Figure 8.7 Average damping ratio of SSWs-engineered multifunctional UHPC

specimen numbers is less than 7.8%, 1.7%, and 4.4%, respectively, for UHPC with 0 vol%, 1 vol%, and 1.5 vol% SSWs, which is within the normal permissible range. Therefore, the experimental data measured by the described preparation, test, and calculation process is uniform and reliable.

In the case when the peak starting point and peak interval number are 3 and 10, respectively, the effect of knock times on the time-domain damping ratio of the same specimen of SSW reinforced UHPC is illustrated in Figure 8.8. It can be seen from Figure 8.8 that the damping ratio of the same specimen changes little with the increase of knock times. The minimum damping ratio of UHPC reinforced with 0 vol%, 1 vol%, and 1.5 vol% SSWs is 0.00351, 0.00402, and 0.00405 under different knock times, while the maximum value is 0.00357, 0.00409, and 0.00412, respectively. The test results verify that the time-domain damping ratio of composites is not affected by knock times.

Based on this analysis, the time-domain damping ratio of composites is achieved on the premise that the peak starting point is 3, the peak interval numbers are 10, the specimen numbers are 3, and the knock times are 4, as shown in Figure 8.9. It is clear from Figure 8.9 that the time-domain damping ratio of UHPC is increased by 18.2% and 19.9% due to the incorporation of 1.0 vol% and 1.5 vol% SSWs, respectively. The values of the time-domain damping ratio for UHPC reinforced with 0 vol%, 1 vol%, and 1.5 vol% SSWs are 0.00346, 0.00409, and 0.00415, respectively. The frequency-domain damping ratio of SSWs-engineered multifunctional UHPC is also plotted in Figure 8.9. As shown in Figure 8.9, the values of the frequency-domain damping ratio for SSWs-engineered multifunctional UHPC are larger than those of the time-domain damping ratio. The addition of 1 vol% and 1.5 vol% SSWs leads to an 8.9% and 12.6% increase, respectively, for the frequency-domain damping ratio of UHPC. It can be concluded that the effect law of SSWs on the damping ratio of UHPC is the same whether the time-domain or frequency-domain calculation method is adopted. The frequency-domain damping ratio of composites is less

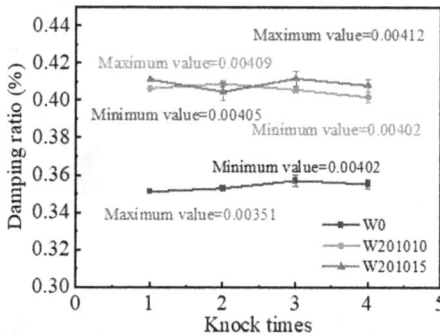

Figure 8.8 Damping ratio of SSWs-engineered multifunctional UHPC under different knock times

representative because it is affected by many factors, such as sampling frequency, frequency resolution, signal frequency, and so on. Therefore, the damping ratio calculated by the time-domain method is taken as the final value for SSWs-engineered multifunctional UHPC.

The enhancement effect of SSWs on the damping ratio of UHPC can be attributed to the following three aspects. (1) Due to the micron diameter, high aspect ratio, and large specific surface area of SSWs, the hydration of cementitious material is promoted, and the compactness of the UHPC matrix is improved. Hence, the viscous resistance between components/phases of SSWs-engineered multifunctional UHPC subjected to dynamic loads increases with increasing SSWs content. (2) Slip and dislocation will occur between UHPC internal components/phases under vibration loads. At this time, SSWs can cross and bridge these slip and dislocation cracks, and hinder the development of cracks, eventually leading to the increase of energy dissipation ability and damping ratio, as shown in Figure 8.10(a). (3)

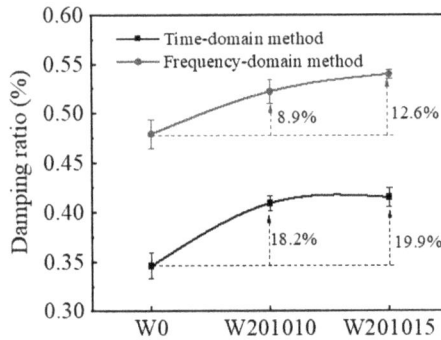

Figure 8.9 Damping ratio of SSWs-engineered multifunctional UHPC obtained by time-domain and frequency-domain method

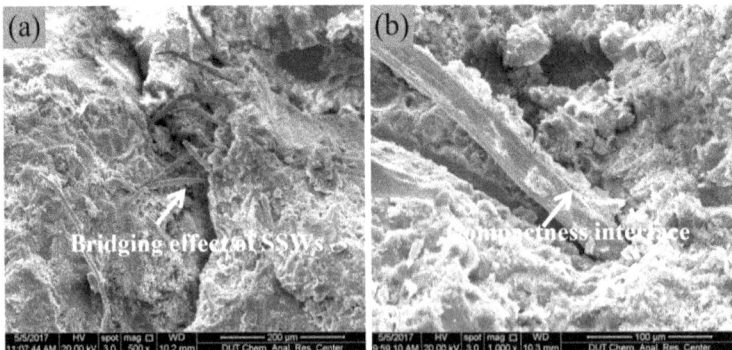

Figure 8.10 SEM images of SSWs-engineered multifunctional UHPC. (a) Bridging effect of SSWs; (b) Interface between SSWs and UHPC matrix

The interface between the SSWs and the UHPC matrix is the main source of energy consumption. Especially, the interfacial bond strength between SSWs and UHPC matrix is enhanced due to the high aspect ratio and large specific surface area of SSWs, as shown in Figure 8.10(b). Therefore, viscous friction resistance is generated on the interface under vibration load, leading to the vibration energy being consumed and the damping ratio of UHPC being improved.

8.4 ELECTROMAGNETIC PROPERTIES OF SSWS-ENGINEERED MULTIFUNCTIONAL UHPC

8.4.1 Electromagnetic wave shielding effectiveness

Powder samples of SSWs-engineered multifunctional UHPC with fineness of 600 μm, 125 μm, and 45 μm were prepared to test the electromagnetic wave shielding effectiveness (SE) using a PNA-X Network analyzer N5244A. The test frequency range was 2–18 GHz. The electromagnetic wave SE was calculated by electromagnetic and S parameters according to Sun et al. [10], Kumar et al. [11], and Almasi-Kashi et al. [12].

The SE of SSWs-engineered multifunctional UHPC calculated on the basis of electromagnetic parameters is demonstrated in Figure 8.11. Figure 8.11 illustrates that the SE of UHPC is closely related to the fineness of the powder sample and the SSWs content. When the powder sample is milled from 600 μm to 45 μm, the maximum SE of UHPC reinforced with 0 vol%, 1 vol%, and 1.5 vol% SSWs is improved by 31.0%, 46.3%, and 60.5%, respectively. This is because the increased fineness of the powder sample leads to enhanced compactness of filling and a relative increase in the proportion of SSWs, resulting in high electromagnetic loss. When the fineness of the powder sample is 45 μm or 125 μm, the SE of UHPC without SSWs increases with frequency, while the SE of UHPC reinforced with 1 vol% and 1.5 vol% SSWs exhibits obvious peak values. The frequency corresponding to the maximum SE for a powder sample with fineness of 45 μm decreases due to the incorporation of SSWs and the increase of SSWs content; it is 18 GHz, 14 GHz and 11.3 GHz for UHPC reinforced with 0 vol%, 1 vol%, and 1.5 vol% SSWs, respectively. The maximum SE of the powder sample with fineness of 45 μm for UHPC without SSWs is 2.85 dB, which increases to 6.65 dB and 7.25 dB for UHPC reinforced with 1 vol% and 1.5 vol% SSWs, indicating a 133.3% and 154.4% increase, respectively. When the fineness of the powder sample is 125 μm, the frequency corresponding to the maximum SE of SSWs-engineered multifunctional UHPC is 18 GHz, 12 GHz and 10 GHz, respectively. Meanwhile, the maximum SE of UHPC reinforced with 1 vol% and 1.5 vol% SSWs is 70% and 104% higher, respectively, than that of UHPC without SSWs. This phenomenon illustrates that the frequency of the consumed electromagnetic wave decreases

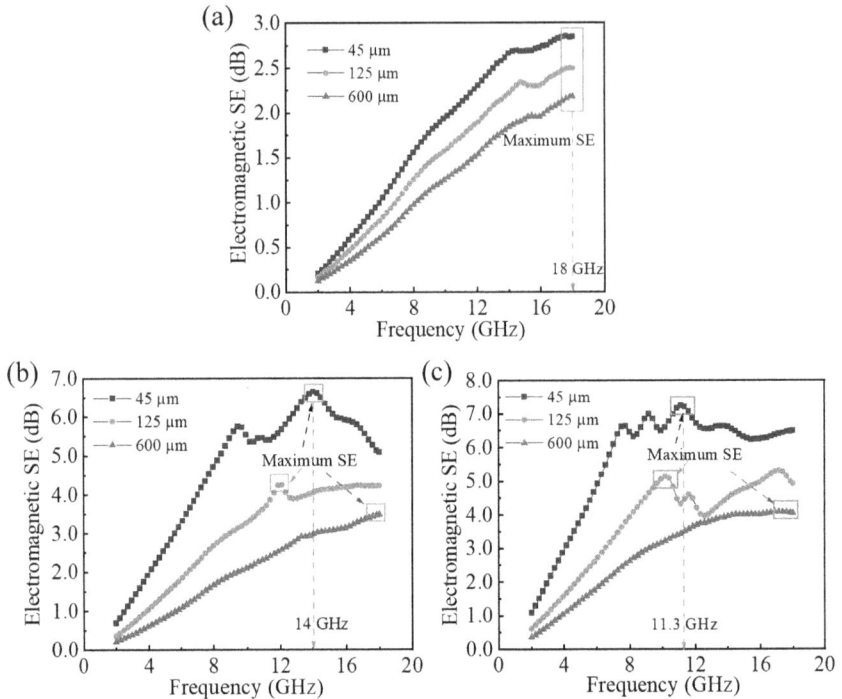

Figure 8.11 SE of SSWs-engineered multifunctional UHPC with different fineness. (a) UHPC without SSWs; (b) UHPC with 1.0 vol% SSWs; (c) UHPC with 1.5 vol% SSWs

and the SE of UHPC increases with increasing SSWs content. As a conductor, SSWs can be coupled with the electromagnetic wave to become electric dipoles, leading to the production of a polarization dissipation current. Hence, the electromagnetic waves are effectively shielded. A high content of SSWs leads to the formation of more polarization dissipation current, resulting in high SE. In addition, the sensitivity of SSWs-engineered multifunctional UHPC in the form of fine powder to electromagnetic frequency can be attributed to the bundling of SSWs. Electromagnetic waves with long wavelength can easily be coupled with bundled SSWs, resulting in a decrease in the frequency corresponding to maximum SE with increasing fineness of the powder sample.

Because the maximum particle size of quartz sand used for UHPC is 840 μm, the powder sample with a fineness of 600 μm is more representative than those with a fineness of 45 μm and 125 μm. As shown in Figure 8.12, when the fineness of the powder sample is 600 μm, the SE of UHPC without SSWs shows an increasing trend as the frequency ranges from 2 GHz to 18 GHz. The variation of SE with frequency can be divided into a linear and

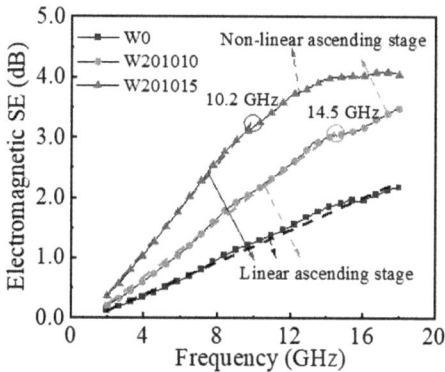

Figure 8.12 SE of SSWs-engineered multifunctional UHPC with fineness of 600 µm

a non-linear stage due to the incorporation of SSWs. Moreover, the linear stage is shortened and the non-linear stage is prolonged as the SSWs content increases. The occurrence of the non-linear stage on SE–frequency curves represents the coupling of SSWs with electromagnetic waves. The coupling frequency is decreased from 14.5 GHz to 10.2 GHz as the SSWs content increases from 1 vol% to 1.5 vol%. Figure 8.12 also demonstrates that compared with UHPC without SSWs, the SE of UHPC reinforced with 1 vol% SSWs increases by 61.5%, 74.3%, 74.2%, 70.4%, 67.5%, 66.7%, 62.5%, 60.0%, and 59.6%, respectively, when the frequency of the electromagnetic wave is 2 GHz, 4 GHz, 6 GHz, 8 GHz, 10 GHz, 12 GHz, 14 GHz, 16 GHz, and 18 GHz. Meanwhile, the addition of 1.5 vol% SSWs leads to an increase of 176.9%, 202.9%, 193.5%, 167.3%, 152.4%, 141.2%, 115.2%, 104.1%, and 85.8% increase, respectively, in the SE of UHPC. It can be concluded that the optimal increment for SE caused by SSWs happens at the frequency of 4 GHz. The maximum SE of SSWs-engineered multifunctional UHPC reaches 4.05 dB when the frequency is 18 GHz. The addition of SSWs can effectively improve the SE of UHPC in the frequency range of 2–18 GHz.

8.4.2 Electromagnetic wave reflectivity

Powder samples of SSWs-engineered multifunctional UHPC with fineness of 600 µm, 125 µm, and 45 µm were prepared to test the electromagnetic wave absorbing performance using a PNA-X Network analyzer N5244A. The test frequency range was 2–18 GHz. The electromagnetic wave reflectivity was calculated by electromagnetic and S parameters according to Sun et al. [10], Kumar et al. [11], and Almasi-Kashi et al. [12]. In addition, the electromagnetic wave reflectivity of plate specimens was directly measured by the bow method, and the test results were compared with those obtained from powder samples.

The electromagnetic wave absorbing performance of powder samples with different degrees of fineness was tested, and the test results were converted into electromagnetic wave reflectivity of plate with different thicknesses, as shown in Figure 8.13. It can be seen from Figure 8.13(a) that when the test results of powder samples of different fineness are converted into results for plate with a thickness of 10 mm, the reflectivity of UHPC without SSWs exhibits two significant peaks at the frequency of 8.4–14.8 GHz and 18 GHz. The reflectivity increases with increasing fineness of the powder sample, and the maximum absolute value of electromagnetic wave reflectivity is 14.6 dB. When the converted plate thickness is 20 mm, the reflectivity of UHPC without SSWs possesses four peaks as the frequency ranges from 2 GHz to 18 GHz. The absolute peak values of reflectivity are 23.7 dB, 30.5 dB, and 28.5 dB when the powder sample fineness is 600 μm, 125 μm, and 45 μm, respectively. The frequency of the electromagnetic wave corresponding to these peak values of reflectivity is 16.3 GHz, 17.0 GHz, and 17.9 GHz, respectively. Figure 8.13(b) shows that the electromagnetic wave reflectivity of UHPC reinforced with 1 vol% SSWs presents two significant peaks at the frequency range of 2–18 GHz when the converted thickness is 10 mm. The absolute peak values of reflectivity

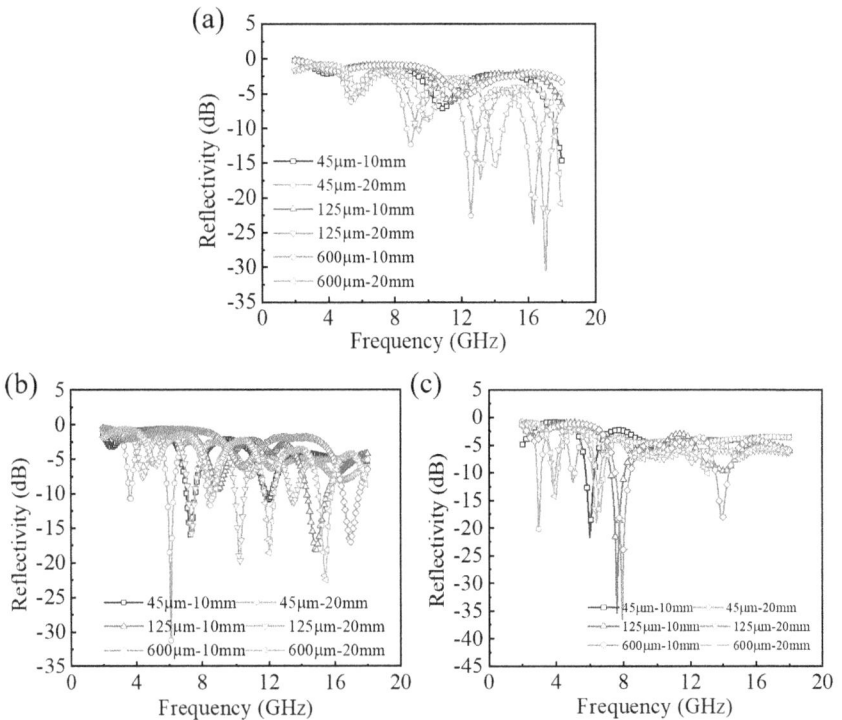

Figure 8.13 Electromagnetic wave reflectivity of SSWs-engineered multifunctional UHPC. (a) UHPC without SSWs; (b) UHPC with 1.0 vol% SSWs; (c) UHPC with 1.5 vol% SSWs

are 6.1 dB and 17.0 dB for powder samples with a fineness of 600 μm, and the corresponding electromagnetic frequencies are 10.4 GHz and 17.0 GHz. When the powder sample fineness is 125 μm, the two absolute peak values of reflectivity are 9.3 dB and 18.2 dB, which appear at the frequency of 9.0 GHz and 14.9 GHz. As the fineness of powder sample increases to 45 μm, the absolute peak values of reflectivity are 15.9 dB and 10.8 dB, occurring when the frequencies are 7.3 GHz and 11.9 GHz. It is clear that the frequency of the electromagnetic wave that can be effectively absorbed becomes lower and the wavelength becomes longer with increasing fineness of the powder sample. When the test results for powder samples are converted into results for plate with a thickness of 20 mm, the electromagnetic wave reflectivity of UHPC reinforced with 1 vol% SSWs has the same change rule as described previously. The absolute peak values of reflectivity for UHPC reinforced with 1 vol% and 1.5 vol% SSWs corresponding to powder sample fineness of 600 μm, 125 μm, and 45 μm are 22.5 dB, 19.7 dB, and 31.2 dB, respectively.

Figure 8.13(c) demonstrates that there are two peaks for electromagnetic wave reflectivity of UHPC reinforced with 1.5 vol% SSWs at the frequency range of 2–18 GHz when the converted thickness is 10 mm. However, unlike UHPC reinforced with 1 vol% SSWs, the absolute peak values of reflectivity increase, and the corresponding frequency of the electromagnetic wave also increases, with decreasing fineness of the powder sample. When the frequency is 7.9 GHz and the fineness of the powder sample is 600 μm, the absolute reflectivity peak value is 36.5 dB. When the test results for the powder sample are converted into results for plate with a thickness of 20 mm, UHPC reinforced with 1.5 vol% SSWs has better absorption performance for low-frequency electromagnetic waves. When the fineness of the powder sample is 45 μm, 125 μm, and 600 μm, the absolute reflectivity peak values are 20.3 dB, 19.2 dB, and 16.6 dB, which appear at the frequency of 3.0 GHz, 7.4 GHz, and 6.6 GHz, respectively.

It can be seen from this analysis that the electromagnetic wave absorbing performance of SSWs-engineered multifunctional UHPC differs with the variation of the fineness of the powder sample and the thickness of the calculation converted plate. Because the maximum particle size of quartz sand is 840 μm, the reflectivity test results obtained from a powder sample with a fineness of 600 μm are compared and analyzed, as demonstrated in Figure 8.14. Figure 8.14 shows that when the converted thickness is 10 mm, the absolute peak value of reflectivity for UHPC without SSW is 5.3 dB at the frequency of 12.2 GHz. UHPC shows two significant reflectivity peaks at the frequency range of 2–18 GHz due to the addition of SSWs. The absolute peak values of reflectivity for UHPC reinforced with 1 vol% and 1.5 vol% SSWs are 17.0 dB and 36.5 dB, respectively, at the frequency of 17.0 GHz and 13.9 GHz. This means that the electromagnetic wave absorbing performance of UHPC is increased by 220.7% and 588.7% due to the

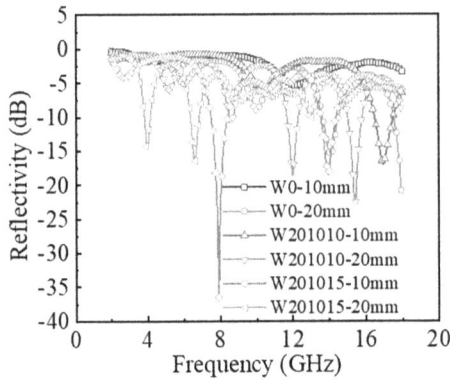

Figure 8.14 Electromagnetic wave reflectivity of SSWs-engineered multifunctional UHPC with fineness of 600 μm

incorporation of 1 vol% and 1.5 vol% SSWs. However, the wavelength of the electromagnetic wave absorbed by UHPC reinforced with SSWs is shorter than that absorbed by UHPC without SSWs. Figure 8.14 also demonstrates that there are four reflectivity peaks at the frequency range of 2–18 GHz for UHPC without SSWs and UHPC reinforced with 1 vol% SSWs when the test results for a powder sample are converted into results of a plate with a thickness of 20 mm. The absolute reflectivity peak values are 4.9 dB, 9.0 dB, 15.3 dB, and 20.8 dB for UHPC without SSWs at the frequency of 6.1 GHz, 10.1 GHz, 14 GHz, and 18 GHz, respectively. However, the absolute reflectivity peaks for UHPC reinforced with 1 vol% SSWs appear at the frequencies of 5.2 GHz, 8.6 GHz, 12.1 GHz, and 15.4 GHz, and the corresponding absolute reflectivity values are 6.0 dB, 10.5 dB, 16.9 dB, and 22.5 dB. When the content of SSWs is 1.5 vol%, UHPC only presents reflectivity peaks at a frequency smaller than 8 GHz. The absolute reflectivity peak values of UHPC reinforced with 1.5 vol% SSWs are 14.4 dB and 16.6 dB, and the corresponding frequencies are 4 GHz and 6.6 GHz. It is clear that the incorporation of SSWs improves the electromagnetic wave absorbing performance of UHPC at low frequency, and the electromagnetic wave frequency that can be effectively absorbed decreases with increasing SSWs content. At a frequency smaller than 11 GHz, the addition of 1.5 vol% SSWs leads to a 193.9% and 82.2% increase in the electromagnetic wave absorbing performance of UHPC as a powder sample are converted into results of plates with thickness of 10 mm and 20 mm, respectively..

The electromagnetic wave absorbing performance of a plate specimen for SSWs-engineered multifunctional UHPC is shown in Figure 8.15. It can be seen from Figure 8.15 that the absolute electromagnetic wave reflectivity of UHPC without SSWs can reach 17.5 dB at the frequency range of 7–8 GHz. The UHPC only presents high electromagnetic wave absorbing performance

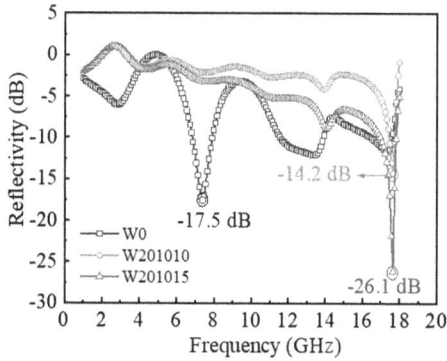

Figure 8.15 Electromagnetic wave reflectivity of SSWs-engineered multifunctional UHPC plate specimen

at the high-frequency range of 17–18 GHz due to the addition of SSWs. In this frequency range, the absolute reflectivity peak values are 14.3 dB and 26.1 dB for UHPC reinforced with 1 vol% and 1.5 vol% SSWs, respectively, which are increased by 22.4% and 125.0% compared with the reflectivity of UHPC without SSWs. The widely and randomly distributed SSWs in UHPC are easily coupled with high-frequency and short electromagnetic waves. Then, an electric dipole and dissipative current are produced to absorb the electromagnetic wave. More electric dipole is generated with increasing SSWs content, resulting in more electromagnetic wave energy being absorbed.

It can be concluded that the electromagnetic wave absorbing performance of SSWs-engineered multifunctional UHPC from a powder sample is very different from that obtained from a plate specimen, which can be attributed to the different bundling condition of SSWs in powder samples and plate specimens. The degree of entanglement of SSWs in powder samples is higher than that in plate, leading to an increase in the effective length of SSWs. Therefore, electromagnetic waves with a long wavelength are easily coupled and dissipated. That is, low-frequency electromagnetic waves can be effectively reflected. However, the randomly distributed SSWs in plate can only be coupled with electromagnetic waves with a short wavelength due to their independent dispersion characteristics and high aspect ratio. Therefore, the test method for the electromagnetic wave absorbing performance of SSWs-engineered multifunctional UHPC should be determined according to the actual structural element types.

8.5 SUMMARY

This chapter introduced the wear resistance, self-damping property, electromagnetic wave shielding effectiveness (SE), and electromagnetic wave reflectivity of SSWs-engineered multifunctional UHPC. The main conclusions can be summarized as follows.

The addition of 1.5 vol% SSWs leads to a 51.8% increment in the wear resistance of UHPC. The time-domain damping ratio of UHPC is increased by 19.9% due to the incorporation of 1.5 vol% SSWs, which is not affected by peak starting point, peak interval numbers, specimen numbers, or knock times. The electromagnetic wave SE of UHPC increases with increasing fineness of the powder sample and increased frequency; it can be improved by 202.9% at the frequency of 4 GHz by the addition of 1.5 vol% SSWs. The SE of SSWs-engineered multifunctional UHPC powder sample with a fineness of 600 μm can reach 4.05 dB at the frequency of 18 GHz. The electromagnetic wave absorbing property of SSWs-engineered multifunctional UHPC is affected by the fineness of the powder sample, frequency, and converted thickness. The electromagnetic wave reflectivity of plate specimens of UHPC reinforced with 1.5 vol% SSWs is 26.1 dB at the high-frequency range of 17–18 GHz.

REFERENCES

1. S. Dong, X. Wang, H. Xu, J. Wang, B. Han. Incorporating super-finer stainless wires to control thermal cracking of concrete structures caused by heat of hydration, *Construction and Building Materials*. 271 (2021) 121896.
2. B. Han, L. Zhang, S. Sun, X. Yu, X. Dong, T. Wu, J. Ou. Electrostatic self-assembly carbon nanotube/nano carbon black composite fillers reinforced cement-based materials with multifunctionality, *Composites – Part A: Applied Science and Manufacturing*. 79 (2015) 103–115.
3. S. Ding, S. Dong, X. Wang, S. Ding, B. Han, J. Ou. Self-heating ultra-high performance concrete with stainless steel wires for active deicing and snow-melting of transportation infrastructures, *Cement and Concrete Composites*. 138 (2023) 105005.
4. S. Dong, D. Zhou, Z.X. Yu, B. Han. Super-fine stainless wires enabled multifunctional and smart reactive powder concrete, *Smart Materials and Structures*. 28(12) (2019) 125009.
5. S. Dong, Y. Wang, A. Ashour, B. Han, J. Ou. Nano/micro-structures and mechanical properties of ultra-high performance concrete incorporating graphene with different lateral sizes, *Composites – Part A: Applied Science and Manufacturing*. 137 (2020) 106011.
6. B. Han, Z. Li, L. Zhang, S. Zeng, X. Yu, B. Han, J. Ou. Reactive powder concrete reinforced with Nano SiO_2-coated TiO_2, *Construction and Building Materials*. 148 (2017) 104–112.
7. B. Felekoğlu, S. Türkel, Y. Altuntaş. Effects of steel fiber reinforcement on surface wear resistance of self-compacting repair mortars, *Cement and Concrete Composites*. 29(5) (2007) 391–396.
8. F. Köksal, O. Gencel, B. Unal, M.Y. Durgun. Durability properties of concrete reinforced with steel-polypropylene hybrid fibers, *Science and Engineering of Composite Materials*. 19(1) (2012) 19–27.

9. Y. Ruan, D. Zhou, S. Sun, X. Wu, X. Yu, J. Hou, X. Dong, B. Han. Self-damping cementitious composites with multi-layer graphene, *Materials Research Express*. 4(7) (2017) 75605.

10. S. Sun, S. Ding, B. Han, S. Dong, X. Yu, D. Zhou, J. Ou. Multi-layer graphene-engineered cementitious composites with multifunctionality/intelligence, *Composites Part B Engineering*. 129 (2017) 221–232.

11. K.B. Kumar, A.K. Thapa, A.K. Yadav. Enhancement of absorption property of one-dimensional ternary periodic structure containing plasma based hyperbolic material for the application of microwave devices, *Journal of Magnetism and Magnetic Materials*. 489 (2019) 165371.

12. M. Almasi-Kashi, S. Alikhanzadeh-Arani, M. Karamzadeh-Jahromi. The role of Sn, Zn, and Cu additions on the microwave absorption properties of Co-Ni alloy nanoparticles, *Materials Research Bulletin*. 118 (2019) 110491.

Chapter 9

Stainless steel wires-engineered multifunctional ultra-high performance concrete incorporating nanofillers

9.1 INTRODUCTION

The contents of Chapters 2–6 have shown that stainless steel wires (SSWs) can strengthen and toughen ultra-high performance concrete (UHPC) mainly by bridging, deflection, and the effect of being pulled off. However, the influence of SSWs on improving the microcracking resistance of UHPC and reducing original flaws is limited [1, 2]. The addition of nanofillers is conducive to refining hydration products, increasing calcium silicate hydrate (C-S-H) gel compactness, and reducing initial flaws in concrete through the nano-core effect, making it possible to enhance the micro-cracking resistance of UHPC and improve the bond strength between SSWs and the UHPC matrix [3–7]. That is to say, nanofillers and SSWs may have a synergistic effect in playing the role of enhancing the crack resistance of matrix and inhibiting the initiation and propagation of cracks, thus endowing UHPC with high first-cracking flexural strength and strain-hardening characteristics as well as high flexural toughness simultaneously [8–10]. In addition, hybrid SSWs and nanofillers are also the best choice for conductive pathway design because they can perform multiscale long- and short-distance conduction effects, respectively [11–14], thus developing self-sensing concrete with high and stable sensitivity. Among various nano-fillers, multi-walled carbon nanotubes (CNTs) with good conductive properties and nano SiO_2-coated TiO_2 (NTs) with good dispersion are selected as representative nanofillers to optimize the properties of SSWs-engineered multifunctional UHPC.

Hence, SSWs-engineered multifunctional UHPC incorporating nano-fillers is described in this chapter. Meanwhile, the mechanical properties (including compressive strength and flexural strength as well as flexural toughness), electrical properties, self-sensing properties under cyclic/mono-tonic compressive load, and underlying mechanisms of SSWs-engineered multifunctional UHPC incorporating nanofillers are introduced. The main contents of this chapter are shown in Figure 9.1.

DOI: 10.1201/9781003276357-9

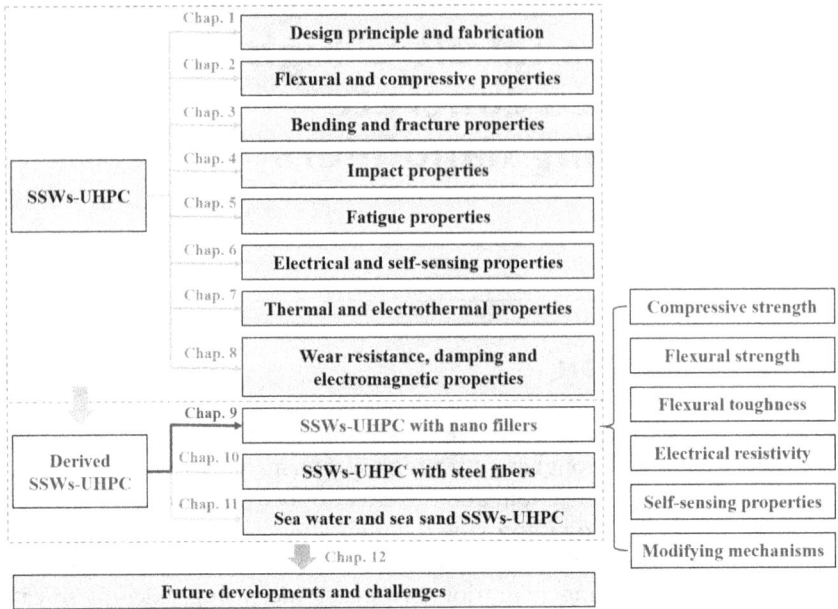

Figure 9.1 Main contents of Chapter 9

9.2 MECHANICAL PROPERTIES OF SSWS-ENGINEERED MULTIFUNCTIONAL UHPC INCORPORATING NANOFILLERS

9.2.1 Compressive/flexural strength and flexural toughness of SSWs-engineered multifunctional UHPC incorporating CNTs

9.2.1.1 Using P•I type of cement

When the cement used is P•I type, the compressive and flexural strength of CNTs and SSWs-engineered multifunctional UHPC are shown in Figure 9.2. Figure 9.2(a) shows that the addition of CNTs in the mass content of 0.25% (C1) and 0.5% (C2) can increase the flexural strength of UHPC by only 1.03 MPa/17.11% and 0.67 MPa/11.13%, respectively, while incorporating 0.6 vol% (S1) and 1.2 vol% SSWs (S2) leads to an increase of 5.42 MPa/90.03% and 7.84 MPa/130.23%, respectively, in flexural strength. The diameter and length of the SSWs used are 20 µm and 10 mm, respectively. The mechanism by which CNTs modify the flexural strength of UHPC mainly results from the nano-core effect, leading to the reduction of original flaws and the increase of microstructural compactness. Meanwhile, SSWs can play a toughening role through inhibiting the initiation and propagation of cracks. When the dosage of SSWs is 0.6 vol%, the incorporation of CNTs has no beneficial

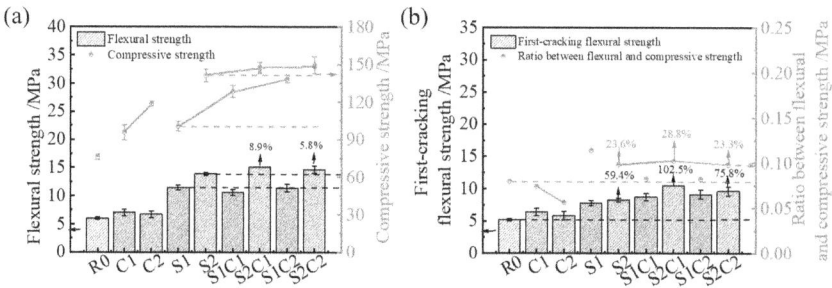

Figure 9.2 Flexural and compressive strength of SSWs-engineered multifunctional UHPC incorporating CNTs (P•I). (a) Flexural and compressive strength; (b) flexural first-cracking strength and the ratio between flexural and compressive strength

effect on flexural strength. However, for 1.2 vol% SSWs-engineered multifunctional UHPC, 1.23 MPa/8.9% and 0.8 MPa/5.8% relative changes can be achieved due to the addition of 0.25% and 0.5% CNTs, respectively, indicating the nano–micro synergistic toughening effect. The flexural strength of UHPC with 1.2 vol% SSWs and 0.25%/0.5% CNTs (S2C1 and S2C2) is 150.7% and 143.5% higher, respectively, than that of UHPC without fillers. Figure 9.2(a) also demonstrates that the compressive strength of UHPC is improved by 19.78 MPa/26.1%, 43.18 MPa/57.0%, 24.69 MPa/32.6%, and 65.38 MPa/86.3% when 0.25% CNTs, 0.5% CNTs, 0.6 vol% SSWs, and 1.2 vol% SSWs is added, respectively. Meanwhile, adding 0.25% and 0.5% CNTs increases the compressive strength of 0.6 vol% SSWs-engineered multifunctional UHPC (S1C1 and S1C2) by 28.12 MPa/28.0% and 38.21 MPa/38.0%, respectively, while the compressive strength of 1.2 vol% SSWs-engineered UHPC is enhanced by only 6.27 MPa/4.4% and 8.50 MPa/6.0%. What is noteworthy is that the limit of proportionality on the linear ascending stage of flexural strain–stress curves (strain gauges were pasted on the mid-span bottom of specimens to test the tensile strain) for composites was determined as flexural first-cracking strength. Figure 9.2(b) shows that adding SSWs or CNTs improves the flexural first-cracking strength of UHPC. When 1.2 vol% SSWs and 0.25% CNTs are incorporated, the maximum relative change for flexural first-cracking strength is 102.5% compared with that of UHPC without fillers and 27.0% higher than that of UHPC with 1.2 vol% SSWs. This results from the reduction of original flaws and the improvement of bond strength between the SSWs and the matrix. It is worth noting that the increase of the ratio between flexural and compressive strength indicates that the enhancement effect of fillers on flexural strength is superior to that on compressive strength, which is also a manifestation of the improved toughness. The ratio between flexural and compressive strength of

UHPC with 1.2 vol% SSWs and 0.25%/0.5% CNTs is increased by 28.8% and 23.3%, respectively, compared with that of UHPC without fillers.

Figure 9.3 shows that the flexural strain–stress curves and load–displacement curves of UHPC with/without CNTs possess only a linear ascending stage (elastic deformation stage); meanwhile, the peak strain and displacement are barely influenced by the adding of CNTs. The non-linear ascending stage (strain-hardening stage) and the slow descending stage appear on the strain–stress curves of UHPC with SSWs due to the crack-bridging effect of SSWs. For CNTs and SSWs-engineered multifunctional UHPC, the shape of flexural strain–stress and load–displacement curves are similar to those of SSWs-engineered multifunctional UHPC, suggesting that SSWs play a leading role during the deformation process of UHPC. When the content of SSWs is 0.6 vol%, the addition of CNTs only increases the peak load and stress of composites. However, the linear ascending stage is extended and the peak strain is improved by 12.5% when 0.25% CNTs is incorporated into UHPC with 1.2 vol% SSWs. Furthermore, there is a more obvious gentle stage before peak stress (the multi-cracking stage, as shown in Figure 9.3(b)), indicating the improvement in toughness.

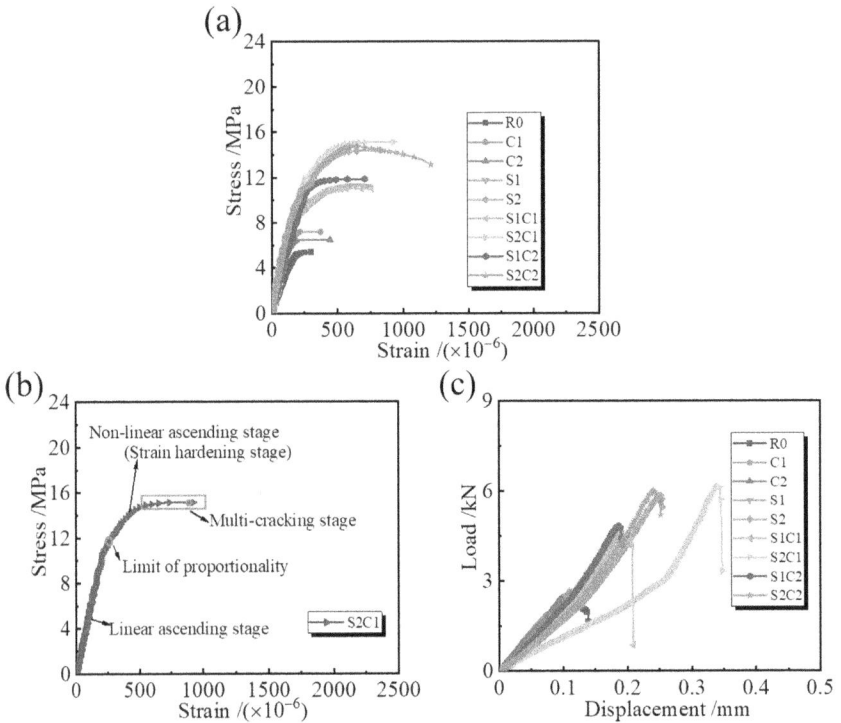

Figure 9.3 Flexural strain–stress and load–deformation curves of SSWs-engineered multifunctional UHPC incorporating CNTs (P•l). (a) Flexural strain–stress curves; (b) flexural strain–stress curve of S2C1; (c) flexural load–displacement curves

Table 9.1 Flexural toughness of SSWs-engineered multifunctional UHPC incorporating CNTs (P•I)

Group	Flexural toughness based on load–displacement curves (J/m²)	Relative changes for flexural toughness based on load–displacement curves (%)	Flexural toughness based on strain–stress curves (J/m³)	Relative changes for flexural toughness based on strain–stress curves (%)
R0	101.0 ± 1.2	–	2743.4 ± 86.6	–
C1	116.9 ± 26.1	15.7	2203.6 ± 181.3	–19.7
C2	96.4 ± 21.3	4.6	2454.6 ± 234.8	–10.5
S1	256.6 ± 11.0	150.1	6617.4 ± 237.2	141.2
S2	358.8 ± 47.6	255.3	9476.6 ± 184.6	245.4
S1C1	232.4 ± 43.9	130.1	6990.6 ± 356.5	154.8
S2C1	417.7 ± 101.8	313.2	11326.6 ± 15.0	312.9
S1C2	265.7 ± 29.3	163.1	6707.6 ± 406.3	144.9
S2C2	377.3 ± 43.6	273.6	6839.0 ± 102.6	149.3

According to Balaguru et al. [15], Banthia and Dubey [16], and Bentur [17] and Japan Society of Civil Engineers (JSCE) standard SF-4 Method [18], the total area of load–displacement curves and stress–strain curves was used to represent the energy absorption ability of composites on the direction of parallel to loading and the flexural-tensile of specimens, called "toughness based on load-displacement curves" and "toughness based on stress–strain curves", respectively. The flexural toughness of UHPC composites based on load–displacement and strain–stress curves is listed in Table 9.1. The addition of CNTs has a less beneficial influence on the toughness of UHPC, while incorporating SSWs brings remarkable increases in composite toughness. Meanwhile, the flexural toughness based on load–displacement and strain–stress curves of UHPC with 1.2 vol% SSWs and 0.25% CNTs is increased by 16.4% and 19.5%, respectively, compared with that of UHPC with 1.2% SSWs. It can therefore be concluded that adding a hybrid of 1.2 vol% SSWs and 0.25% CNTs is conductive to the enhancement of toughness, with the result that CNTs can effectively reduce original flaws at the nano scale, and SSWs can inhibit the initiation and propagation of microcracks at the micro scale.

9.2.1.2 Using P•O type of cement

When the cement used is P•O type, the flexural and compressive strength of CNTs and SSWs-engineered multifunctional UHPC are shown in Figure 9.4(a). The flexural strength of UHPC with 0.25% CNTs and that with 1.2 vol% SSWs is 1.60 MPa/13.5% and 6.64 MPa/56.0% higher, respectively,

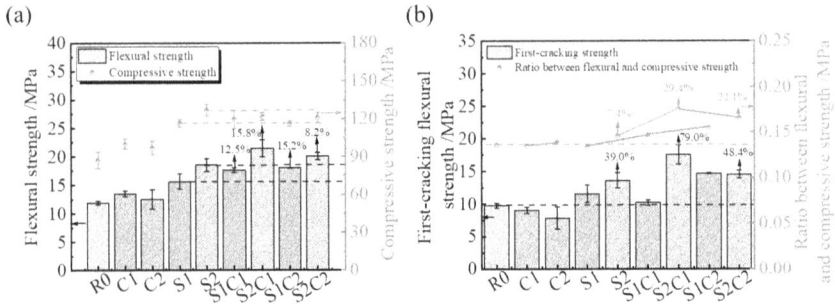

Figure 9.4 Flexural and compressive strength of SSWs-engineered multifunctional UHPC incorporating CNTs (P•O). (a) Flexural and compressive strength; (b) flexural first-cracking strength and the ratio between flexural and compressive strength

than that of UHPC without fillers. Due to the addition of 0.25% and 0.5% CNTs, the flexural strength of UHPC with 1.2 vol% SSWs is increased by 2.93 MPa/15.8% and 1.52 MPa/8.2%, respectively. Meanwhile, the relative changes for UHPC with 0.6 vol% SSWs can reach 1.96 MPa/12.5% and 2.38 MPa/15.2%, respectively. The flexural strength of UHPC with 1.2 vol% SSWs and 0.25%/0.50% CNTs is 80.7% and 68.8% higher, respectively, than that of UHPC without fillers. The synergistic enhancing effect of SSWs and CNTs on the flexural strength of UHPC is much greater than that of normal steel fibers and CNTs on the tensile strength of concrete [19]. Comparative analysis between Figure 9.2(a) and Figure 9.4(a) shows that adding a hybrid of 0.6 vol% SSWs and CNTs also has a synergistic toughening effect on UHPC when the cement used is P•O type. The content of C_3A in the P•O type of cement is high, leading to high early hydration heat in composites. Meanwhile, due to the high thermal conductivity, the addition of CNTs can effectively transfer hydration heat, decrease temperature stress, and reduce original microcracks, resulting in a high synergistic toughening effect with SSWs. Figure 9.4(a) also shows that the strengthening effect of SSWs on the compressive strength of UHPC is better than that of CNTs. Meanwhile, the addition of CNTs has no significant effect on the compressive strength of SSWs-engineered UHPC. Figure 9.4(b) shows that UHPC with 1.2 vol% SSWs and 0.25% CNTs possesses the maximum flexural first-cracking strength, 79.0% and 28.8% higher than that of UHPC without fillers and UHPC with 1.2% SSWs, respectively, due to the improvement effect of CNTs on the UHPC matrix structure. It can also be found that adding hybrid SSWs and CNTs endows UHPC with a higher ratio between flexural and compressive strength than that incorporating SSWs or CNTs only, also indicating the synergistic toughening effect of micro- and nanofillers. The ratio between flexural and compressive strength for UHPC with 1.2 vol% SSWs and 0.25%/0.5% CNTs is increased by 29.4% and 22.1%, respectively, compared with that of UHPC without fillers.

The flexural strain–stress and load–displacement curves of CNTs and SSWs-engineered multifunctional UHPC fabricated with the P•O type of cement are shown in Figure 9.5. It can be seen that the flexural strain–stress curves for UHPC with 1.2 vol% SSWs, UHPC with 1.2 vol% SSWs and 0.25% CNTs, and UHPC with 1.2 vol% SSWs and 0.5% CNTs all possess an obvious non-linear ascending stage (strain-hardening stage). Compared with other composites, the addition of 0.25% CNTs endows the curves of UHPC with 1.2 vol% SSWs with the slowest descending stage (strain-softening stage), indicating that the hybrid SSWs and CNTs can effectively bridge cracks and then be pulled off, as shown in Figure 9.5(b). This is different from the behavior of UHPC filled with 1.2 vol% SSWs and 0.25% CNTs and fabricated with the P•I type of cement. It can be attributed to the nano-core effect of CNTs in composites fabricated with the P•I type of

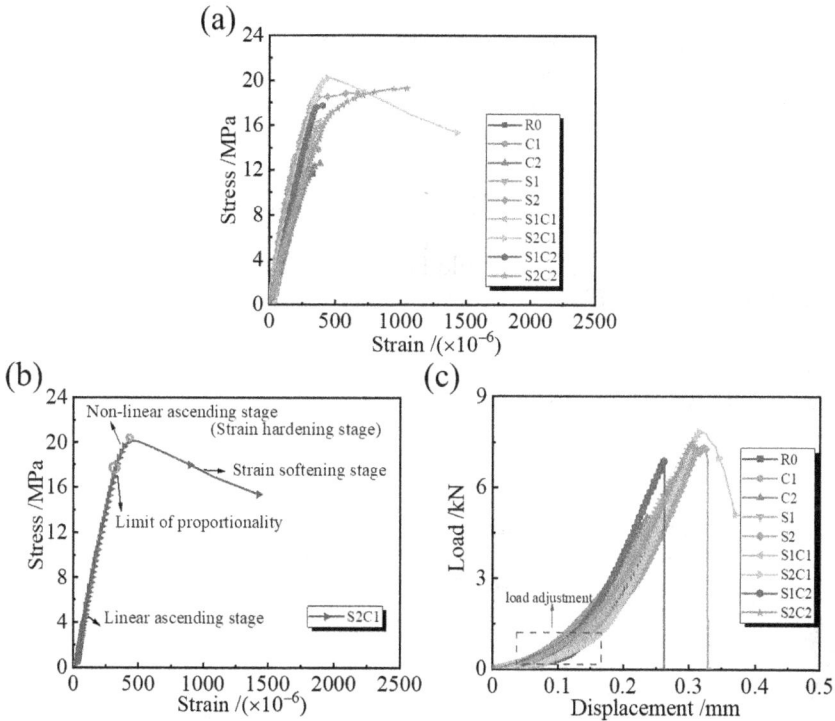

Figure 9.5 Flexural strain–stress and load–deformation curves of SSWs-engineered multifunctional UHPC incorporating CNTs (P•O). (a) Flexural strain–stress curves; (b) Flexural strain–stress curve of S2C1; (c) Flexural load–displacement curves

cement being dominated by the nucleation effect, such that the enhancement effect on interfacial bond strength between the SSWs and the matrix is positively correlated with the number of nanofillers. When the P·O type of cement is used, CNTs can greatly enhance the interfacial bond strength by reducing the original cracks of the matrix, resulting in more SSWs being pulled off when failure happens (this can be proved by the calculation results for SSWs efficiency in Section 9.4). Hence, the strain–stress curve of UHPC with 0.25% CNTs and 1.2 vol% SSWs fabricated with the P·O 42.5 type of cement has a significant descending stage. Figure 9.5(c) shows that there is a slow descending stage on the bending load–displacement curve of UHPC with 1.2 vol% SSWs and 0.25% CNTs, which is consistent with flexural stress–strain curves. Table 9.2 shows that the flexural toughness of UHPC with SSWs is higher than that of UHPC with CNTs. Due to the addition of 0.25% and 0.5% CNTs, the flexural toughness based on load–displacement curves of UHPC with 1.2 vol% SSWs is improved by 19.8% and 18.1%, respectively. Meanwhile, the relative changes in flexural toughness based on strain–stress curves can reach 133.7% and 57.1%. That is to say, no matter what kind of cement is used, 1.2 vol% SSWs and CNTs have a synergistic enhancement effect on the toughness of UHPC. SSWs can effectively inhibit the initiation and propagation of cracks through bridging, debonding, and the effect of being pulled off. The addition of CNTs can enhance crack resistance and reduce the original cracks of concrete matrix by the nano-core effect [20, 21]. Hybrid SSWs and CNTs work together to improve the first-cracking flexural strength and toughness of UHPC at the nano and micro scale. It should be noted that the toughening effect of hybrid SSWs and CNTs is more pronounced when the P•O type of cement

Table 9.2 Flexural toughness of SSWs-engineered multifunctional incorporating CNTs UHPC (P•O)

Group	Flexural toughness based on load–displacement curves (J/m^2)	Relative changes for flexural toughness based on load–displacement curves (%)	Flexural toughness based on strain–stress curves (J/m^3)	Relative changes for flexural toughness based on strain–stress curves (%)
R0	167.3 ± 9.5	–	3152.3 ± 95.7	–
C1	271.4 ± 34.4	62.3	2843.4 ± 99.3	−9.8
C2	213.0 ± 25.5	27.3	2813.0 ± 382.3	−10.8
S1	326.5 ± 0.6	95.1	3729.9 ± 317.5	18.3
S2	427.8 ± 68.1	155.7	9728.7 ± 610.1	208.6
S1C1	326.1 ± 12.1	94.9	3774.0 ± 92.3	19.7
S2C1	512.3 ± 109.7	206.2	22733.5 ± 1570	621.2
S1C2	322.9 ± 38.1	93.1	4108.4 ± 16.0	30.3
S2C2	505.4 ± 30.7	202.1	15276.0 ± 473.1	384.6

is used, resulting from the nano-core effect of CNTs being closely related to the properties of the matrix.

9.2.2 Compressive/flexural strength and flexural toughness of SSWs-engineered multifunctional UHPC incorporating NTs

9.2.2.1 Using P•I type of cement

The flexural and compressive strength of NTs and SSWs-engineered mul-tifunctional UHPC fabricated with the P•I type of cement are shown in Figure 9.6(a). The addition of 0.6 vol% and 1.2 vol% SSWs leads to an increase of 5.42 MPa/90.03% and 7.84 MPa/130.0%, respectively, in the flexural strength of UHPC. Incorporating 1.5% and 3.0% NTs (mass content of cement, marked as T1 and T2) does not have a favorable effect on the flexural strength of UHPC with 0.6 vol% SSWs (S1T1 and S1T2), while it increases the flexural strength of UHPC with 1.2 vol% SSWs by 2.49 MPa/18.0% and 1.87 MPa/13.5%, respectively. This indicates that adding 1.2 vol% SSWs and NTs (S2T1 and S2T2) together has a multiscale synergistic enhancement effect on the flexural strength of UHPC. Relative changes of 171.6%/10.33 MPa and 161.3%/9.71 MPa, respectively, in the flexural strength of UHPC with 1.2 vol% SSWs can be obtained compared with that of UHPC without fillers due to the addition of 1.5% and 3.0% NTs. The content of C_3S in the P•I type of cement is high, indicating that the early hydration reaction speed is fast. Due to the small size, large amounts, large specific surface area, and surface energy of NTs, the nano-core effect is dominated by the nucleation effect, causing the deposition of hydration products, increasing the amount and density of C-S-H gels, and reducing crystal orientation. Meanwhile, the $Ca(OH)_2$ content in the UHPC matrix

(a)   (b)

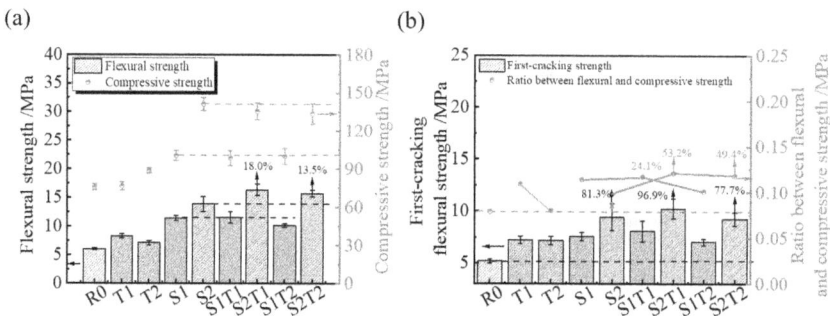

Figure 9.6 Flexural and compressive strength of SSWs-engineered multifunctional UHPC incorporating NTs (P•I). (a) Flexural and compressive strength; (b) Flexural first-cracking strength and the ratio between flexural and compressive strength

is high, and the SiO_2 coating the surface of TiO_2 can react with $Ca(OH)_2$ to further improve the structural compactness of C-S-H gels. Therefore, the addition of NTs is conducive to the enhancement of the crack resistance of the matrix. The multiscale synergistic effect of SSWs and NTs endows UHPC with high flexural strength. Figure 9.6(a) also shows that the compressive strength of UHPC with 0.6 vol% SSWs and 1.2 vol% SSWs is increased by 24.69 MPa/32.6% and 65.38 MPa/86.3%, respectively, compared with that of UHPC without fillers. At the same dosage of SSWs, the difference in compressive strength between UHPC with SSWs and UHPC with SSWs and NTs is within 5.5%, indicating that NTs have no obvious influence on the compressive strength of UHPC with SSWs and that SSWs play a critical role in the compressive strength of UHPC. As shown in Figure 9.6(b), the addition of SSWs or NTs can significantly improve the first-cracking flexural strength of UHPC, resulting from the improvement of structural compactness and crack resistance of the matrix. The maximum relative change of 96.9% can be obtained for the first-cracking flexural strength of UHPC when the hybrid of 1.2 vol% SSWs and 1.5% NTs is added, which is comparable to that achieved by incorporating 1.2 vol% SSWs and 0.25% CNTs. Figure 9.6(b) also demonstrates that the ratio between flexural and compressive strength for UHPC with 0.6 vol% SSWs and 1.2 vol% SSWs is 44.3% and 24.1% higher, respectively, than that of UHPC without fillers. Meanwhile, incorporating 1.5% and 3.0% NTs leads to an increase of 23.5% and 20.4%, respectively, for the ratio between the flexural and compressive strength of UHPC with 1.2 vol% SSWs, representing the enhancement of toughness.

Figure 9.7(a) shows that the addition of SSWs endows the bending strain–stress curves of UHPC with a long linear ascending stage and a gentle development stage before peak load. The shape of the curve for UHPC with

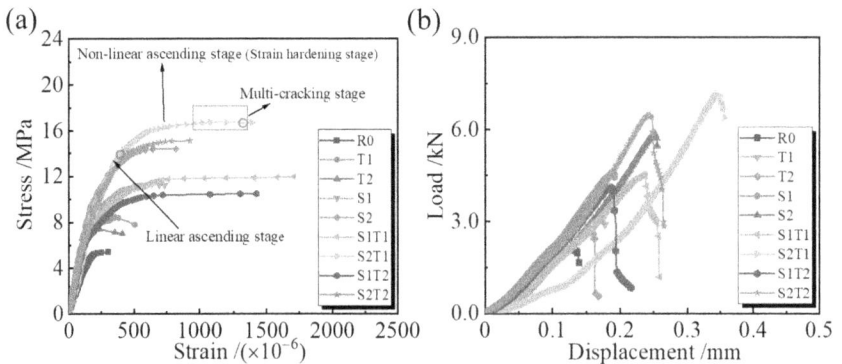

Figure 9.7 Flexural strain–stress and load–displacement curves of SSWs-engineered multifunctional UHPC incorporating NTs (P•I). (a) Flexural strain–stress curves; (b) Flexural load–displacement curves

SSWs and NTs depends mainly on the content of SSWs. The peak strain of UHPC with 1.2 vol% SSWs is increased by 34.8% due to the incorporation of 1.5% NTs. Compared with Figure 9.3(a), the trend toward the *x*-axis of the non-linear ascending stage on bending strain–stress curves of UHPC with SSWs and NTs is weakened because CNTs can effectively bridge cracks to inhibit the propagation of cracks in addition to the nanocore effect. However, the flexural stress–strain curves of UHPC with SSWs and NTs have a longer gentle development stage before peak load compared with that of UHPC with SSWs and CNTs, resulting from the wide distribution of NTs. Figure 9.7(b) demonstrates that the flexural load–displacement curves of UHPC with SSWs and NTs possess high peak displacement and a slow descending stage. Meanwhile, the peak displacement on the load–displacement curve for UHPC with 1.2 vol% SSWs and 1.5% NTs is 37.2% higher than that for UHPC with 1.2 vol% SSWs, resulting from the improvement of matrix structural compactness and the increase of bond strength between the SSWs and the UHPC matrix.

Table 9.3 shows that incorporating SSWs endows UHPC with higher flexural toughness compared with that endowed by NTs. Relative changes of 82.8% and 46.1% can be achieved for flexural toughness based on load–displacement curves of UHPC with 1.2 vol% SSWs due to the addition of 1.5% and 3.0% NTs, respectively. Meanwhile, the flexural toughness based on strain–stress curves of UHPC with 0.6 vol% SSWs is increased by 172.4% and 99.4%, and that of UHPC with 1.2 vol% SSWs is improved by 104.2% and 17.5%, respectively. It can be concluded that 1.2 vol% SSWs and 1.5% NTs manifest an outstanding synergistic toughening effect, even

Table 9.3 Flexural toughness of SSWs-engineered multifunctional UHPC incorporating NTs (P•I)

Group	Flexural toughness based on load–displacement curves (J/m^2)	Relative changes for flexural toughness based on load–displacement curves (%)	Flexural toughness based on strain–stress curves (J/m^3)	Relative changes for flexural toughness based on strain–stress curves (%)
R0	101.0 ± 1.2	–	2743.4 ± 86.6	–
T1	149.9 ± 1.6	48.4	2277.5 ± 96.0	−17.0
T2	121.0 ± 10.8	19.8	791.9 ± 42.4	–
S1	252.6 ± 11.0	150.1	6617.4 ± 237.2	141.2
S2	358.8 ± 47.6	255.2	9476.6 ± 184.6	245.4
S1T1	347.0 ± 60.0	243.5	18028.0 ± 1562.2	557.2
S2T1	572.1 ± 16.5	466.4	19348.7 ± 1124.2	605.3
S1T2	232.2 ± 27.7	129.9	13194.5 ± 405.0	380.9
S2T2	477.5 ± 22.8	372.8	11131.9 ± 431.7	305.8

better than the hybrid incorporation of SSWs and CNTs, due to the high content and extensive distribution of NTs. Adding hybrid SSWs and NTs can enhance the nano- and microstructure of UHPC to reduce original flaws and improve the crack resistance of the matrix. Meanwhile, the core-shell elements of NTs and SSWs together restrict the initiation and stable propagation of cracks. When cracks enter the unstable propagation stage, SSWs can bridge the cracks until they are pulled off.

9.2.2.2 Using P•O type of cement

When the cement used is the P•O type, the enhancement effect on the strengths of UHPC caused by the addition of SSWs is still far superior to that caused by the addition of NTs, as shown in Figure 9.8(a). However, the synergistic enhancement effect of SSWs and NTs is not remarkable. When the content of NTs is 1.5%, the flexural strength of UHPC with 1.2 vol% SSWs is increased by only 1.06 MPa/5.7%. The relative change of flexural strength for NTs and SSWs-engineered multifunctional UHPC fabricated with the P•O type of cement is far below than that for composites fabricated with the P•I type of cement. This can be attributed to a reduction in $Ca(OH)_2$ crystal content in the UHPC matrix due to the high content of super-fine mineral admixture in P•O cement, such that nano SiO_2 coating the surface of NTs cannot effectively participate in chemical reactions and limits the nano-core effect of NTs. The change rule of compressive strength for UHPC with SSWs and NTs is the same as that of flexural strength. Figure 9.8(b) shows that the first-cracking flexural strength and the ratio between flexural and compressive strength of UHPC with 1.2 vol% SSWs and 1.5% NTs is 89.7% and 13.2% higher, respectively, than that of UHPC without fillers. These relative changes are smaller than for UHPC fabricated with the P•I type of cement. This illustrates that the chemical reaction

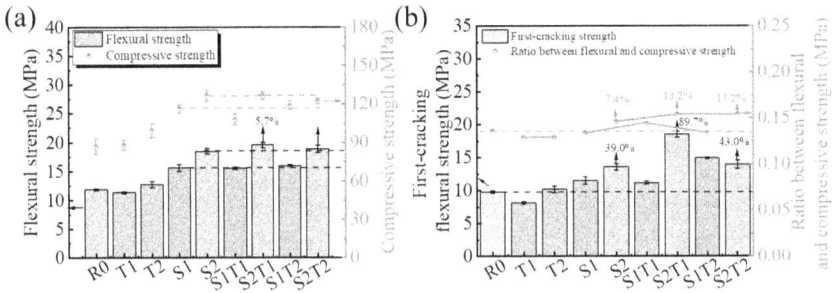

Figure 9.8 Flexural and compressive strength of SSWs-engineered multifunctional UHPC incorporating NTs (P•O). (a) Flexural and compressive strength; (b) Flexural first-cracking strength and the ratio between flexural and compressive strength

between SiO_2 and $Ca(OH)_2$ plays an important role in the nano-core effect of TiO_2 to form more effective core-shell elements.

Figure 9.9(a) shows that the flexural strain–stress curves of UHPC with 1.2 vol% SSWs have three stages: the linear ascending stage (elastic ascending stage), the non-linear ascending stage (strain-hardening stage), and the slow descending stage (strain-softening stage). Meanwhile, with increasing NTs content, the non-linear ascending and descending stage are extended. Incorporating 3.0% NTs increases the peak strain of UHPC with 1.2 vol% SSWs by 81.3%, due to the fact that the widely distributed NTs can also effectively enhance the crack resistance of the matrix. It can be found by comparison between Figure 9.7(a) and Figure 9.9(a) that the non-linear ascending stage and the stationary development stage before peak stress are obviously shortened due to the use of the P•O type of cement. Meanwhile, the descending stage of the strain–stress curve for UHPC with 1.2 vol% SSWs and 1.5% NTs is also shorter than that of UHPC with 1.2 vol% SSWs and 0.25% CNTs, as shown in Figure 9.9(b). Figure 9.9(c) shows that the addition of 1.5% and 3.0% NTs leads to an increase of 7.4% and 8.2%,

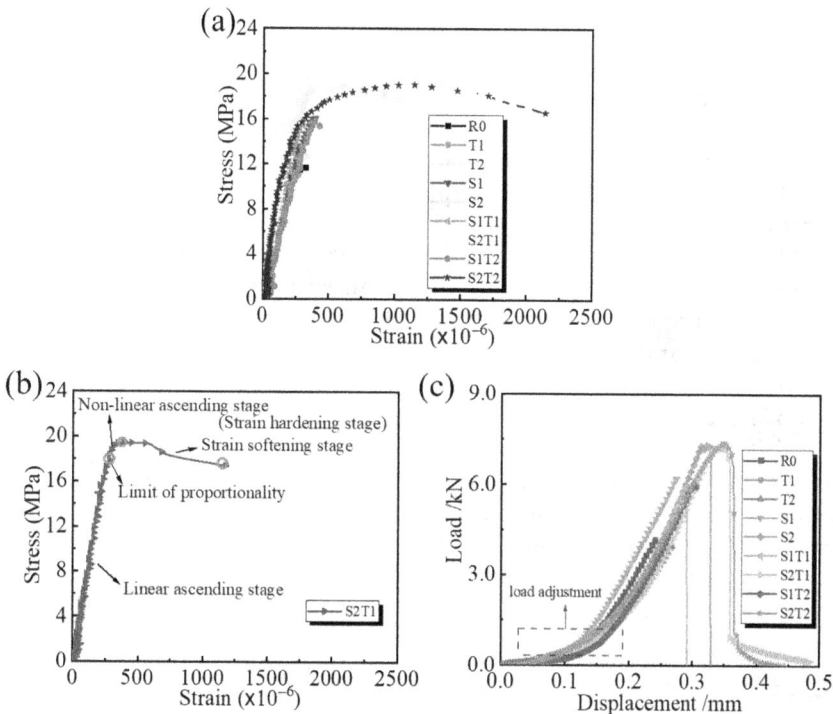

Figure 9.9 Flexural strain–stress and bending load–displacement curves of SSWs-engineered multifunctional UHPC incorporating NTs (P•O). (a) Flexural strain–stress curves; (b) Flexural strain–stress curve of S2CI; (c) Bending load–displacement curves

Table 9.4 Bending toughness of SSWs-engineered multifunctional UHPC incorporating NTs (P•O)

Group	Flexural toughness based on load–displacement curves (J/m²)	Relative changes for flexural toughness based on load–displacement curves (%)	Flexural toughness based on strain–stress curves (J/m³)	Relative changes for flexural toughness based on strain–stress curves (%)
R0	167.3 ± 9.5	–	3152.3 ± 95.7	–
T1	160.2 ± 13.3	−4.2	1788.1 ± 26.7	–
T2	190.9 ± 12.5	14.1	2433.4 ± 100.7	–
S1	326.5 ± 0.6	95.1	3729.9 ± 317.5	18.3
S2	427.8 ± 68.1	155.7	9728.7 ± 610.1	208.6
S1T1	361.1 ± 102.3	115.8	3630.7 ± 49.2	15.2
S2T1	590.0 ± 32.6	252.6	19217.5 ± 451.9	509.6
S1T2	296.5 ± 21.6	77.2	3884.5 ± 29.1	23.2
S2T2	583.6 ± 57.4	248.8	17801.0 ± 594.6	464.7

respectively, for peak displacement of UHPC with 1.2 vol% SSWs. The load–displacement curves of UHPC with 1.2 vol% SSWs and NTs have the same three stages as the strain–stress curves, representing the toughening effect of SSWs and NTs.

Table 9.4 shows that the flexural toughness based on load–displacement curves for UHPC with 1.2 vol% SSWs and 1.5%/3.0%NTs is 62.1% and 59.8% higher, respectively, than that of UHPC with 1.2 vol% SSWs. These relative changes are higher than those caused by the addition of CNTs, mainly because NTs are more beneficial in improving structural compactness. The flexural toughness based on strain–stress curves of UHPC with 1.2 vol% SSWs is increased by 97.5% and 83.0% due to the incorporation of 1.5% and 3.0% NTs, respectively, lower than the relative changes caused by CNTs. This can be attributed to CNTs having a more significant inhibiting effect on crack propagation. Meanwhile, these relative changes are comparable to those of composites fabricated with the P•I type of cement, but the flexural failure modes are different (as shown in Figures 9.7(a) and 9.9(a)). The toughness results reveal again that 1.2 vol% SSWs and NTs possess a synergistic toughening effect.

9.2.3 Flexural toughening calculation models

It can be seen from these test results that when incorporating hybrid SSWs and CNTs/NTs into UHPC, the load will be totally transferred to the SSWs after the concrete matrix cracking until they are pulled off. Due to the high specific surface area of SSWs, some of the CNTs/NTs tend to aggregate on the surface of the SSWs to play the role of the nano-core effect, enhancing the bond strength between the UHPC matrix and the SSWs. Therefore, the synergistic toughening effect of SSWs and CNTs/NTs can be calculated by modified fiber reinforced concrete theory and composite theory.

9.2.3.1 Calculation model of three-point flexural strength

The volume fraction of CNTs (V_{CNT}) can be obtained from Eq. (9.1):

$$V_{CNT} = \frac{m}{\rho V} \qquad (9.1)$$

where m is the total mass of CNTs in UHPC specimens, ρ is the density of CNTs, and V is the specimen volume of UHPC. Therefore, the volume fraction of CNTs corresponding to the mass fraction of 0.25% and 0.5% is 0.093 vol% and 0.186 vol%, respectively. Similarly, for NTs with mass dosage of 1.5% and 3.0%, the volume fraction (V_{NT}) is 0.275 vol% and 0.550 vol%, respectively. The volume fraction of SSWs (V_{SW}) is set as 0.5vol%, 1.0vol%, and 1.5vol%.

The numbers of CNTs (N_{CNT}), NTs (N_{NT}), and SSWs (N_{SW}) in the UHPC specimen are calculated by Eqs (9.2)–(9.4). The calculation results are listed in Table 9.5.

$$N_{CNT} = \frac{V_{CNT}V}{\frac{1}{4}\pi l_c d_c^2} \qquad (9.2)$$

$$N_{NT} = \frac{V_{NT}V}{\frac{4}{3}\pi R_t^3} \qquad (9.3)$$

$$N_{SW} = \frac{V_{SW}V}{\frac{1}{4}\pi l_w d_w^2} \qquad (9.4)$$

where d_c is the outer diameter of CNTs (< 8 nm), l_c is the length of CNTs (0.5–2 μm), R_t is the diameter of NTs (20 nm), d_w is the diameter of SSWs (20 μm), and l_w is the length of SSWs (10 mm).

Assuming that all CNTs/NTs were absorbed on the surface of SSWs, the equivalent diameter of SSWs can be obtained from Eqs (9.5) and (9.6):

$$d_{SW-CNT} = \sqrt{\frac{N_{CNT}}{10N_{SW}} l_c d_c^2 + d_w} \qquad (9.5)$$

Table 9.5 Flexural strength calculation parameters of SSWs-engineered multifunctional UHPC incorporating CNTs/NTs

Group	Numbers	Specimens	Equivalent diameter (μm)	Spacing of SSWs (mm)	Spacing coefficient (S')	FSS (mm²/mm³)	η_l P•I	η_l P•O
S1	4.892×10^5	S1	20.00	3.56	–	1.20	0.814	0.857
S2	9.783×10^5	S2	20.00	2.52	–	2.40	0.726	0.780
C1	3.791×10^{15}	S1C1	27.87	4.97	1.39	0.86	0.802	0.871
		S2C1	25.50	3.21	1.28	1.88	0.743	0.804
C2	7.582×10^{15}	S1C2	31.13	5.55	1.56	0.77	0.814	0.873
		S2C2	27.87	3.51	1.39	1.72	0.737	0.793
T1	2.109×10^{16}	S1T1	33.50	5.97	1.67	0.72	0.815	0.856
		S2T1	29.57	3.73	1.47	1.62	0.758	0.789
T2	4.204×10^{16}	S1T2	39.10	6.97	1.96	0.61	0.795	0.810
		S2T2	33.40	4.21	1.67	1.44	0.751	0.783

$$d_{\text{SW-NT}} = \sqrt{\frac{N_{\text{NT}}}{10 N_{\text{SW}}} R_t^3} + d_w \tag{9.6}$$

The calculation formula of the spacing ($\overline{S}_w$) between SSWs is shown as Eq. (9.7) [22]:

$$\overline{S}_w = 13.8 d_w \sqrt{\frac{1}{V_{\text{SW}}}} \tag{9.7}$$

It can be seen from Eq. (9.7) that the spacing between SSWs is proportional to diameter. The corresponding spacing coefficient (S') for SSWs in UHPC with CNTs/NTs is defined as the ratio between the equivalent diameter and the actual diameter of SSWs. Meanwhile, the specific surface area of SSWs (FSS_{SW}) in SSWs-engineered multifunctional UHPC can be obtained from Eq. (9.8) [23]:

$$FSS_{\text{SW}} = \frac{4}{d_w} V_{\text{SW}} \tag{9.8}$$

Correspondingly, the specific surface area of SSWs in UHPC with SSWs and CNTs/NTs ($FSS_{\text{SW-CNT}}$ and $FSS_{\text{SW-NT}}$) can be calculated using Eqs (9.9) and (9.10), as follows:

$$FSS_{\text{SW-CNT}} = \frac{4}{d_{\text{sw-CNT}}} V_{\text{SW}} \tag{9.9}$$

$$FSS_{\text{SW-NT}} = \frac{4}{2 d_{\text{sw-NT}}} V_{\text{SW}} \tag{9.10}$$

The spacing coefficient (S') and specific surface area (FSS) of SSWs in different UHPC matrixes are displayed in Table 9.5.

According to composite theory, the tensile strength of UHPC with fillers can be calculated using Eq. (9.11) [17]:

$$\sigma_{\text{fc}} = \sigma_f (1 - V_{\text{SW}}) + \eta_0 \eta_1 \sigma_f^u V_{\text{SW}} \tag{9.11}$$

where σ_{fc} is the tensile strength of UHPC with fillers, and σ_f is the tensile strength of UHPC without fillers. On the basis of existing investigations [24, 25], Eq. (9.11) is modified to the following Eq. (9.12) considering the influence of the spacing coefficient of SSWs, the specific surface area of SSWs, the volume fraction of SSWs, and the relationships between flexural and tensile strength.

$$\sigma_{bc} = \sigma_f\left(1 - V_{SW}\right) + 2.44S'\eta_0\eta_1\sigma_f^u V_{SW}\exp\left(-10FSS \times V_{SW}\right) \tag{9.12}$$

where σ_{bc} is the flexural strength of UHPC with fillers, σ_b is the flexural strength of UHPC without fillers, η_0 is the orientation coefficient of SSWs (0.4574 in this study), σ_f^u is the tensile strength of SSWs (780 MPa), and η_1 is the length coefficient of SSWs without considering the effect of nanofillers, which can be calculated using Eq. (9.13) [26]:

$$\eta_1 = \frac{2\sigma_{bt}}{2\sigma_{bt} + \eta_0 V_{SW}\sigma_f^u} \tag{9.13}$$

where σ_{bt} is the measured flexural strength of composites. The corresponding values of η_1 are listed in Table 9.5.

It should be noted that rather than being absorbed on the surface of SSWs, NTs are more easily dispersed in UHPC matrix due to the SiO_2 coated on the surface of TiO_2, such that the calculation results of equivalent diameter for SSWs in UHPC with NTs are much larger than the actual value. Therefore, the effect of the spacing coefficient (S') should not be considered in the calculation of flexural strength for UHPC with SSWs and NTs. The errors (calculated by Eq. [9.14]) between the calculated and the measured flexural strengths for CNTs/NTs and SSWs-engineered UHPC are displayed in Table 9.6.

$$\Delta = \left|\frac{\sigma_{bt} - \sigma_{bc}}{\sigma_{bt}}\right| \times 100\% \tag{9.14}$$

where σ_{bc} is the theoretical value of flexural strength, and σ_{bt} is the experimental value of flexural strength for UHPC. Table 9.6 shows that except for UHPC with 1.2 vol% SSWs and 3.0% NTs fabricated with the P•O type of cement, Eq. (9.12) can be used to accurately describe the flexural strength of UHPC with hybrid micro- and nanofillers. The flexural strength shows an increasing trend with the increase of SSWs' equivalent diameter and the decrease of specific surface area.

9.2.3.2 Calculation of SSWs' efficiency

Assuming that SSWs are pulled off at half their length, Eq. (9.15) is established:

$$\pi d_w \frac{l_{SW}^{crit}}{2}\tau_{SW} = \frac{\pi}{4}d_w^2\sigma_f^u \tag{9.15}$$

Table 9.6 Theoretical and experimental values of flexural strength of UHPC composites with different types of cement

Cement type	Group	Theoretical value (MPa)	Experimental value (MPa)	Δ (%)
P·I42.5	SI	10.84	11.44	5.24
	S2	13.77	13.86	0.65
	SICI	11.52	10.59	8.78
	S2CI	13.83	15.09	8.35
	SIC2	12.28	11.39	7.81
	S2C2	14.66	14.66	0
	SITI	10.06	11.54	12.80
	S2TI	15.52	16.35	5.07
	SIT2	9.99	10.1	1.09
	S2T2	16.98	15.73	7.95
P·O42.5R	SI	16.65	15.62	6.60
	S2	19.55	18.50	5.77
	SICI	17.33	17.58	1.42
	S2CI	19.61	21.43	8.49
	SIC2	18.11	18.00	0.67
	S2C2	20.43	20.02	2.05
	SITI	16.15	15.50	4.19
	S2TI	21.87	19.56	11.81
	SIT2	16.18	16.00	1.13
	S2T2	23.32	18.86	23.65

where l_{sw}^{crit} is the critical length of SSWs, which can be calculated using Eq. (9.16) [26]. Other symbols have the same meaning as in previous equations.

$$l_{sw}^{crit} = \frac{2\eta_0 V_{SW}\sigma_f^u l_{SW}}{2\sigma_{bt} + \eta_0 V_{SW}\sigma_f^u} \tag{9.16}$$

Then, the bond strength τ_{SW} between SSWs and UHPC matrix can be expressed by Eq. (9.17):

$$\tau_{SW} = \frac{d_{SW}\sigma_f^u}{2l_{sw}^{crit}} \tag{9.17}$$

It can be seen that the bond strength is proportional to the diameter of SSWs at a certain critical length. The bond strength $\tau_{SW\text{-}CNT/NT}$ in CNTs/

NTs and SSWs-engineered multifunctional UHPC can be obtained according to the equivalent diameter of SSWs, as listed in Table 9.7.

The energy consumed by a single SSW at failure ($\bar{W}_{SW,P}$) is expressed by Eq. (9.18) [27]:

$$\bar{W}_{SW,P} = \frac{1}{192} \frac{\pi d_w^4 \left(\sigma_f^u\right)^3}{l_{SW}\tau_{sw}^2} \tag{9.18}$$

Based on the values of equivalent diameter of SSWs and the bond strength between SSWs and the UHPC matrix, the energy values of a single SSW in UHPC without/with CNTs/NTs $\bar{W}_{SW\text{-}CNT/NT,P}$) can be obtained, as demonstrated in Table 9.7. After that, the numbers of effective SSWs in UHPC (N_{SW}') are obtained according to the experimental relative changes in flexural toughness ($E_{SW,I}$, integration differences of flexural load–displacement curves).

$$N_{SW}' = \frac{E_{SW,i}}{\bar{W}_{SW,P}} \tag{9.19}$$

Table 9.7 Calculation parameters of SSWs' efficiency for UHPC composites

Cement type	Group	Bond strength (MPa)	Flexural energy of a single SSW (×10⁻³ J)	Experimental flexural toughness relative changes (J)	N_{SW}' (×10³)
P·I42.5	SI	2.10	0.028	0.243	8.783
	S2	1.43	0.061	0.364	5.953
	SICI	2.74	0.062	0.210	3.367
	S2CI	1.93	0.088	0.507	5.758
	SIC2	3.26	0.068	0.264	3.875
	S2C2	2.07	0.109	0.442	4.856
	SITI	3.53	0.078	0.390	5.000
	S2TI	2.37	0.106	0.750	7.089
	SIT2	3.71	0.132	0.210	1.590
	S2T2	2.61	0.141	0.602	4.248
P·O42.5R	SI	2.72	0.017	0.255	15.190
	S2	1.77	0.039	0.417	10.540
	SICI	4.19	0.026	0.250	8.460
	S2CI	2.53	0.051	0.550	10.700
	SIC2	4.80	0.031	0.249	8.030
	S2C2	2.63	0.068	0.540	7.940
	SITI	3.53	0.048	0.310	6.480
	S2TI	2.37	0.080	0.676	8.450
	SIT2	3.71	0.061	0.207	3.388
	S2T2	2.61	0.107	0.666	6.220

Similarly, the numbers of effective SSWs in UHPC with CNTs/NTs are calculated and listed in Table 9.7.

Table 9.7 shows that the bond strength between SSWs and the UHPC matrix increases with increasing CNTs/NTs dosage. When the cement used is the P•I type, the enhancement effect on the bond strength of NTs is better than that of CNTs. The numbers of effective SSWs show a decreasing trend with the increase of bond strength. However, the numbers of effective SSWs in UHPC with 1.2 vol% SSWs and 1.5% NTs are 19.1% higher than those in UHPC with 1.2 vol% SSWs. Meanwhile, the bond strength in the former matrix is increased by 44.8% compared with that in the latter matrix. This is consistent with UHPC with 1.2 vol% SSWs and 1.5% NTs showing the highest flexural strength and toughness, and the flexural failure mode having a tendency to multi-cracking. Due to the existence of SiO_2 coating the surface of TiO_2, NTs can be widely distributed in the UHPC matrix to perform the nano-core effect, thus enhancing structural compactness and reducing original flaws. Meanwhile, SSWs can effectively inhibit the initiation and propagation of cracks to improve strengths and toughness. When the cement used is the P•O type, UHPC with SSWs and CNTs has a higher bond strength and greater numbers of effective SSWs than UHPC with SSWs and NTs at the same dosage of SSWs. The reduction of $Ca(OH)_2$ content in the matrix is not conducive to the dispersion of NTs, while CNTs can also bridge cracks except for the nano-core effect. The numbers of effective SSWs for UHPC with 1.2 vol% SSWs and 0.25% CNTs are almost identical to that for UHPC with 1.2 vol% SSWs. Meanwhile, for other composites, the numbers of effective SSWs are reduced due to the addition of nanofillers. This is in agreement with the previous conclusions that UHPC with 1.2 vol% SSWs and 0.25% CNTs possesses the highest flexural strength.

9.2.4 Flexural toughening mechanisms

It can be seen from experimental results and theoretical calculations that the flexural failure modes of CNTs/NTs and SSWs-engineered multifunctional UHPC are different from those of other concrete, as shown in Figure 9.10. The flexural strain–stress curve of conventional steel fiber (SF) reinforced concrete has an obvious strain-softening stage [28], i.e. the addition of normal steel fiber mainly endows concrete with post-cracking ductility. Although the strain–stress curve of glass fiber reinforced concrete (GFRC) possesses a multi-cracking stage, and the peak stress is increased, [17], the typical brittle failure finally occurs. The strain–stress curve of commonly used UHPC and SSWs-engineered multifunctional UHPC includes three stages: an elastic deformation stage, a strain-hardening stage, and a strain-softening stage; therefore, the composites have high first-cracking strength, peak strength, and toughness. When CNTs/NTs are incorporated

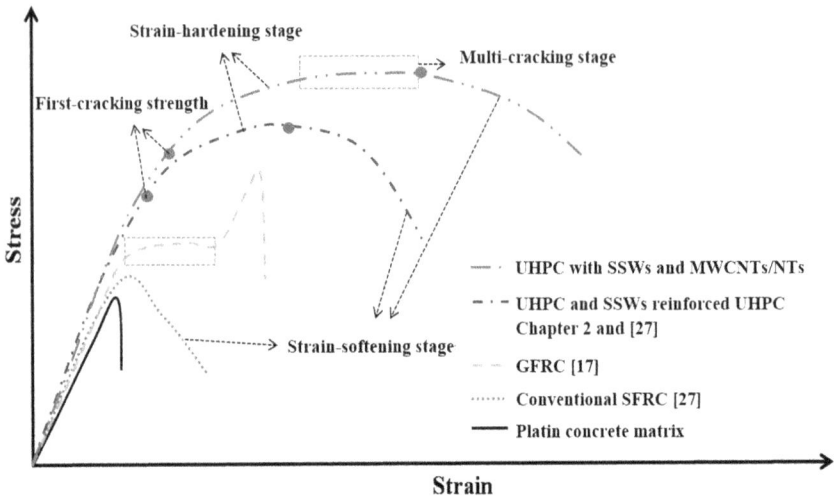

Figure 9.10 Flexural-tensile strain–stress curves of different concretes

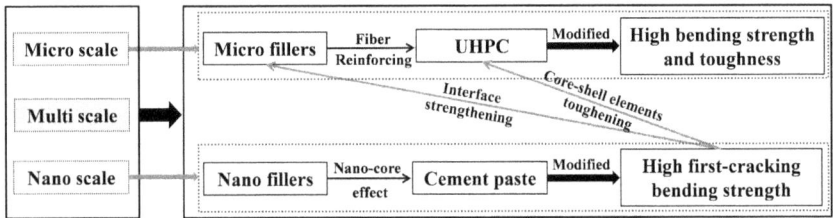

Figure 9.11 Multiscale modification mechanisms of micro- and nanofillers in UHPC

into SSWs-engineered multifunctional UHPC, the strain–stress curve still consists of these three stages. However, the value of the flexural first-cracking strength is increased, and the strain-hardening stage is prolonged. Furthermore, the gentle stage before peak stress indicates that there is a tendency to multi-cracking, which is consistent with the improvement of the bond strength between SSWs and the UHPC matrix and the efficiency of SSWs.

The multiscale synergistic modification mechanisms of SSWs and CNTs/ NTs on the performance of UHPC are summarized in Figure 9.11. At the nano scale, the addition of CNTs/NTs can increase the polymerization of C-S-H gels and decrease the CH crystal size through the nano-core effect, leading to the reduction of original flaws and the improvement of structural compactness [20, 29, 30]. Due to the small size and wide distribution of nanofillers, the core-shell elements are close to each other, which improves the first-cracking bending strength of UHPC. Meanwhile, nanofillers can aggregate on the surface of SSWs to enhance the bond strength between SSWs and the UHPC matrix [21]. At the micro scale, SSWs can form a widely distributed network in UHPC at a low volume fraction, effectively preventing the initiation and propagation of microcracks by bridging, deflecting, and stripping effects. This results in the increase of peak strengths and toughness. It is particularly important that the core-shell elements governed by the nano-core effect can also inhibit the emergence and development of cracks through a pinning effect, leading to the extension of the strain-hardening stage and even the appearance of a multi-cracking stage before peak stress on strain–stress curves. This can be proved by the fact that the bending tensile strain–stress curve of UHPC with 1.2 vol% SSWs and 1.5% NTs has a longer multi-cracking stage than that of UHPC with 1.2 vol% SSWs and 0.25% CNTs, as shown in Figures 9.3(a) and 9.7(a). The nano-core effect and micro-fiber reinforcing effect can promote each other to improve the first-cracking strength and toughness of UHPC simultaneously, indicating the multiscale synergistic effect of micro- and nanofillers.

9.3 ELECTRICAL AND SELF-SENSING PROPERTIES OF SSWS-ENGINEERED MULTIFUNCTIONAL UHPC INCORPORATING NANOFILLERS

9.3.1 Electrical resistivity

Due to the low content of SSWs in this part (0.15 vol% and 0.30 vol%), the compressive strength of CNTs/NTs and SSWs-engineered multifunctional UHPC at the curing age of 28 days is lower than 120 MPa, but the compressive strength of these composites at the curing age of 72 days can reach 120 MPa. Hence, these composites are also called UHPC. Figure 9.12 shows that the direct current (DC) resistivity of UHPC shows little change as the curing age increases from 3 days to 28 days. P•O 42.5R type of cement was used in

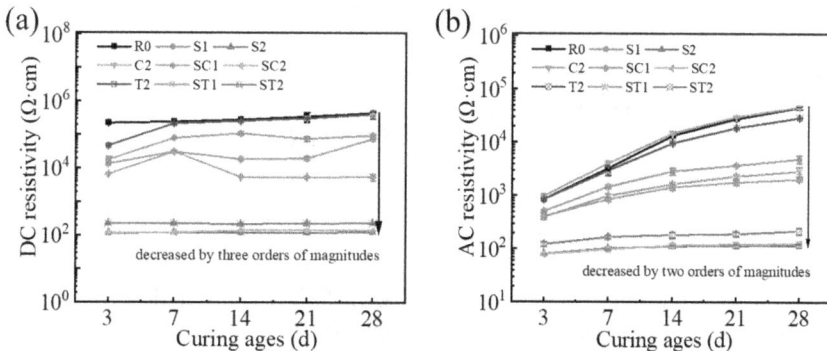

Figure 9.12 (a) DC and (b) 100-Hz AC resistivity of SSWs-engineered multifunctional UHPC incorporating CNTs/NTs

this part. Adding 0.5% CNTs (C2) or 3% NTs (T2) in mass content of cement can only reduce the DC resistivity of UHPC at the curing age of 3 days. This is mainly related to the nano-core effect of fillers to form more C-S-H gels at early ages. When the curing age is larger than 3 days, the DC resistivity of UHPC with 0.5% CNTs or 3% NTs is almost the same as that of UHPC without fillers, and this can be attributed to the low content of CNTs and the semiconductor characteristics of NTs. With the increase of SSWs content, the DC resistivity of UHPC presents a declining trend. Decrements of one and three orders of magnitude, respectively, can be obtained for DC resistivity when 0.15 vol% (S1) and 0.3 vol% SSWs (S2) are added, indicating that 0.15 vol% and 0.3 vol% SSWs are all in the percolation threshold zone. The former is located in the middle stage of the percolation threshold zone, and the latter is nearing the end of the percolation threshold zone. At the curing age of 3 days, the incorporation of CNTs and NTs decreases the DC resistivity of UHPC with 0.15 vol% SSWs (the corresponding UHPCs are marked as SC1 and ST1) by 63.0% and 24.3%, respectively. Furthermore, the decrements are 94.2% and 21.0%, respectively, at the curing age of 28 days, showing the cooperative conductive effect of micro-fibers and nanofillers. Due to the excellent conductivity, the synergistic conduction effect of SSWs and CNTs is better than that of SSWs and NTs. Adding CNTs has little influence on the DC resistivity of UHPC with 0.3 vol% SSWs (the corresponding UHPC is marked as SC2), indicating that SSWs have formed a widely distributed conductive network in the UHPC. However, the incorporation of NTs increases the DC resistivity of UHPC with 0.3 vol% SSWs (the corresponding UHPC is marked as ST2) by 85.4%, because NTs can permeate into the space of SSWs to interrupt the overlapping network.

Figure 9.12(b) shows that at the frequency of 100 Hz, the alternating current (AC) resistivity of UHPC with fillers increases with the increase of curing age, but the increments show a decreasing tendency with the conductivity improvement of UHPC. At the curing age of 28 days, the addition of 0.5% CNTs has no influence on the AC resistivity of UHPC; in contrast, the AC resistivity of UHPC with 3.0% NTs is 36.3% lower than that of UHPC without fillers. This phenomenon is consistent with the conclusion of DC resistivity. The AC resistivity of UHPC is reduced by one and two orders of magnitude, respectively, due to the addition of 0.15 vol% and 0.3 vol% SSWs. Meanwhile, the incorporation of 0.5% CNTs or 3% NTs enables the AC resistivity of UHPC with 0.15 vol% SSWs to decrease by 58.2% and 41.0%, respectively, representing the synergistic conduction effect of micro-fibers and nanofillers. However, the AC resistivity of UHPC with 0.3 vol% SSWs is increased by 7.8% and 88.0%, respectively. This again shows that the overlapped conductive network of SSWs plays a decisive role in the resistivity of composites with 0.3 vol% SSWs and is hardly affected by the addition of CNTs. The pathway composed of hybrid SSWs and NTs is

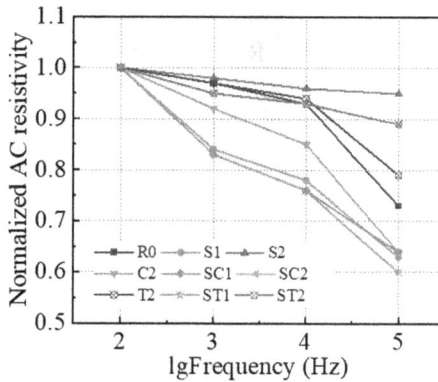

Figure 9.13 Normalized AC resistivity of SSWs-engineered multifunctional UHPC incorporating CNTs/NTs at the curing age of 28 days

less conductive compared with that formed by overlapped SSWs due to the semiconductor characteristics of NTs.

The normalized AC resistivity of UHPC at the curing age of 28 days is demonstrated in Figure 9.13. It can be seen that the AC resistivity of UHPC with 0.3 vol% SSWs is only slightly affected by frequency, also verifying that the SSWs have overlapped with each other to form a conductive pathway. Due to the increase of micro-capacitance elements in the conductive pathway formed by hydration products, the normalized AC resistivity of CNTs/NTs and SSWs-engineered UHPC is smaller than that of UHPC with SSWs alone. The AC resistivity of UHPC with NTs is more prone to be influenced by frequency compared with that of UHPC with CNTs. It can therefore be concluded that when the cement matrix occupies a high proportion of the conductive path, the resistivity is more susceptible to frequency.

According to these test results, the conductive pathway of UHPC composites can be established, as shown in Figure 9.14. It is worth noting that the capacitance of composites (Q) was not measured in this study, and it was only used to characterize the composition of the conductive pathway. It can be seen that the conductive pathway of UHPC without fillers includes three parts, pore solution, cement matrix, and electrodes (Figure 9.14(a)) and is dominated by ion conduction. When NTs/CNTs/0.15 vol% SSWs are added to UHPC, pore solution, cement matrix, CNTs/NTs/SSWs, and electrodes are connected in series to form the conductive pathway of composites, as demonstrated in Figure 9.14(b). The ion conduction and tunneling effect work together to contribute to the conductive properties. As shown in Figure 9.14(c), the conductive pathway of UHPC with hybrid 0.3 vol% SSWs and CNTs/NTs is mainly composed of pore solution, CNTs/NTs, and SSWs together.

(a)

Pore solution
resistance   Cement matrix   EIS of electrode

(b)

Pore solution   Cement matrix   MWCNTs/   EIS of electrode
resistance   NTs/SSWs

(c)

Pore solution   MWCNTs/   SSWs and   EIS of electrode
resistance   NTs and   cement matrix
                    cement matrix

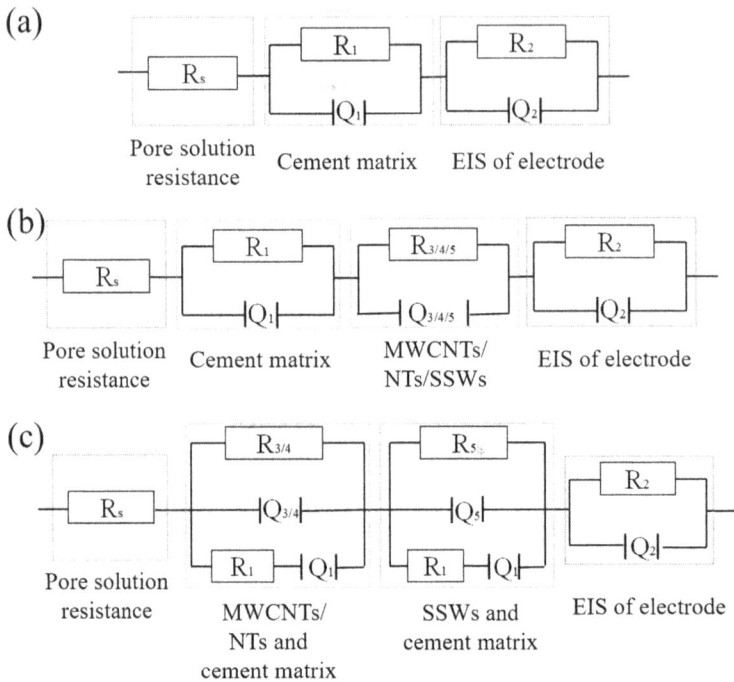

Figure 9.14 Conductive pathway of SSWs-engineered multifunctional UHPC incorporating CNTs/NTs. (a) UHPC without fillers; (b) UHPC with 0.5% CNTs/3.0% NTs/0.15 vol% SSWs; (c) UHPC with hybrid 3.0 vol% SSWs and CNTs/NTs R_s is the pore solution resistance, R_1 represents the charge diffusion impedance of the cement matrix, R_2 indicates the contact impedance between the electrode and the cement matrix, R_3 represents the resistance of CNTs, R_4 is the resistance of NTs, and R_5 is the resistance of SSWs. EIS means electrochemical impedance spectroscopy. Q_1, Q_2, Q_3, Q_4, and Q_5 are the double-layer capacitance formed by hydration of C-S-H gels and the interfacial double-layer capacitance between the electrode and the cement matrix, between CNTs and the cement matrix, between NTs and the cement matrix, and between SSWs and the cement matrix, respectively

9.3.2 Self-sensing properties

9.3.2.1 Under cyclic compressive load

Experimental results show that all UHPC composites have stable and repetitive self-sensing properties under cyclic compressive loading. R0, ST1, and ST2 are taken as examples to analyze the stress/strain–fractional change in resistivity (FCR) time history relationships, as shown in Figure 9.15. It can be seen from Figure 9.15 that with the increase of stress/strain, FCR displays a decreasing trend; on the contrary, FCR increases with decreasing stress/strain, showing significant piezoresistive properties. For UHPC without fillers, the absolute value of FCR is only 0.54% under the cyclic

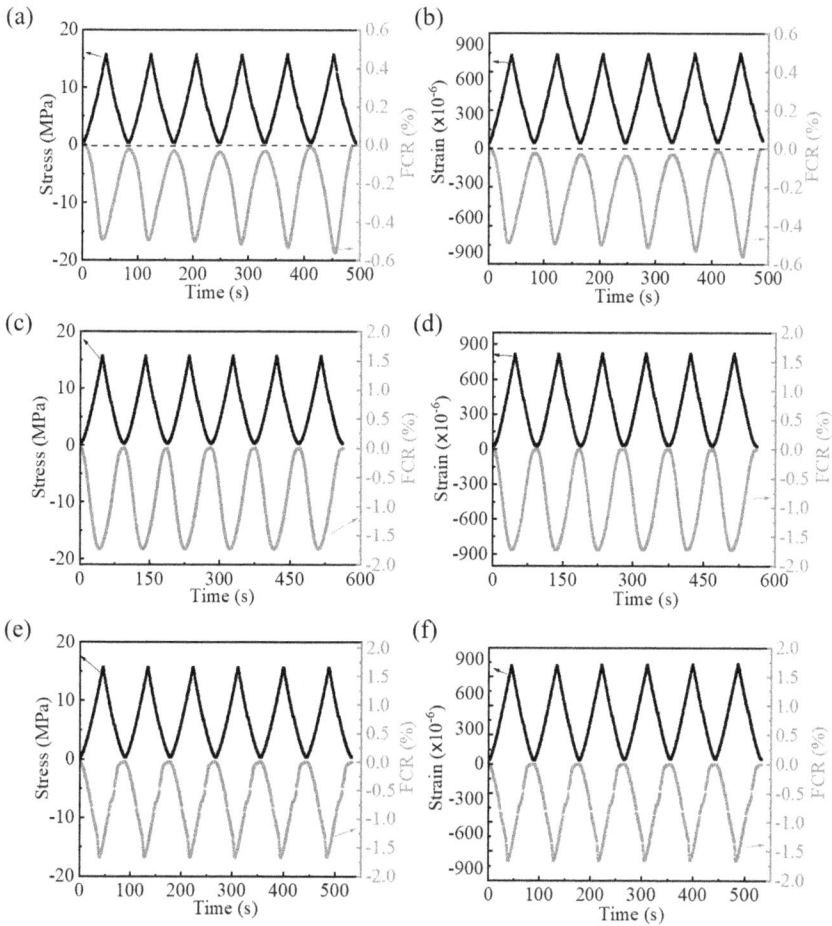

Figure 9.15 Stress/strain–FCR time history curves of SSWs-engineered multifunctional UHPC incorporating CNTs/NTs under cyclic compressive load. (a) R0, stress–FCR; (b) R0, strain–FCR; (c) ST1, stress–FCR; (d) ST1, strain–FCR; (e) ST2, stress–FCR; (f) ST2, strain–FCR

compressive load of 25 kN. The FCR mainly results from the distance change of capacitor plates caused by the elastic deformation of hydration products. When hybrid 0.15 vol% SSWs and 3.0% NTs are added to composites, the absolute value of FCR can reach 1.68%, 95.3% higher than that of UHPC with 0.15 vol% SSWs alone. Meanwhile, the absolute value of FCR for UHPC with 0.3 vol% SSWs is increased by 22.1% due to the addition of 3.0% NTs. The addition of 0.5% CNTs increases the absolute value of FCR for UHPC with 0.15 vol% SSWs by 59.3% and decreases the absolute value of FCR for UHPC with 0.3 vol% SSWs by 11.8%. This indicates that when the content of functional fillers is in the middle of the percolation threshold zone, the synergistic effect of multiscale conductive fillers on the self-sensing property is more obvious.

The stress/strain sensitivity of UHPC composites refers to the fractional change in electrical resistivity per unit stress/strain. The stress/strain sensitivity of UHPC under cyclic compressive loading is listed in Table 9.8. It can be found that the strain/stress sensitivity of UHPC increases with increasing SSWs content, and 83.5% and 173.7% increments, respectively, can be obtained for strain sensitivity due to the addition of 0.15 vol% and 0.30 vol% SSWs. Meanwhile, the strain sensitivity of UHPC with 0.5% CNTs/3.0% NTs is increased by 56.4% and 54.6%, respectively, compared with that of UHPC without fillers. The similar modification effect can be attributed to the low content of CNTs and the semiconductor characteristics of NTs. When hybrid 0.15 vol% SSWs and 0.5% CNTs are added, the strain sensitivity of UHPC is improved by 145.1% compared with that of UHPC without fillers. Furthermore, the strain sensitivity is 56.7% and 33.6%, respectively, higher than that of composites with 0.5% CNTs and composites with 0.15 vol% SSWs, again verifying the synergistic effect of micro-fibers and nanofillers. Although the strain sensitivity of UHPC with 0.3 vol% SSWs and 0.5% CNTs is 62.0% higher than that of UHPC with 0.5% CNTs, it is decreased by 7.4% compared with that of UHPC with 0.3 vol% SSWs, verifying that SSWs have overlapped with each other when the dosage is 0.3 vol%. The strain sensitivity of UHPC with 0.15 vol% SSWs and 3.0% NTs is increased by 65.7% and 96.6% compared with that of composites with 0.15 vol% SSWs and composites with 3.0% NTs, respectively. That is to say, the synergistic effect of 0.15 vol% SSWs and NTs is better than that of 0.15 vol% SSWs and CNTs, mainly resulting from the high content of NTs. The strain sensitivity of UHPC with 0.30 vol% SSWs and 3.0% NTs is 92.3% and 8.6% higher that of composites with 3.0% NTs and composites with 0.30 vol% SSWs, respectively. This is caused by the semiconductor characteristics of NTs. UHPC with 0.15 vol% SSWs and

Table 9.8 Stress/strain sensitivity of SSWs-engineered multifunctional UHPC incorporating CNTs/NTs under cyclic compressive loading

Specimens	|FCR| (%)	Stress sensitivity (%/MPa)	Strain sensitivity
R0	0.54 ± 0.02	0.035 ± 0.001	6.77 ± 0.33
C2	0.82 ± 0.04	0.052 ± 0.003	10.6 ± 0.7
T2	0.93 ± 0.03	0.06 ± 0.002	10.5 ± 0.5
S1	0.86 ± 0.04	0.055 ± 0.003	12.4 ± 0.6
S2	1.36 ± 0.02	0.087 ± 0.01	18.5 ± 1.0
SC1	1.37 ± 0.12	0.089 ± 0.008	16.6 ± 0.9
SC2	1.20 ± 0.09	0.077 ± 0.006	17.2 ± 0.9
ST1	1.68 ± 0.13	0.108 ± 0.008	20.6 ± 1.1
ST2	1.66 ± 0.12	0.107 ± 0.008	20.1 ± 0.7

3.0% NTs has the optimal self-sensing properties under cyclic compressive loading.

It is worth noting that the strain sensitivity of UHPC in the elastic regime shown in Table 9.8 is lower than that in Wen and Chung [31], and this phenomenon may be attributable to the fact that the matrix in Wen and Chung [31] is cement paste and cement mortar. Compared with cement paste, the use of fine aggregate reduces the elastic deformation and blocks the formation of the conductive pathway.

9.3.2.2 Under monotonic compressive loading

The stress/strain–FCR curves of CNTs/NTs and SSWs-engineered multifunctional UHPC under monotonic compressive load are shown in Figure 9.16. Meanwhile, the maximum absolute values of FCR and the corresponding strain sensitivity are listed in Table 9.9. It can be seen from Figure 9.16(a) that with the increase of stress/strain, the FCR of UHPC without fillers and with 0.5% CNTs/3.0% NTs presents a linear decreasing trend. Due to the good conductivity of CNTs, the maximum absolute value of FCR for UHPC with 0.5% CNTs is increased by 44.1% compared with that of UHPC without fillers. The FCR of UHPC with 3.0% NTs is

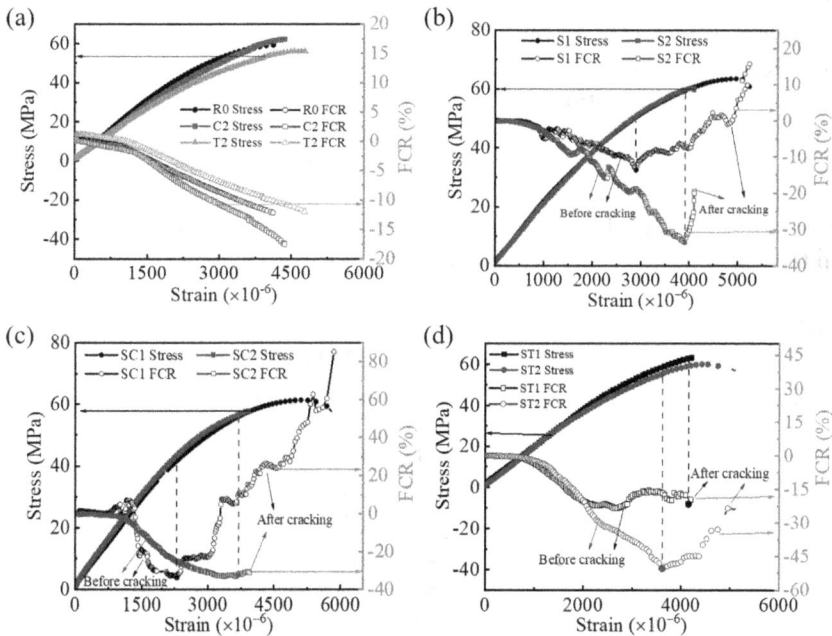

Figure 9.16 Stress/strain–FCR curves of SSWs-engineered multifunctional UHPC incorporating CNTs/NTs under monotonic compressive load. (a) R0, C2, and T2; (b) S1 and S2; (c) SC1 and SC2; (d) ST1 and ST2

close to that of UHPC without fillers and is mainly determined by cement matrix deformation. The FCR of UHPC with SSWs first decreases and then increases with the increase of stress/strain, as shown in Figure 9.16(b), presenting in-situ monitoring properties. The inflection points of FCR can be considered as the sign of initiation of microcracks. The maximum absolute value of FCR for UHPC with 0.30 vol% SSWs is 175.3% higher than that of UHPC without fillers. As displayed in Figure 9.16(c) and (d), the FCR of UHPC with hybrid SSWs and CNTs/NTs has the same change trend as that of UHPC with SSWs. The addition of CNTs increases the maximum absolute value of FCR of UHPC with 0.15 vol% SSWs by 137.9% but has little effect on the FCR absolute value of UHPC with 0.30 vol% SSWs. The maximum absolute value of FCR of UHPC with 0.15 vol% SSWs and 0.30 vol% SSWs is increased by 61.3% and 51.2%, respectively, due to the addition of 3.0% NTs.

It can be found in Table 9.9 that the strain sensitivity of UHPC with 0.5% CNTs is 36.0% higher than that of UHPC without fillers due to the excellent electrical conductivity of CNTs. In contrast, incorporating 3.0% NTs has no beneficial effect on the strain sensitivity. Adding 0.15 vol% and 0.3 vol% SSWs increases the strain sensitivity of UHPC by 35.3% and 191.5%, respectively. The strain sensitivity of UHPC with hybrid 0.15 vol% SSWs and 0.5% CNTs is 379.9%, 252.9%, and 254.7% higher than that of UHPC without fillers, UHPC with 0.5% CNTs, and UHPC with 0.15 vol% SSWs, respectively. This is an impressive display of the synergistic effect of micro-fibers and nanofillers and also verifies that fillers with content in the percolation threshold zone can endow UHPC with high sensitivity. When hybrid 0.3 vol% SSWs and 0.5% CNTs are added, the strain sensitivity of UHPC is 230.2%, 142.8%, and 13.3.0% higher than that of UHPC without fillers, UHPC with 0.5% CNTs, and UHPC with 0.3 vol% SSWs, respectively. This indicates that 0.3 vol% SSWs plays a decisive role in the self-sensing properties under monotonic loading. The increments are

Table 9.9 FCR and strain sensitivity of SSWs-engineered multifunctional UHPC incorporating CNTs/NTs under monotonic compressive load

Group	|FCR| (%)	Strain sensitivity corresponding to maximum |FCR|
R0	12.16 ± 1.02	29.5 ± 1.3
C2	17.52 ± 1.17	40.1 ± 2.0
T2	11.98 ± 0.87	25.1 ± 1.0
S1	13.60 ± 2.01	39.9 ± 1.6
S2	33.47 ± 1.54	86.0 ± 3.1
SC1	32.36 ± 1.72	142 ± 5
SC2	32.86 ± 2.02	97.4 ± 3.5
ST1	21.93 ± 1.13	52.6 ± 1.6
ST2	50.62 ± 2.69	140 ± 6

78.4%, 109.9%, and 31.9% for the strain sensitivity of UHPC with 0.15 vol% SSWs and 3.0% NTs, compared with UHPC without fillers, UHPC with 3.0% NTs, and UHPC with 0.15 vol% SSWs, respectively. Meanwhile, 374.7%, 458.5%, and 62.8% increments, respectively, can be obtained for UHPC with 0.30 vol% SSWs and 3.0% NTs. That is to say, 3.0% NTs have an excellent synergistic effect with 0.3 vol% SSWs for self-sensing properties under monotonic compressive loading, resulting from the different sensing mechanisms of hybrid fillers.

9.3.3 Synergistic conductive mechanisms

On the basis of the preceding test results, Figure 9.17 is plotted to reveal the conductivity and self-sensing mechanisms of CNTs/NTs and SSWs-engineered multifunctional UHPC. As shown in Figure 9.17, ionic conduction plays the decisive role in the conductive pathway of UHPC with 0.5% CNTs/3.0% NTs, and the filler content is at the beginning of the percolation threshold zone. The self-sensing properties of UHPC under cyclic/monotonic compressive load mainly come from the change of capacitance, indicating that the change of FCR is small under low load amplitude. Because CNTs/NTs can be regarded as capacitance plates, the capacitance plate distance is shortened and the relative dielectric constant is decreased under compressive loading, leading to the decrease of resistivity. Although the content of NTs is larger than that of CNTs, the strain sensitivity of UHPC with NTs is comparable to that of UHPC with CNTs, which can be

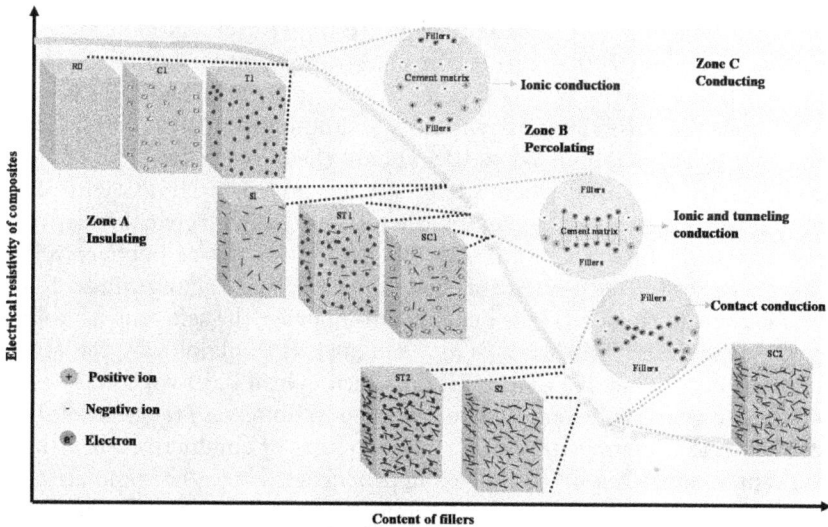

Figure 9.17 Conductivity and self-sensing mechanisms of SSWs-engineered multifunctional UHPC incorporating CNTs/NTs

attributed to the high conductivity of CNTs (100 S/cm). Under monotonic compressive load, the change of intrinsic resistivity of CNTs/NTs makes an important contribution to the self-sensing properties of UHPC. This can be verified by the phenomenon that the maximum absolute value of FCR for UHPC with CNTs is higher than that for UHPC with NTs, as shown in Table 9.9.

The filler contents of UHPC with 0.15 vol% SSWs, UHPC with hybrid 0.15 vol% SSWs and CNTs, and UHPC with hybrid 0.15 vol% SSWs and NTs are in the middle of the percolation threshold zone. The change of capacitance and tunneling resistance between fillers work together to contribute to the self-sensing properties of UHPC composites under cyclic compressive load, such that strain sensitivity is positively related to the content of fillers. This is why the strain sensitivity of UHPC with hybrid 0.15 vol% SSWs and NTs is higher than that of UHPC with hybrid 0.15 vol% SSWs and CNTs. Under monotonic compressive load, the modification mechanisms for self-sensing properties include three aspects: the change of capacitance, the change of tunneling resistance between fillers, and the change of intrinsic resistivity of fillers, with the result that the maximum absolute value of FCR for UHPC with hybrid 0.15 vol% SSWs and CNTs is significantly higher than that of UHPC with hybrid 0.15 vol% SSWs and NTs. This means that the sensitivity of UHPC can be improved through adjusting filler content and hybrids using fillers with different conductivity.

The contact state between fillers is the dominant factor in the self-sensing properties of UHPC with 0.30 vol% SSWs. When hybrid 0.5% CNTs and 0.30 vol% SSWs are added to composites, the conductive pathway is more complete and is not easily changed under cyclic/monotonic compressive load. This is the reason why the strain sensitivity of UHPC with 0.30 vol% SSWs and CNTs is smaller than that of UHPC with 0.30 vol% SSWs alone. Because the resistivity of NTs (about 6.0×10^3 $\Omega \cdot cm$) is larger than that of SSWs (about 7.1×10^{-5} $\Omega \cdot cm$), the existing conductive pathway is partly destroyed when 3.0% NTs are added to UHPC with 0.30 vol% SSWs. This can be proved by the test results of electrical resistivity, as shown in Figures 9.12 and 9.13. The suitable distance between SSWs makes it easier for the electrical resistivity of UHPC to change under loading, i.e. the addition of NTs is beneficial to improve the self-sensing ability of UHPC with 0.30 vol% SSWs alone. This also explains why the maximum absolute value of FCR for UHPC with hybrid 0.30 vol% SSWs and NTs under monotonic loading can reach up to 50.62%. Previous research has shown that composites with a high content of conductive fillers have excellent strength but low self-sensing properties due to the complete conductive path [12]. The addition of semiconductor fillers has the potential to solve the contradiction between mechanical performance and damage monitoring.

According to the theory of general effective media (GEM), the resistivity of CNTs/NTs and SSWs-engineered multifunctional UHPC is closely related to the content, conductivity, shape, and distribution of each phase of the media [32]. Assumed that the conductive fillers are uniformly distributed in the UHPC, the electrical resistance of each conductive path may be reduced to a series of the resistance of fillers and the resistance of the cement matrix between fillers, as shown in Eq. (9.20):

$$R_i = (\eta_i - 1)R_m + \eta_i R_C \approx \eta_i (R_m + R_C) \tag{9.20}$$

where R_i is the electrical resistance of each conductive path, ηI_i is the amount of conductive fillers in each pathway, R_m is the electrical resistance of the cement matrix between fillers, and R_C is the electrical resistivity of the fillers. As the resistivity of the fillers used in this chapter is much lower than that of the cement matrix, the expression of Eq. (9.20) can be simplified as Eq. (9.21):

$$R_i \approx \eta_i' R_m \tag{9.21}$$

where $\eta_i' > \eta_i$, η_i' is the effective amounts of fillers considering the relationship between fillers and cement matrix resistance. The electrical resistance of UHPC can be approximately considered to be composed of the resistance of numbers of conductive paths in parallel.

Based on the tunneling theory proposed by Simmons and Unterkofler [33], the electrical resistivity of the cement matrix between fillers can be expressed using Eq. (9.22):

$$R_m = \frac{8\pi h d}{3\gamma a^2 e^2} \exp(\gamma d) \tag{9.22}$$

$$\gamma = \frac{4\pi^2 \sqrt{2m\varphi}}{h} \tag{9.23}$$

where h is the Planck constant, d is the minimum distance between adjacent conductive fillers, a^2 is the effective section area, e is the electron charge, m is the mass of an electron, and φ is the barrier height between adjacent conductive fillers. According to the derivation process of Han et al. [34], the relationships between FCR and compressive stress/strain before cracking are expressed as Eqs (9.24) and (9.25) under monotonic compressive loading.

$$FCR = (A\sigma + B\sigma^2) \times 100\% \tag{9.24}$$

Table 9.10 Strain/stress–FCR curve fitting parameters for hybrid CNTs/NTs and SSWs-engineered multifunctional UHPC incorporating CNTs/NTs under monotonic compressive load (as shown in Figure 9.16 in elastic deformation stage)

Group		A	B	R^2
Strain–FCR	R0	−0.00121	-5.121×10^{-7}	0.98694
	C2	−0.00169	-5.582×10^{-7}	0.99483
	T2	−0.00055	-4.787×10^{-7}	0.96344
	S1	−0.00129	-9.788×10^{-7}	0.98651
	S2	−0.00199	-9.788×10^{-7}	0.99537
	SC1	−0.01355	-1.441×10^{-5}	0.84545
	SC2	−0.00273	-2.741×10^{-6}	0.97104
	ST1	−0.00220	-3.17×10^{-6}	0.97287
	ST2	−0.00160	-3.697×10^{-6}	0.98479

Group		A'	B'	R^2
Stress–FCR	R0	−0.00337	−0.00385	0.99178
	C2	−0.00476	−0.00413	0.99815
	T2	−0.10031	−0.00543	0.98765
	S1	−0.00300	−0.00444	0.98316
	S2	−0.06824	−0.00982	0.99297
	SC1	−0.85107	−0.04408	0.85344
	SC2	−0.14852	−0.01521	0.98829
	ST1	−0.02874	−0.01139	0.97923
	ST2	−0.22300	−0.02053	0.99417

$$\text{FCR} = \left(A'\varepsilon + B'\varepsilon^2 \right) \times 100\% \qquad (9.25)$$

where σ is the compressive stress, ε is the compressive strain, and A, B, A' and B' are parameters. The corresponding fitting results are displayed in Table 9.10. The self-sensing properties of UHPC with CNTs/NTs and SSWs can be accurately explained by GEM and tunneling theory. When the UHPC with hybrid CNTs/NTs and SSWs is used as a sensor, the relationships between stress/strain and FCR established in this chapter can play the role of calibration curves, providing guidance for practical application of composites.

9.4 SUMMARY

In this chapter, the mechanical properties (including compressive/flexural strength, flexural strain–stress/load–displacement curves, and flexural toughness), electrical resistivity, and self-sensing properties under cyclic and monotonic compressive load of SSWs-engineered multifunctional UHPC incorporating nanofillers were introduced. Meanwhile, the flexural toughening calculation models and the modification mechanisms for these properties were presented. The main conclusions are as follows:

1) The toughening effect of hybrid SSWs and NTs on UHPC is better than that of SSWs and CNTs when the P•I type of cement is used. Relative changes of 171.6%, 572.1%, 605.3%, 96.9%, and 53.2%, respectively, can be achieved for flexural strength, flexural toughness based on load–displacement curves, flexural toughness based on strain-stress curves, first-cracking flexural strength, and the ratio between flexural and compressive strength of UHPC due to the addition of 1.2 vol% SSWs and 1.5 wt% NTs. When the P•O type of cement is used, these parameters for UHPC incorporating hybrid 1.2 vol% SSWs and 0.25% CNTs are 80.7%, 512.3%, 621.2%, 79.0%, and 29.4% higher, respectively, than that of UHPC without fillers.

2) The calculation model of flexural strength established on the basis of composite theory can accurately describe the synergistic toughening effect of SSWs and CNTs/NTs. The numbers of effective SSWs and the bond strength between SSWs and UHPC for UHPC with 1.2 vol% SSWs and 1.5% NTs are 19.1% and 44.8% higher, respectively, than that of UHPC with 1.2 vol% SSWs. At the nano scale, the addition of CNTs/NTs reduces original flaws and improves bond strength between SSWs and the UHPC matrix through the nano-core effect. At the micro scale, the widely distributed core-shell elements and SSWs work together to inhibit the initiation and propagation of microcracks by pinning and bridging effects, endowing UHPC with high bending strength and toughness.

3) It is shown that 0.3 vol% SSWs have overlapped to form a complete conductive network in UHPC, leading to the electrical resistivity being decreased by three orders of magnitude and being barely affected by the addition of CNTs. However, incorporating NTs leads to an increase of 85.4% and 88.0%, respectively, for the DC and AC electrical resistivity of UHPC with 0.3 vol% SSWs.

4) Under cyclic compressive load, the strain sensitivity of UHPC with 0.15 vol% SSWs and 3.0% NTs is 65.7% higher than that of UHPC with 0.15 vol% SSWs. The change of capacitance and tunneling resistance between fillers work together to contribute to the sensing properties of composites incorporating NTs. The maximum absolute value of FCR for UHPC with 0.15 vol% SSWs and 0.5% CNTs under monotonic compressive load is increased by 179.0% compared with that of UHPC with 0.15 vol% SSWs, and the change of the intrinsic resistivity of fillers plays an important role in this process. The maximum

absolute value of FCR reaches 50.6% under monotonic compressive load when hybrid 0.3 vol% SSWs and 3.0% NTs are added, 51.6% higher than that of UHPC with 0.3 vol% SSWs. This results from the semiconductor characteristics of NTs and the increased proportion of the interfacial layer in the conductive pathway caused by the nano-core effect of NTs.

REFERENCES

1. S. Dong, W. Zhang, D. Wang, B. Han. Bending toughness and calculation model of ultra-high performance concrete with hybrid micro- and nanofillers, *Journal of Materials in Civil Engineering.* 33(8) (2021) 4021201.
2. S. Dong, W. Zhang, D. Wang, X. Wang, B. Han. Modifying self-sensing cement-based composites through multiscale composition, *Measurement Science and Technology.* 32(7) (2021) 074002.
3. L. Li, Q. Zheng, B. Han, J. Ou. Fatigue behaviors of graphene reinforcing concrete composites under compression, *International Journal of Fatigue.* 151 (2021) 106354.
4. L. Li, X. Wang, H. Du, B. Han. Comparison of compressive fatigue performance of cementitious composites with different types of carbon nanotube, *International Journal of Fatigue.* 165 (2022) 107178.
5. B. Han, L. Zhang, J. Ou. *Smart and Multifunctional Concrete Toward Sustainable Infrastructures,* Springer, 2017.
6. B. Han, L. Zhang, S. Sun, X. Yu, X. Dong, T. Wu, J. Ou. Electrostatic self-assembly carbon nanotube/nano carbon black composite fillers reinforced cement-based materials with multifunctionality, *Composites Part A: Applied Science and Manufacturing.* 79 (2015) 103–115.
7. X. Wang, S. Ding, A. Ashour, L. Qiu, Y. Wang, B. Han, J. Ou. Improving bond of fiber-reinforced polymer bars with concrete through incorporating nanomaterials, *Composites Part B Engineering.* 239 (2022) 109960.
8. J. Wang, S. Dong, S. Pang, X. Yu, B. Han, J. Ou. Tailoring anti-impact properties of ultra-high performance concrete by incorporating functionalized carbon nanotubes, *Engineering.* 18 (2022) 232–245.
9. S. Ding, Y. Xiang, Y. Ni, V. Thakur, X. Wang, B. Han, J. Ou. In-situ synthesizing carbon nanotubes on cement to develop self-sensing cementitious composites for smart high-speed rail infrastructures, *Nano Today.* 43 (2022) 101438.
10. S. Ding, X. Wang, L. Qiu, Y. Ni, X. Dong, Y. Cui, A. Ashour, B. Han, J. Ou. Self-sensing cementitious composites with hierarchical carbon fiber-carbon nanotube composite fillers for crack development monitoring of a maglev girder, *Small.* 19(9) (2023) 2206258.
11. J. Luo, Z. Duan, T. Zhao, Q. Li. Hybrid effect of carbon fiber on piezoresistivity of carbon nanotube cement-based composite, *Advanced in Materials Research.* 143–144 (2010) 639–643.
12. F. Azhari, N. Banthia. Carbon fiber/nanotube cement-based sensors for structural health monitoring, Structural Health Monitoring, 2009: From System Integration to Autonomous Systems-Proceedings of the 7th International Workshop on Structural Health Monitoring. IWSHM 2009, (2009) 1013–1020.

13. I. You, D. Yoo, S. Kim, M. Kim, G. Zi. Electrical and self-sensing properties of ultra-high-performance fiber-reinforced concrete with carbon nanotubes, *Sensors.* 17(11) (2017) 2481–2499.

14. S. Ding, S. Dong, A. Ashour, B. Han. Development of sensing concrete: Principles, properties and its applications, *Journal of Applied Physics.* 126(24) (2019) 241101.

15. P. Balaguru, R. Narahari, M. Patel. Flexural toughness of steel fiber reinforced concrete, *ACI Materials Journal.* 89(6) (1992) 541–546.

16. N. Banthia, A. Dubey. Measurement of flexural toughness of fiber reinforced concrete using a novel technique-part 2: Performance of various composites, *ACI Materials Journal.* 97(1) (2000) 3–11.

17. S. Bentur. *Mindess, Fibre Reinforced Cementitious Composites* (2nd Edition), Taylor and Francis, 2007.

18. JSCE. Test method for bending strength and bending toughness of steel fiber reinforced concrete. In Standard Specification for Concrete Structures. Test Methods and Specifications; JSCE: Tokyo, Japan

19. Z. Hajar, D. Lecointre, A. Simon, J. Petitjean. Design and construction of the world first ultra-high performance concrete road bridges, Proceedings of the International Symposium on Ultra High Performance Concrete, Kassel, Germany (2004) 39–48.

20. B. Han, L. Zhang, S. Zeng, S. Dong, X. Yu, Y. Yang, J. Ou. Nano-core effect in nano-engineered cementitious composites, *Composites Part A: Applied Science and Manufacturing.* 95 (2017) 100–109.

21. X. Wang, S. Dong, A. Ashour, W. Zhang, B. Han. Effect and mechanisms of nanomaterials on interface between aggregates and cement mortars, *Construction and Building Materials.* 240 (2020) 117942.

22. J.P. Romualdi, J.A. Mandel. Tensile strength of concrete affected by uniformly distributed and closely spaced short lengths of wire reinforcement, *ACI Structural Journal.* 61(6) (1964) 657–672.

23. H. Krenchel. *Fibre Reinforcement, Copenhagen: Akademisk Forlay,* 1964.

24. D.J. Hannant. The effect of post cracking ductility on the flexural strength of fiber cement and fiber concrete, A Neville (ed) Fiber Reinforced Cement and Concrete, Proceedings of the RILEM Symposium, Construction Press Ltd. Lancaster. 2 (1975) 499–508.

25. S.P. Shah, V.B. Rangan. Fibre reinforced concrete properties, *Journal of the American Concrete Institute.* 68(2) (1971) 126–135.

26. S. Dong, D. Zhou, A. Ashour, B. Han, J. Ou. Flexural toughness and calculation model of super-fine stainless wire reinforced reactive powder concrete, *Cement and Concrete Composites.* 104 (2019) 103367.

27. M.R. Pigott. Theoretical estimation of fracture of fibrous composites, *Journal of Materials Science.* 5(8) (1970) 669–675.

28. D. Kim, S. Park, G. Ryu, K. Koh. Comparative flexural behavior of hybrid ultra high performance fiber reinforced concrete with different macro fibers, *Construction and Building Materials.* 25(11) (2011) 4144–4155.

29. A. Hawreen, B. Alexandre, G. Mafalda. Mechanical behavior and transport properties of cementitious composites reinforced with carbon nanotubes, *ASCE Journal of Materials in Civil Engineering.* 30(10) (2018) 04018257.

30. B. Han, Z. Li, L. Zhang, S. Zeng, X. Yu, B. Han, J. Ou. Reactive powder concrete reinforced with Nano SiO_2-coated TiO_2, *Construction and Building Materials.* 148 (2017) 104–112.

31. S. Wen, D.D.L. Chung. A comparative study of steel- and carbon-fibre cement as piezoresistive strain sensors, *Advances in Cement Research*. 15(3) (2003) 119–128.
32. D. Stauffer, A. Aharony. *Introduction to Percolation Theory*, Taylor and Francis, 1985.
33. J.G. Simmons, G.J. Unterkofler. Potential barrier shape determination in tunnel junctions, *Journal of Applied Physics*. 34(6) (1963) 1828–1830.
34. B. Han, S. Ding, X. Yu. Intrinsic self-sensing concrete and structures: A review, *Measurement*. 59 (2015) 110–128.

Chapter 10

Stainless steel wires-engineered multifunctional ultra-high performance concrete incorporating steel fibers

10.1 INTRODUCTION

Steel fibers (SFs) with a diameter larger than 0.16 mm can be used to improve the toughness of conventional ultra-high performance concrete (UHPC) by exerting a bridge effect across macroscopic cracks, but cannot modify the electrically conductive properties and cannot endow UHPC with self-sensing characteristics [1–5]. Meanwhile, the initial microcracks under loading propagate rapidly, and the multiple-cracking failure mode under flexural/tensile load is still not available in UHPC when the total SFs content is 2.0 vol% [6–9]. This can be attributed to the fact that SFs have no significant modification effect on the UHPC matrix and cannot form a widely distributed network at low content levels [7, 10, 11]. A high content of fibers seriously reduces the workability, increases the density, and raises the cost of UHPC [12–15]. An important way to realize the sustainable development of UHPC would be to endow it with multiple-cracking properties at a low content level of SFs and improve its conductivity and self-sensing properties [2].

The research results in Chapters 2–9 show that stainless steel wires (SSWs) with a micro diameter are arranged in a widely distributed overlapping network in UHPC at low contents, with the result that the microstructural compactness of UHPC matrix is improved, the microcracks in UHPC matrix induced by hydration heat temperature gradient, shrinkage, and initial load can be effectively controlled, and the conductivity and self-sensing properties of UHPC are improved. Furthermore, the formation of a complete network of SSWs is expected to exert a favorable influence on the fibers' distribution. SSWs have potential to simultaneously enhance the UHPC matrix and form a conductive network synergistically with SFs to optimize the flexural properties of UHPC and endow UHPC with high electrically conductive and self-sensing properties.

Hence, SSWs-engineered multifunctional UHPC incorporating SFs is presented in this chapter, and the flexural (including flexural strain–stress curves, flexural strength, flexural toughness, complementary energy), electrical, and self-sensing properties of SSWs-engineered multifunctional UHPC incorporating SFs are introduced. In addition, the synergistic

DOI: 10.1201/9781003276357-10

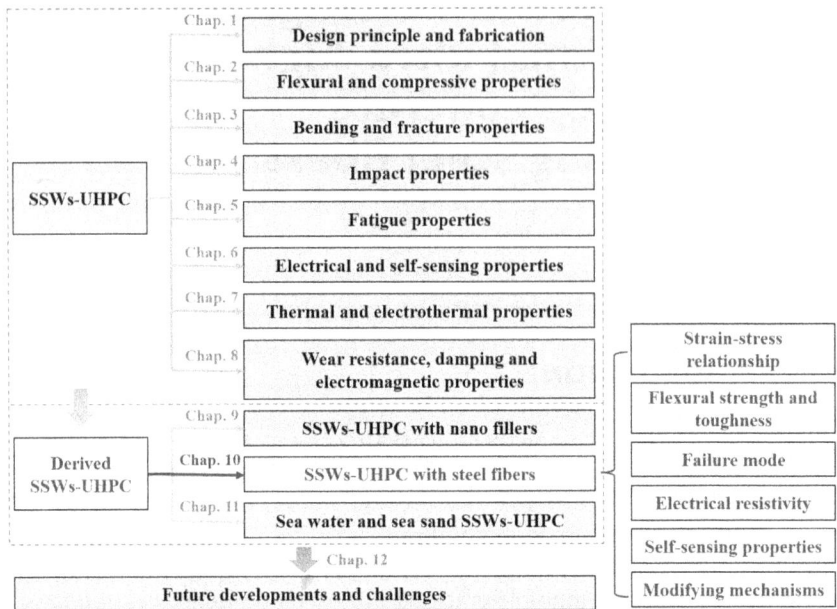

Figure 10.1 Main contents of Chapter 10

modification mechanisms of SFs and SSWs on the flexural cracking process and the self-sensing properties of UHPC are presented. The diameter of SSWs used in this chapter is 20 μm and the length is 10 mm. The main contents of this chapter are shown in Figure 10.1.

10.2 FLEXURAL PROPERTIES OF SSWS-ENGINEERED MULTIFUNCTIONAL UHPC INCORPORATING STEEL FIBERS

10.2.1 Flexural strain–stress curves

The flexural strain–stress curves before initial macrocracking were used to characterize the cracking resistance of the UHPC matrix. The shading in Figure 10.2 represents the range of curves obtained by different specimens with the same mixing ratio. Figure 10.2(a) shows that there are two stages, the elastic stage and the propagation stage, of macrocracks on the flexural strain–stress curves of SSWs-engineered multifunctional UHPC (marked W0, W0.2, and W0.4 according to the volume fraction of SSWs). As demonstrated in Figure 10.2(b) and (c), three stages on the flexural strain–stress curves of UHPC with SFs alone (marked as S1.4, S1.6, S1.8, S2.0, S2.2, and S3.0 according to the volume fraction of SFs) or hybrid SFs and SSWs-engineered multifunctional UHPC are visible: the elastic stage, the

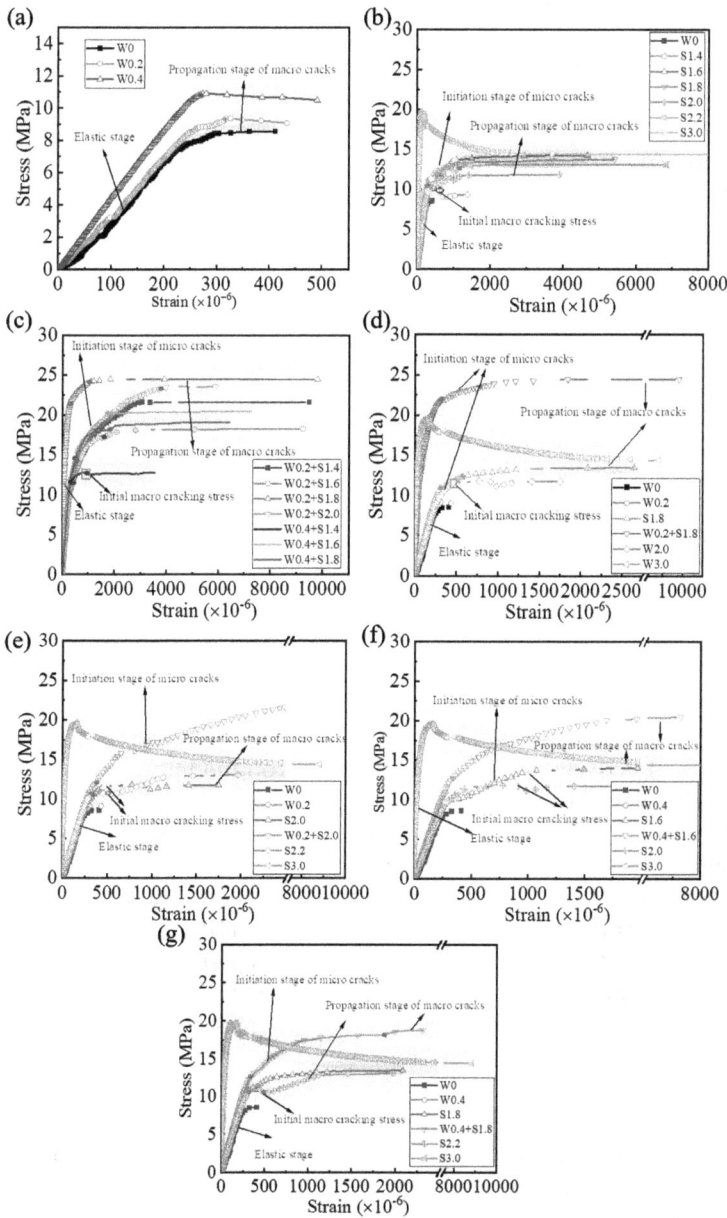

Figure 10.2 Flexural strain–stress curves of SSWs-engineered multifunctional UHPC incorporating SFs. (a) Curves for UHPC with mono SSWs alone; (b) Curves for UHPC with SFs alone; (c) Curves for UHPC with hybrid SSWs and SFs; (d) Curves for UHPC with hybrid 0.2 vol% SSWs and 1.8 vol% SFs; (e) Curves for UHPC with hybrid 0.2 vol% SSWs and 2.0 vol% SFs; (f) Curves for UHPC with hybrid 0.4 vol% SSWs and 1.6 vol% SFs; (g) Curves for UHPC with hybrid 0.4 vol% SSWs and 1.8 vol% SFs

initiation stage of microcracks, and the propagation stage of macrocracks. Figure 10.2(b) shows that when the content of SFs is less than 2.2 vol%, the slope of the initiation stage of microcracks first increases and then decreases with increasing SFs content. This indicates that the resistance to microcrack formation first increases and then decreases, resulting from the uneven dispersion of SFs and the large pores introduced in the UHPC matrix. For UHPC with 3.0 vol% SFs, the absence of the initiation stage of microcracks verifies that the resistance of SFs to microcrack generation is very limited due to the large space between them. Meanwhile, adding SSWs is conducive to increasing the slope of the elastic stage, while SFs have little effect on the elastic stage slope. UHPC with hybrid 0.2 vol% SSWs and SFs has a higher elastic stage slope compared with that of UHPC with 0.2 vol% SSWs alone. However, hybrid reinforcements do not have significant enhancement effect on the elastic stage slope of UHPC with 0.4 vol% SSWs alone. It is worthwhile to note that the elastic stage slope for UHPC with 3.0 vol% SFs and with hybrid 0.2 vol% SSWs/1.8 vol% SFs is especially higher than that for other UHPC composites. The increase of elastic slope represents an increase in the flexural-tensile modulus of composites, and it mainly results from the integrity and extensiveness of the enhancing network in UHPC. This also shows that the modification effect of hybrid 0.2 vol% SSWs and 1.8 vol% SFs on the flexural-tensile modulus is comparable to that of 3.0 vol% SFs alone. In addition, the incorporation of SFs lengthens the elastic stage of UHPC to varying degrees, but the modification effect of hybrid SSWs and SFs on the elastic stage is far superior to that of SFs alone, indicating that SSWs can effectively improve the resistance to random microcrack generation. As displayed in Figure 10.2(c)–(g), the flexural strain–stress curves of UHPC with hybrid SSWs and SFs also possess a significantly longer initiation stage of microcracks and higher initial macrocracking stress compared with those of UHPC with SFs alone at the same total volume fraction.

As the flexural load–displacement curves of UHPC develop from the linear ascending stage to the non-linear ascending stage, there is a turning point, representing the formation of macrocracks. The flexural load corresponding to this turning point is the macrocracking load, and the corresponding stress is the initial macrocracking stress displayed in Figure 10.2. The initial macrocracking stresses on flexural stress–strain curves were labeled and compared. The initial macrocracking stress for UHPC with single and hybrid reinforcements is demonstrated in Figure 10.3(a). It can be seen that with increasing SFs content, the initial macrocracking stress of UHPC first increases and then decreases due to the homogeneity of composites. When the content of SFs is 1.4 vol%, 1.6 vol%, 1.8 vol%, and 2.0 vol%, the initial macrocracking stress is increased by 103%, 9%, 69%, and 101%, respectively, due to the addition of 0.2 vol% SSWs. When 0.4 vol% SSWs are added, 21%, 23%, and 30% increments can be obtained for the initial macrocracking stress of UHPC with 1.4 vol%, 1.6 vol%, and 1.8

(a)   (b)

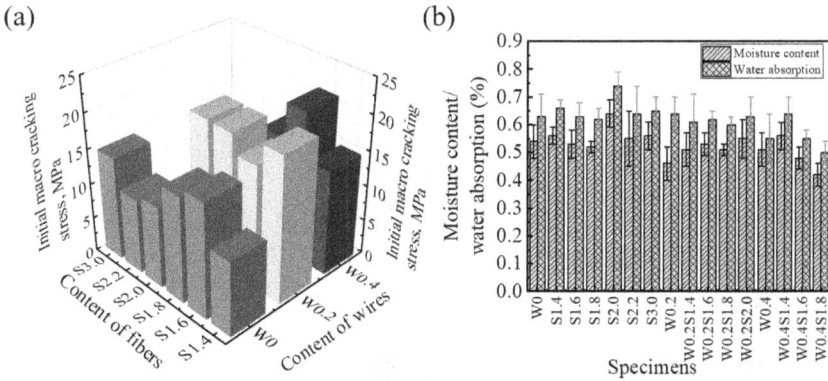

Figure 10.3 Initial macrocracking stress and moisture content and water absorption of SSWs-engineered multifunctional UHPC incorporating SFs. (a) Initial macrocracking stress; (b) Moisture content and water absorption

vol% SFs, respectively. The modification effect of hybrid SFs and SSWs on initial macrocracking stress is closely related to the influence of SSWs on the dispersion and orientation of SFs, and this will be analyzed in Section 10.3. The initial macrocracking stress of UHPC with hybrid 0.2 vol% SSWs/1.8 vol% SFs and hybrid 0.4 vol% SSWs/1.6 vol% SFs is 107%/30% and 73%/9% higher, respectively, than that of UHPC with 2.0 vol%/3.0 vol% SFs alone. The initial macrocracking stress for UHPC with hybrid 0.2 vol% SSWs/1.8 vol% SFs and hybrid 0.4 vol% SSWs/1.6 vol% SFs is increased by 102%/69% and 79%/49%, respectively, compared with that presented in the references of Niu et al. [16] and Yoo et al. [17] for UHPC with hybrid reinforcements using SFs of different lengths.

The extension of the elastic/initiation stage of microcracks and the increase of initial macrocracking stress is closely related to the modification effect of SSWs on structural compactness and their hindering effect on microcrack initiation and convergence. These can be proved by the test results of moisture content/water absorption and the optical microscopic images of SSWs in UHPC, respectively. As shown in Figure 10.3(b), when the content of SFs is larger than 1.8 vol%, the moisture content and water absorption of UHPC increase significantly. This result can be ascribed to the high content and uneven dispersion of SFs. The addition of 0.2 vol% and 0.4 vol% SSWs decreases the moisture content of UHPC without SFs by 15% and 6%, respectively. The water absorption of UHPC with 0.4 vol% SSWs is 13% lower than that of UHPC without SSWs and SFs, representing the effect of SSWs on refining pores. Meanwhile, the moisture content/water absorption of UHPC with 1.4 vol%, 1.6 vol%, 1.8 vol%, and 2.0 vol% SFs is reduced by 9%/8%, 0%/2%, 2%/3%, and 14%/15%, respectively, due to the incorporation of 0.2 vol% SSWs. The reduction caused by

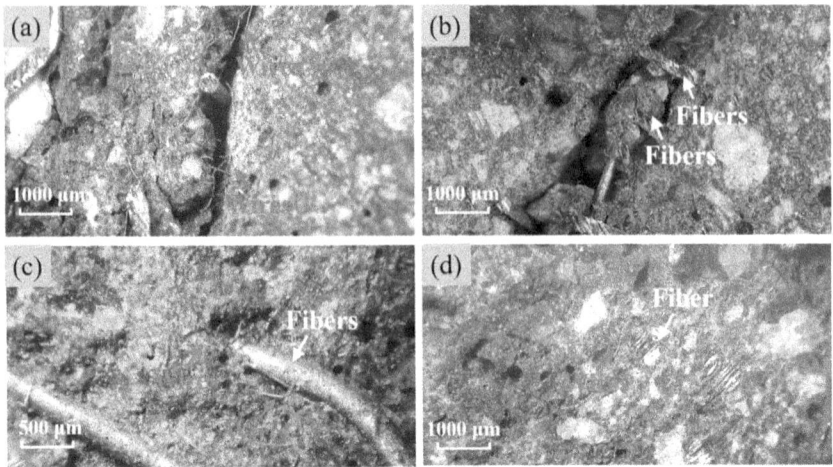

Figure 10.4 Optical microscopic images of SWs-engineered multifunctional UHPC incorporating SFs. (a) Delaying effect of SSWs on crack coalescence; (b) Hindering effect of SSWs on crack initiation; (c) Effect of SSWs on SFs settlement; (d) Effect of SSWs on SFs aggregation

the addition of 0.4 vol% SSWs to UHPC with 1.4 vol%, 1.6 vol%, and 1.8 vol% SFs is 0%/3.0%, 9.4%/12.7%, and 19%/19%, respectively. The modification effect of SSWs on macroscopic defects of the matrix for UHPC with a high content of SFs is more pronounced.

Figure 10.4 shows that SSWs can prevent microcrack initiation at earlier stages of flexural load, delay the coalescence of microcracks, and improve the ability of the UHPC matrix to resist cracking. Meanwhile, the settlement and aggregation of SFs can be reduced by adding SSWs due to their high flexibility, thus improving structural compactness and enhancing the cracking resistance of the concrete matrix.

10.2.2 Optimization effect of SSWs on the dispersion and orientation of SFs

Optical microscopy was used to observe the distribution state of SFs and SSWs in UHPC. Cross-sectional images of specimens (with an area of 40 mm × 40 mm) near the main cracks of failure specimens after flexural failure were used to obtain the dispersion and orientation factors of SFs, according to Liu et al. [18] and Meng and Khayat [19]. In order to avoid the boundary effect, the 5-mm edges of the cross section were not included in the range of the image, as shown in Figure 10.5(a). The cross section with an area of 30 mm × 30 mm was divided into five parts from top to bottom, and the images of each part were collected by a high-definition camera. The resolution of the images was high enough to identify fibers with a diameter

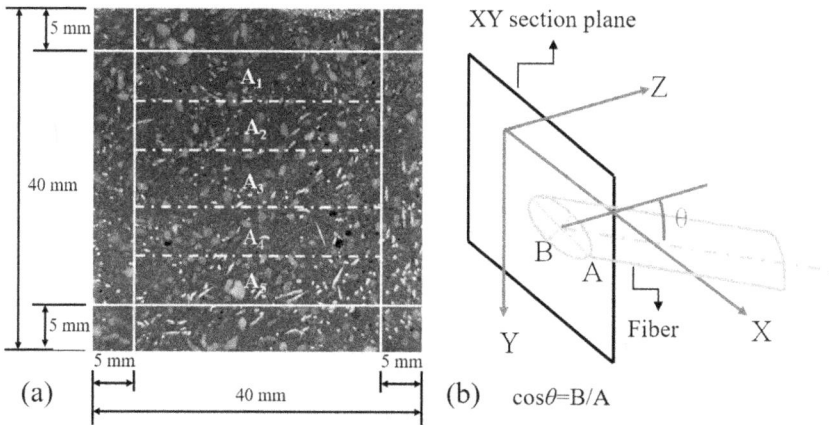

Figure 10.5 Dispersion and orientation factor calculation for SFs. (a) Mesoscale steel fiber statistical region; (b) Orientation factor

of 0.2 mm; then, Image-Pro Plus was used to process the acquired images and obtain the fiber area in the five parts through four processing steps: image binarization (which allows the SFs to be distinguished from the concrete matrix according to the brightness), image segmentation, morphologic processing, and image cleaning [18, 19]. The dispersion factor (D) for SFs was characterized by Eq. (10.1) [18]:

$$D = e^{-\sqrt{\frac{\sum_{i=1}^{5}(A_s - A_i)}{5}}}$$ (10.1)

where A_i represents fiber area for each part in one image, and A_s is the sum of SFs area for one image. In this image processing, the cross section parallel to the loading direction was defined as the XY cutting section plane (as shown in Figure 10.5(b)), and the cross section parallel to the flexural-tensile direction was defined as the Z cutting plane, so SFs distributed in the XY plane and inclined to the Z plane were elliptic. Based on the research results of Meng and Khayat [19], Zak et al. [20], and Lee et al. [21], the orientation factor ($\cos\theta$, $0 \leq \theta \leq \pi/2$) of SFs in the Z direction can be calculated from the quotient between the minor axis (B) and major axis (A) lengths of the fiber elliptic. The average value of the orientation factors of all SFs in a cross section was used as the final orientation factor of this cross section. For each specimen, 3–5 pieces of cross section were measured; there were 9–15 values for fiber distribution and the orientation factor of each mix proportion, and the average value was taken as the final value.

As shown in Table 10.1, the dispersion factor of SFs with a content of less than 2.2 vol% in UHPC presents a declining trend with the increase

Table 10.1 Dispersion/orientation factor of SFs and interfacial bond strength in SSWs-engineered multifunctional UHPC incorporating SFs

| | Dispersion factor of SFs | | | | | |
	S1.4	S1.6	S1.8	S2.0	S2.2	S3.0
W0	0.90 ± 0.01	0.82 ± 0.07	0.77 ± 0.05	0.79 ± 0.05	0.77 ± 0.05	0.83 ± 0.02
W0.2	0.83 ± 0.03	0.81 ± 0.01	0.81 ± 0.02	0.85 ± 0.01	–	–
W0.4	0.85 ± 0.02	0.85 ± 0.06	0.83 ± 0.03	–	–	–
Orientation factor of SFs						
W0	0.47 ± 0.08	0.49 ± 0.08	0.58 ± 0.08	0.46 ± 0.09	0.46 ± 0.03	0.46 ± 0.04
W0.2	0.61 ± 0.07	0.56 ± 0.04	0.56 ± 0.10	0.47 ± 0.09	–	–
W0.4	0.60 ± 0.10	0.53 ± 0.12	0.55 ± 0.07	–	–	–
Interfacial bond strength between SFs and UHPC matrix						
W0	7.40	9.08	6.30	5.96	6.50	5.59
W0.2	10.76	9.39	9.33	10.61	–	–
W0.4	7.41	10.95	8.50	–	–	–

of SFs content due to uneven dispersion. The large dispersion factor for UHPC with 3.0 vol% SFs mainly comes from the high content of SFs. The increments in the dispersion factor of SFs caused by 0.2 vol% SSWs for UHPC with 1.8 vol% and 2.0 vol% SFs are 5% and 8%, respectively. The incorporation of 0.4 vol% SSWs increases the dispersion factor of UHPC with 1.6 vol% and 1.8 vol% SFs by 3% and 8%, respectively. It can be seen that adding SSWs is particularly beneficial to modify the dispersion state of high-content SFs in UHPC, and the modification effect improves with increasing SSWs content.

It is worth noting that the orientation factors of SFs expressed in Table 10.1 have a similar change pattern to the initial macrocracking stress of composites. The values of the orientation factor for UHPC with SFs alone are close to that obtained by Romualdi and Mandel [7]. When the content of SFs is less than 1.8 vol%, the orientation factor increases with increasing SFs content. The orientation factors of SFs in UHPC with 2.0 vol%, 2.2 vol%, and 3.0 vol% SFs are almost equivalent, although they are smaller than that for UHPC with 1.8 vol% SFs. The orientation factor for UHPC with a low content of SFs is mainly affected by aggregates, while that with a high content of SFs is also influenced by interaction between SFs. Incorporating 0.2 vol% SSWs improves the orientation factor of SFs in UHPC with 1.4 vol% and 1.6 vol% SFs by 29% and 15%, respectively. Meanwhile, 26% and 8% increments can be obtained when 0.4 vol% SSWs are added. However, the orientation factor of SFs in UHPC is not improved when the SFs content is larger than 1.8 vol%. It can therefore be concluded from Figure 10.3(a) and Table 10.1 that the influence of SSWs on the dispersion factor of SFs plays a leading role in the initial macrocracking stress of UHPC with a high content of SFs, while the impact of SSWs on the orientation factor of SFs occupies the main position in the initial macrocracking stress of UHPC with a low content of SFs. The modification effect of SSWs on the dispersion and orientation of SFs mainly results from the hindering function of SSWs on the settlement and aggregation of SFs.

These results for the dispersion and orientation of SFs directly affect the structural compactness of hardened UHPC and consequently, the interfacial bond strength between the SFs and the concrete matrix. The interfacial bond strength between the SFs and the UHPC matrix (τ_{fu}) was calculated using Eq. (10.2) [8, 10]:

$$\tau_{fu} = \frac{\sigma_f d}{4\alpha\lambda_c L V_f} \tag{10.2}$$

where σ_f is the flexural strength, α is the orientation factor of fibers, and λ_c is the length and bond strength factor, equal to 1 in this study. The calculated interfacial bond strengths between SFs and the concrete matrix for hybrid SFs and SSWs-engineered multifunctional UHPC are reported in Table 10.1. It can be observed that with increasing SFs content, bond strengths between SFs and the concrete matrix for UHPC with SFs alone first increase and then decrease, relating to the increase of macroscopic defects caused by a high content of SFs. Incorporating 0.2 vol% SSWs increases the bond strengths of UHPC with 1.4 vol%, 1.6 vol%, 1.8 vol%, and 2.0 vol% SFs by 45%, 3%, 48%, and 78%, respectively. Meanwhile, the bond strengths of 1.4 vol%, 1.6 vol%, and 1.8 vol% SFs reinforced UHPC are enhanced by 0.1%, 21%, and 35%, respectively, due to the addition of 0.4 vol% SSWs. The values of calculated interfacial bond strengths are similar to that presented in Shannag et al. [22]. Meanwhile, the increase of bond strength caused by the addition of SSWs is aligned with the results for SFs hybrid systems (with a diameter of 0.19 mm and different lengths) [23, 24].

10.2.3 Flexural strength and toughness

It can be seen from Figure 10.6(a) and (b) that 47%, 114%, 99%, 65%, 98%, and 132% increments are achieved for the flexural strength of UHPC when 1.4 vol%, 1.6 vol%, 1.8 vol%, 2.0 vol%, 2.2 vol%, and 3.0 vol% SFs are added. When the SFs content is lower than 2.0 vol%, the flexural strength of UHPC first increases and then decreases with increasing SFs content. This is consistent with the conclusion of Al-Mwanes and Agayari [25]. Incorporating 0.2 vol% and 0.4 vol% SSWs improves the flexural strength of UHPC by 19% and 28%, respectively. Due to the modification effect of SSWs on the concrete matrix, the dispersion/orientation of SFs, and the interfacial bond strength between SFs and the matrix, the flexural strengths of hybrid SFs and SSWs-engineered multifunctional UHPC are higher than that of 3.0 vol% SFs reinforced UHPC, except for UHPC with hybrid 0.4 vol% SSWs and 1.4 vol% SFs. The addition of 0.2 vol% SSWs improves the flexural strength of UHPC with 1.4 vol%, 1.6 vol%, 1.8 vol%,

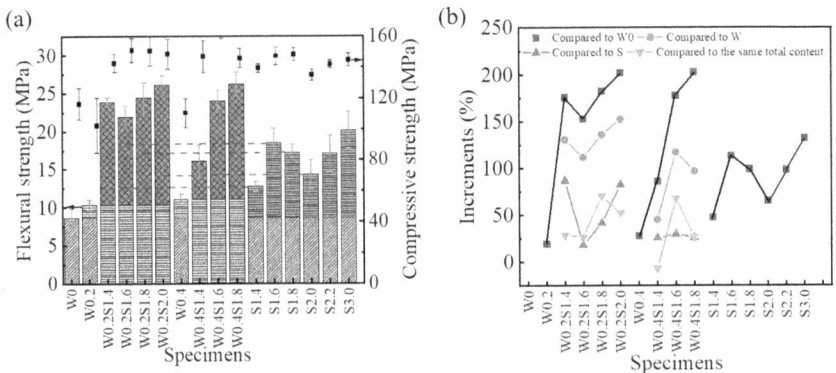

Figure 10.6 Flexural/compressive strength and toughness of SSWs-engineered multifunctional UHPC incorporating SFs. (a) Flexural and compressive strength; (b) Increments of flexural strength

and 2.0 vol% SFs alone by 87%, 18%, 42%, and 82%, respectively. The incorporation of 0.4 vol% SSWs results in 26%, 30%, and 26% increments, respectively, for the flexural strength of UHPC with 1.4 vol%, 1.6 vol%, and 1.8 vol% SFs alone. Using hybrid 0.2 vol% SSWs and 2.0 vol% SFs or 0.4 vol% SSWs and 1.8 vol% SFs increases the flexural strength of UHPC by 52% and 27%, respectively, compared with that of UHPC with 2.2 vol% SFs alone; it can reach 26.1 MPa and 21.8 MPa, respectively. The flexural strength of UHPC with hybrid 0.2 vol% SSWs/1.8 vol% SFs (24.4 MPa) and with hybrid 0.4 vol% SSWs/1.6 vol% SFs (24.0 MPa) is 71% and 68% higher, respectively, than that of UHPC with 2.0 vol% SFs alone, 22% and 21% higher, respectively, than that of UHPC with hybrid 0.5 vol% medium length (10 mm) SFs and 1.5 vol% long SFs (20 mm) [16], and 6% and 4% higher, respectively, than that of UHPC with 4.0 vol% SFs (diameter of 0.2 mm and length of 13 mm) [17]. Meanwhile, these two values of flexural strength are comparable to that of UHPC with hybrid 1.0 vol% hooked SFs and 1.0 vol% SFs (26.5 MPa) [17] and that of UHPC with hybrid 16 wt% SFs and 0.8 wt% PVA fibers (23.9 MPa) [26].

The compressive strength of UHPC with hybrid SSWs and SFs can also be found in Figure 10.6(a). It can be observed that the compressive strength of UHPC with SFs alone first increases and then decreases with increasing content of SFs, and 27% increments can be obtained when 1.8 vol% SFs are incorporated. Adding 0.2 vol% SSWs has only a small impact on the compressive strength of UHPC with 1.4 vol%, 1.6 vol% and 1.8 vol% SFs, but can increase the compressive strength of UHPC with 2.0 vol% SFs by 10%. The compressive strength of UHPC with hybrid 0.4 vol% SSWs and 1.6 vol% SFs is 12% higher than that of UHPC with 1.6 vol% SFs.

The flexural toughness index ($\bar{v}_b$) was evaluated by the following Eq. (10.3) based on the specification of JSCE SF-4.

$$\bar{v}_b = \frac{T_b}{d_{tb}} \frac{l}{bh^2} \tag{10.3}$$

where T_b is the integration area of flexural load–displacement curves up to $l/150$ of mid-span vertical displacement, l is the distance between bearing points (100 mm), d_{tb} is the set value $l/150$ of mid-span vertical displacement, and b and h are the width and height of specimens (40 mm). Furthermore, the complementary energy (G_{com}) of UHPC matrix in different composites was expressed as the difference between the product of peak load and peak displacement and the integral of the peak flexural load–displacement curves [27].

As shown in Figure 10.7(a), when the SFs content is less than 1.8 vol%, the flexural toughness index presents an increasing trend with the increase of SFs content. However, the flexural toughness index of UHPC with 2.0 vol% SFs is 23% lower than that of UHPC with 1.8 vol% SFs, indicating

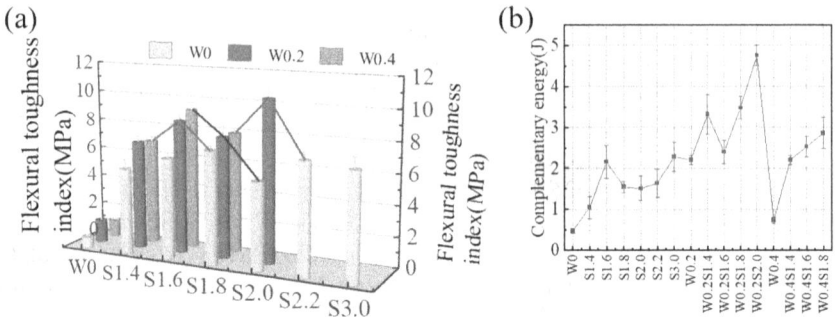

Figure 10.7 Flexural toughness index and complementary energy of SSWs-engineered multifunctional UHPC incorporating SFs. (a) Flexural toughness index; (b) Complementary energy

that the dispersion state and volume fraction of SFs contribute to the flexural toughness index together. Adding 0.2 vol% and 0.4 vol% SSWs alone increases the flexural toughness index of UHPC by 96% and 47%, respectively. The flexural toughness index of UHPC with 1.4 vol%/1.6 vol%/1.8 vol%/2.0 vol% SFs is improved by 26%, 33%, 8%, and 86%, respectively, when 0.2 vol% SSWs are added. Meanwhile, 23%, 41%, and 7% increments are obtained for UHPC with 1.4 vol%/1.6 vol%/1.8 vol% SFs due to the addition of 0.4 vol% SSWs. Increments of 39% and 63%, respectively, are achieved for the flexural toughness index when 0.2 vol% SSWs/1.8 vol% SFs and 0.4 vol% SSWs/1.6 vol% SFs are used compared with that of UHPC with 2.0 vol% SFs alone, and the increasements are 13% and 13% when compared with that of UHPC with 3.0 vol% SFs alone. The flexural toughness index of UHPC with hybrid 0.2 vol% SSWs/2.0 vol% SFs and 0.4 vol% SSWs/1.8 vol% SFs is 45%/27% and 51%/32% higher, respectively, than that of UHPC with 2.2 vol%/3.0 vol% SFs alone.

Figure 10.7(b) shows that when 0.2 vol% SSWs are added, the complementary energy of UHPC with 1.4 vol%/1.6 vol%/1.8 vol%/2.0 vol% SFs alone is enhanced by 216%, 11%, 125%, and 215%, respectively. The complementary energy of UHPC with hybrid 0.4 vol% SSWs and 1.4 vol%/1.6 vol%/1.8 vol% SFs is 111%, 17%, and 85% higher, respectively, than that of UHPC with SFs alone at the same contents. When the SFs content is low, the improvement of complementary energy caused by the addition of SSWs can be mainly ascribed to the hindering effect of SSWs on the initiation and convergence of microcracks and the supplementary effect of SSWs on the SFs' toughening network. When the SFs content is higher than 1.8 vol%, the modification effect of SSWs on the dispersion and orientation of SFs plays an important role in enabling UHPC to have high complementary energy. Increments of 55% and 43% can be obtained for the complementary energy of UHPC with hybrid 0.2 vol%

SSWs/1.6 vol% SFs and 0.4 vol% SSWs/1.4 vol% SFs compared with that of UHPC with 1.8 vol% SFs alone. The complementary energy of UHPC with hybrid 0.2 vol% SSWs/1.8 vol% SFs and 0.4 vol% SSWs/1.6 vol% SFs is 131% and 68% higher, respectively, than that of UHPC with 2.0 vol% SFs alone. Furthermore, using hybrid 0.2 vol% SSWs/2.0 vol% SFs and 0.4 vol% SSWs/1.8 vol% SFs causes the complementary energy of UHPC to increase by 190%/74% and 109%/25%, respectively, compared with UHPC with 2.2 vol% and 3.0 vol% SFs alone.

10.2.4 Multiple-cracking flexural failure

As demonstrated in Figure 10.8(a), the flexural load–displacement curves of UHPC with SSWs alone contain only two stages, the linear ascending stage before initial macrocracking and the vertical descending stage, and the corresponding specimens break into two sections instantly after reaching peak load. This is because the ability of SSWs to bridge macroscopic cracks is relatively weak. Figure 10.8(b) illustrates that there is no multiple-cracking stage in the flexural load–displacement curves of UHPC with SFs alone at less than 3.0 vol%, and the corresponding flexural load–displacement curves of composites contain three stages: the linear ascending stage before initial macrocracking, the fiber reinforcement stage, and the descending stage. This is consistent with the results of Tjiptobroto and Hansen [28]. There is a multiple-cracking stage in the flexural load–displacement curves of UHPC with 3.0 vol% SFs, benefitting from the bridging effect of widely distributed SFs on macrocracks. Figure 10.8(c)–(f) show that the flexural load–displacement curves of hybrid SFs and SSWs-engineered multifunctional UHPC have a longer linear ascending stage compared with that of UHPC with SFs alone at the same total content. The transition from the initial macrocracking stress to the SFs reinforcement stage in the flexural load–displacement curves of hybrid SFs and SSWs-engineered multifunctional UHPC becomes smoother and smoother as the SFs content increases, until the SFs reinforcement stage is transformed into the multi-cracking stage. Adding 0.2 vol% SSWs enables the flexural load–displacement curves of UHPC with 1.8 vol% and 2.0 vol% SFs to possess a multiple-cracking stage, and there is already a multiple-cracking stage in the flexural load–displacement curves of UHPC with 1.6 vol% and 1.8 vol% SFs due to the incorporation of 0.4 vol% SSWs. The coalescence of microcracks into macrocracks is hindered, and additional small cracks then form, resulting from the modification effect of SSWs on the concrete matrix and the dispersion/orientation of SFs, and the synergistic effect of hybrid SFs and SSWs on the initiation and propagation of micro- and macro-cracks. It should be noted that the descending stages of these curves have a high slope due to the enhancement effect of SSWs on the bond strength between fibers and matrix.

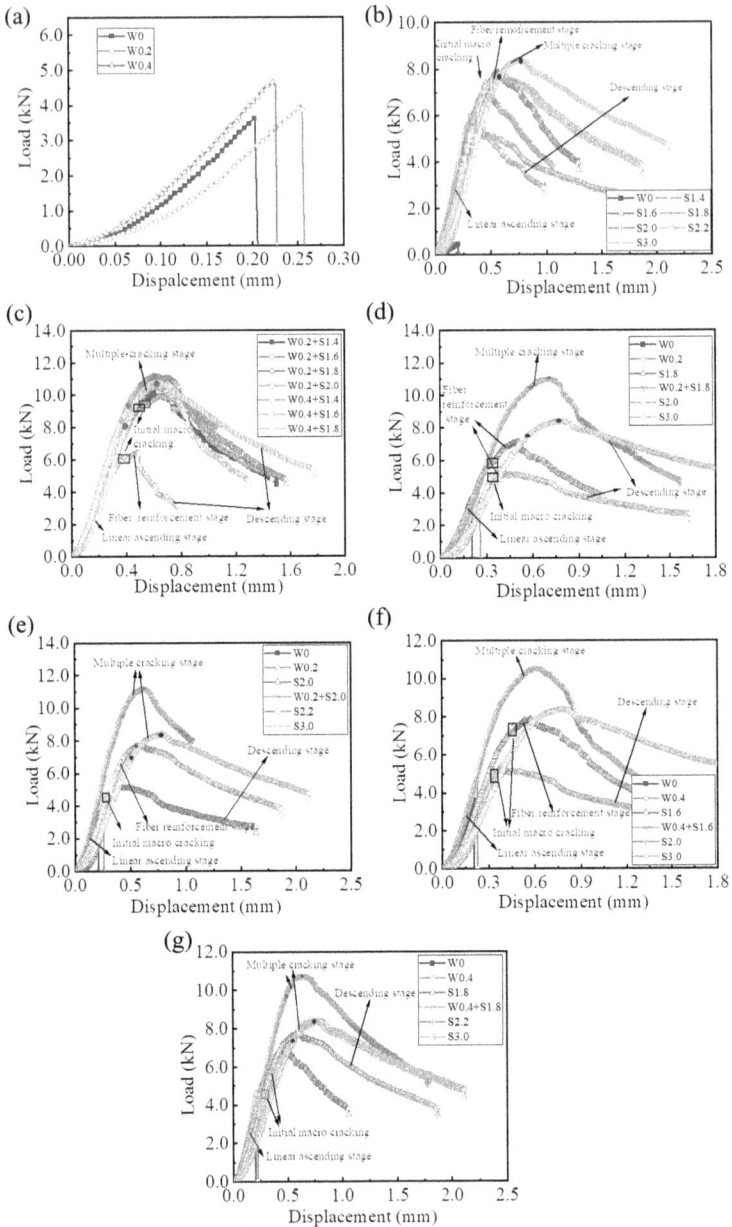

Figure 10.8 Flexural load–displacement curves of SSWs-engineered multifunctional UHPC incorporating SFs. (a) Curves of UHPC with wires alone; (b) Curves of UHPC with SFs alone; (c) Curves of hybrid SFs and SSWs-engineered UHPC; (d) Curves of UHPC with hybrid 0.2 vol% SSWs and 1.8 vol% SFs; (e) Curves of UHPC with hybrid 0.2 vol% SSWs and 2.0 vol% SFs; (f) Curves of UHPC with hybrid 0.4 vol% SSWs and 1.6 vol% SFs; (g) Curves of UHPC with hybrid 0.4 vol% SSWs and 1.8 vol% SFs

Because of the size of specimens, the multiple-cracking failure state cannot be visually observed. Hence, it is necessary to theoretically calculate crack numbers to verify the multiple cracking failure mode of UHPC. If the energy for the first cracking (E_f) is larger than that required for the formation of one additional microcrack (E_a), the multiple-cracking failure mode under flexural loading for UHPC will occur. The process of additional microcracks is repeated until the total cracking energy of (nE_a) becomes larger than the first cracking energy (E_f). Then, the crack numbers (n) after flexural failure are obtained by the expression of E_f/E_a. The first cracking energy (E_f) was obtained using Eqs (10.4)–(10.7) according to Wei et al. [29]:

$$E_f = \Delta U_{f\text{-mc}} + \Delta U_{fr} + \Delta U_{db} \tag{10.4}$$

$$\Delta U_{f\text{-mc}} = \frac{V_{ef}}{E_f}\left[\frac{7}{48}\frac{\tau_{fu}^2}{d^2} - \left(E_f\varepsilon_{mu}\right)^2\frac{l_f}{4}\right] \tag{10.5}$$

$$\Delta U_{fr} = \frac{V_{ef}\tau_{fu}l_f}{4d}\left[\frac{\tau_{fu}l_f}{3E_f d} - \frac{1}{2}\varepsilon_{mu}\right] \tag{10.6}$$

$$U_{db} = \frac{V_{ef}l_f G_{II}}{d} \tag{10.7}$$

where $\Delta U_{f\text{-mc}}$ is the strain energy of crack-bridging SFs; ΔU_{fr} is the slipping fractional energy between SFs and the concrete matrix; U_{db} is the debonding energy consumed by the elastic bond between SFs and the concrete matrix; V_{ef} is the effective volume fraction of SFs, equal to the product of SFs' volume fraction and orientation factor in the flexural-tensile direction; E_f is the elastic modulus of SFs (200 GPa); τ_{fu} is the bond strength between SFs and the concrete matrix; ε_{mu} is the limit elastic strain on the elastic stage of flexural-tensile strain–stress curves; d and l_f are the diameter and length of SFs; and G_{II} is the second mode fracture energy in a shearing fracture process of the SFs–matrix interface [10]. Due to the compact microstructure of composites, the values of second mode fracture energy (G_{II}) are the equivalent of the fracture energy (G_m) of the UHPC matrix [9]. Meanwhile, the values of fracture energy (G_m) are closely related to the complementary energy (G_{com}) of UHPC and have the following relationships with complementary energy:

$$G_{II} = 2\Delta a G_{com} \tag{10.8}$$

where Δa is the allowable width of main cracks, and the value is set as 0.05 mm [9]. The values of complementary energy (G_{com}) have been obtained according to flexural load–displacement curves, as mentioned before. The energy for forming one additional microcrack (E_a) can be denoted by Eqs (10.9)–(10.11) [7, 29]:

$$E_a \approx G_m V_m + \Delta U_{f\text{-mu}} - U_m \tag{10.9}$$

$$\Delta U_{f\text{-mu}} = \frac{1}{24} E_f V_{ef} l_f \frac{E_m V_m}{E_f V_f} \left(18 + 7 \frac{E_m V_m}{E_f V_f} \right) \varepsilon_{mu}^2 \tag{10.10}$$

$$\Delta U_m = \frac{1}{24} E_m V_m l_f \varepsilon_{mu}^2 \tag{10.11}$$

where E_m is the elastic modulus of the concrete matrix, and the value is determined as 40 GPa according to Bentur and Mindess [10], V_m is the matrix volume fraction, $\Delta U_{f\text{-mu}}$ is the bridging SFs' strain energy, and ΔU_m is the decrease in value of matrix strain energy as cracking occurs.

As shown in Table 10.2, there is no multiple-cracking failure mode for 2.0 vol% SFs reinforced UHPC, and this is consistent with the result of Xu et al. [30]. Multiple-cracking failure mode occurs for UHPC with 3.0 vol% SFs alone under flexural loading, and the number of cracks is 3.9. Incorporating 0.2 vol% SSWs enables the crack numbers for 1.8 vol% and 2.0 vol% SFs reinforced UHPC to reach 1.7 and 2.0, respectively; meanwhile, UHPC with hybrid 0.4 vol% SSWs and 1.6 vol% SFs also possesses a tendency to multiple cracking. The calculation results for crack numbers are in good agreement with the experimental phenomena presented in Table 10.2. Table 10.2 also shows that the change rule of first cracking energy (E_f) for UHPC with SFs alone is consistent with that of initial macrocracking stress. The first cracking energy for hybrid SFs and SSWs-engineered multifunctional UHPC is influenced by both initial macrocracking stress and interfacial bond strength. Incorporating 0.2 vol% SSWs increases the first cracking energy for UHPC with 1.4 vol%, 1.6 vol%, 1.8 vol%, and 2.0 vol% SFs alone by 212%, 62%, 48%, and 164%, respectively. Meanwhile, 105%, 38%, and 74% increments can be obtained for UHPC with 1.4 vol%, 1.6 vol%, and 1.8 vol% SFs alone due to the addition of 0.4 vol% SSWs. What calls for special attention is that the existence of a complete toughening network in UHPC causes the influence area of the first cracking energy to enlarge, which in turn, reduces the energy required for additional cracks. This is the reason for the emergence of multiple-cracking failure mode and the embodiment of high toughness. It can be seen that the additional energy for one more crack (E_a) is reduced by 40% and 47%, respectively, for UHPC with 1.8 vol% and 2.0 vol% SFs alone when 0.2 vol% SSWs are added.

Table 10.2 Multiple-cracking energy and crack numbers for SSWs-engineered multifunctional UHPC incorporating SFs

Group	$\varepsilon_{mu} \times 10^{-6})$	$\Delta U_{f\text{-mc}}$ (J/m^2)	ΔU_{fr} (J/m^2)	U_{db} (J/m^2)	E_f (J/m^2)	$\Delta U_{f\text{-mu}}$ (J/m^2)	ΔU_m (J/m^2)	E_a (J/m^2)	E_f/E_a
S1.4	261.2	14.3	7.0	28.3	49.5	73.3	1.1	101.7	0.49
S1.6	302.4	25.4	12.5	68.7	106.7	90.5	1.7	118.3	0.90
S1.8	341.8	15.9	7.2	66.3	89.5	125.1	2.5	152.1	0.59
S2.0	447.5	11.9	4.9	56.7	73.6	155.3	4.8	179.9	0.41
S2.2	318.7	16.5	7.6	67.6	91.7	72.8	2.7	99.4	0.92
S3.0	138.6	17.1	8.8	128.3	154.2	10.8	6.8	39.2	3.93
W0.2S1.4	382.4	38.7	18.9	115.0	172.6	201.8	2.4	228.9	0.76
W0.2S1.6	407.7	30.7	14.6	87.4	132.6	188.3	3.2	214.6	0.62
W0.2S1.8	281.5	34.6	17.3	142.3	194.2	81.1	1.7	108.9	1.78
W0.2S2.0	346.8	41.9	20.7	183.0	245.7	95.5	2.9	122.1	2.01
W0.4S1.4	296.8	17.9	8.6	74.9	101.3	119.0	1.5	147.1	0.69
W0.4S1.6	325.3	39.9	20.0	86.8	146.7	112.8	2.0	140.4	1.05
W0.4S1.8	343.7	27.7	13.2	114.5	155.5	118.4	25	145.3	1.07

The images of the cracked specimens for UHPC are shown in Figure 10.9. As shown in Figure 10.9(a) and (c), the width of cracks decreases and the bending degree of cracks increases on the flexural-tensile bottom surface with increasing content of SFs. Meanwhile, Figure 10.9(b) and (d) show that the height of cracks on the flexural-tensile side surface displays a downward trend when the SFs content increases from 2.0 vol% to 3.0 vol%. This phenomenon indicates that a high content of SFs possesses a strong ability to hinder crack propagation. Figure 10.9(e)–(l) demonstrate that the cracked specimens of hybrid SFs and SSWs-engineered UHPC have typical multiple-cracking patterns, and the crack height on the flexural-tensile side surface is significantly reduced. Limited by specimen size, the failure specimens did not show multiple independent cracks as described in Dong et al. [31]. The reason for the multiple-cracking tendency can be attributed to the widely distributed SSWs and SFs working together to form a more

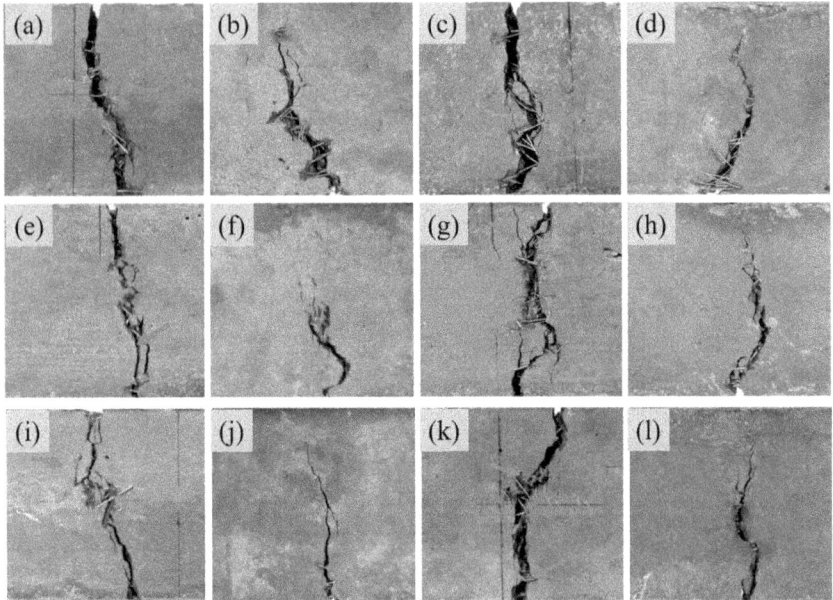

Figure 10.9 Crack images of the cracked specimens under flexure. (a) S2.0 flexural-tensile bottom surface; (b) S2.0 flexural-tensile side surface; (c) S3.0 flexural-tensile bottom surface; (d) S3.0 flexural-tensile side surface; (e) W0.2S1.8 flexural-tensile bottom surface; (f) W0.2S1.8 flexural-tensile side surface; (g) W0.2S2.0 flexural-tensile bottom surface; (h) W0.2S2.0 flexural-tensile side surface; (i) W0.4S1.6 flexural-tensile bottom surface; (j) W0.4S1.6 flexural-tensile side surface; (k) W0.4S1.8 flexural-tensile bottom surface; (l) W0.4S1.8 flexural-tensile side surface S2.0 and S3.0 represent UHPC with 2.0 vol%/3.0 vol% SFs alone, W0.2S1.8 and W0.2S2.0 represent UHPC with 0.2 vol% SSWs and 1.8 vol%/2.0 vol% SFs, and W0.4S1.6 and W0.4S1.8 represent UHPC with hybrid 0.4 vol% SSWs and 1.6 vol%/1.8 vol% SFs

complete cracking resistance network. In particular, the effective hindering effect of SSWs on the initiation and propagation of microcracks remarkably prolongs the development footprint of cracks. The cracked failure state of hybrid SFs and SSWs-engineered UHPC is consistent with the calculated results in Table 10.1

The flexural load–displacement curves of hybrid SFs and SSWs-engineered multifunctional UHPC can be divided into four categories according to the characteristics of the cracking process, as shown inFigure 10.10. A type I curve occurs for UHPC with SFs alone at less than 2.2 vol%, and it comprises three stages: the elastic stage, the SFs reinforcement stage, and the descending stage. With the increase of flexural load, the main crack is formed rapidly, and the load is transferred to the SFs. SFs are mainly pulled out of failure after peak flexural load, leading to the small slope of the descending stage. UHPC with type I curves has low initial macrocracking stress and flexural strength/toughness, resulting from the poor crack resistance of the matrix, which causes the SFs not to be fully utilized. The flexural load–displacement curves of UHPC with 3.0 vol% SFs alone belong to

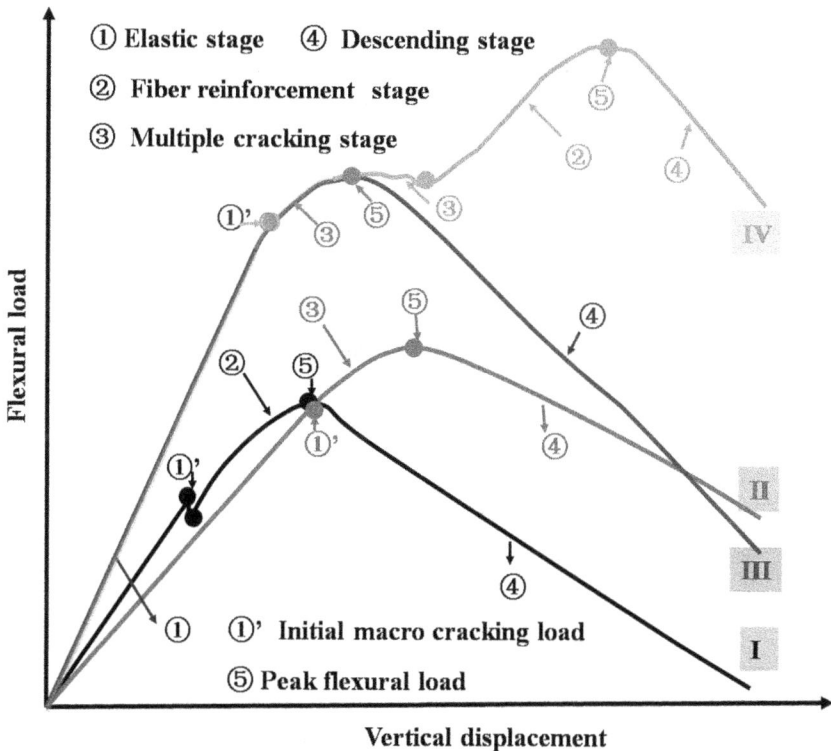

Figure 10.10 Typical flexural load–displacement curves for hybrid SFs and SSWs-engineered multifunctional UHPC incorporating SFs

type II and consist of three stages: the elastic stage, the multiple-cracking stage, and the descending stage. The increase of initial macrocracking stress and the formation of the multiple-cracking stage benefits from the high elastic modulus and high content of SFs. Due to the low interfacial bond strength between the SFs and the matrix, SFs are mainly pulled out after peak flexural load, further reducing the slope of the descending stage. The flexural strength and toughness of UHPC with type II curves need to be further improved due to the defect in the matrix caused by the uneven dispersion of SFs. SSWs can increase the microstructural compactness of the UHPC matrix to increase the initial macrocracking load, and inhibit the initiation, propagation, and convergence of microcracks into macrocracks together with SFs, thus leading to the existence of a multiple-cracking stage in flexural load–displacement curves, i.e. type III curves in Figure 10.9. UHPC with hybrid 0.2 vol% SSWs/1.8 vol% SFs, 0.2 vol% SSWs/2.0 vol% SFs, 0.4 vol% SSWs/1.6 vol% SFs, and 0.4 vol% SSWs/1.8 vol% SFs has this type III of flexural load–displacement curves. However, due to the modification effect of SSWs on the interfacial bond strength between SFs and the matrix, a descending stage with a large slope occurs after the multiple-cracking stage in type III curves, representing the SFs being pulled off. This indicates that a UHPC matrix containing SSWs can be used in conjunction with higher–tensile strength SFs to coordinate their best performance. If SFs with higher tensile strength and SSWs are incorporated together, UHPC has the potential to possess a type IV cracking process under flexural load and obtain better flexural strength and toughness as well as a more pronounced multiple-cracking failure state. This provides the possibility to prepare highly ductile UHPC for special applications.

10.2.5 Optimization results

As shown in Figure 10.11(a), 0.2 vol% SSWs play a more significant role in improving the initial macrocracking stress, interfacial bond strength, and

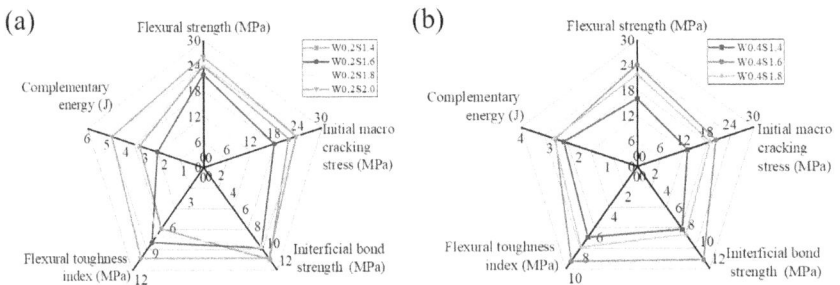

Figure 10.11 Properties of SSWs-engineered multifunctional UHPC incorporating SFs. (a) Optimization results of 0.2 vol% SSWs; (b) Optimization results of 0.4 vol% SSWs

flexural strength of UHPC with 1.4 vol%, 1.6 vol%, and 1.8 vol% SFs. Using hybrid 0.2 vol% SSWs and 2.0 vol% SFs endows UHPC with both excellent strength and toughness. Figure 10.11(b) shows that the improvement effect of hybrid 0.4 vol% SSWs and 1.4 vol% SFs on flexural toughness index and complementary energy is better than the effect on initial macrocracking stress and flexural strength. With increasing SFs content, the modification effect of hybrid 0.4 vol% SSWs and SFs on the flexural strength and toughness index of UHPC behaves better and better. The flexural toughness index of UHPC with 0.4 vol% SSWs and 1.6 vol% SFs is better (17% higher) than that of UHPC with 0.2 vol% SSWs and 1.8 vol% SFs, but the initial macrocracking stress, interfacial bond strength, and flexural strength for the latter are higher than for the former. When the total content of SSWs and SFs is 2.2 vol%, UHPC with 0.2 vol% SSWs and 2.0 vol% SFs possesses higher initial macrocracking stress (26%), interfacial bond strength (33%), flexural strength (20%), flexural toughness index (34%), and complementary energy (70%) than UHPC with 0.4 vol% SSWs and 1.8 vol% SFs.

The mechanisms by which SSWs modify the flexural strength and toughness of UHPC with SFs can be illustrated by Figure 10.12. As demonstrated in Figure 10.12(a), SSWs with a micro diameter can improve the structural compactness and the cracking resistance of the UHPC matrix by the filling effect at low content. Meanwhile, the addition of SSWs can modify the

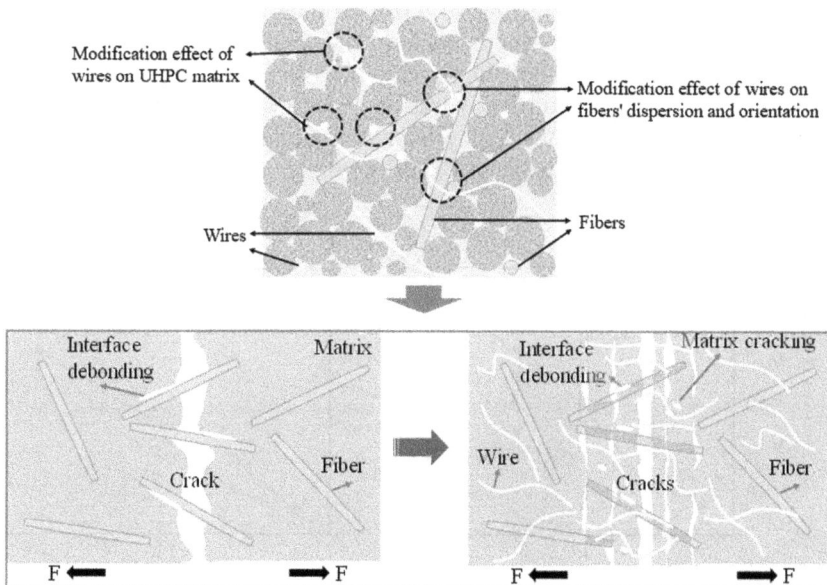

Figure 10.12 Modification mechanisms of SSWs-engineered multifunctional UHPC incorporating SFs

dispersion and orientation state of SFs in UHPC through the hindrance effect, reducing the defects caused by a high content of SFs and also enhancing the compactness of the matrix. This is helpful for the improvement of SFs' efficiency. The failure cracking of UHPC with SFs less than 3.0 vol% alone or with SSWs alone is dominated by a single crack, and no new cracks develop once the main crack opens, as shown in Figure 10.12(b). For UHPC with hybrid SSWs and SFs, the growth of the main crack is accompanied by the formation of additional small cracks. The presence of SSWs provides a system with sufficiently small spacing to control the initiation and propagation of microcracks at the early stage of flexural loading and impedes the coalescence of microcracks into bigger cracks, leading to a smooth transition from microcracks to the first macrocrack, i.e. the occurrence of multiple-cracking failure.

10.3 ELECTRICAL AND SELF-SENSING PROPERTIES OF SSWS-ENGINEERED MULTIFUNCTIONAL UHPC INCORPORATING STEEL FIBERS

10.3.1 Electrical properties

The direct current (DC) resistivity of UHPC with SSWs alone and with SFs at different curing ages is presented in Figure 10.13. As can be seen from Figure 10.13(a), the DC resistivity of the composites without fillers continuously increases at curing ages from 3 days to 28 days. Meanwhile, the increment of resistivity increases first and then decreases as the curing age increases, indicating that the development of electrical performance is affected by the progress of the hydration reaction [32]. It should be noted that the addition of SSWs can effectively enhance the conductivity of UHPC [33]. The DC resistivity of UHPC with 0.2 vol% and 0.4 vol% SSWs

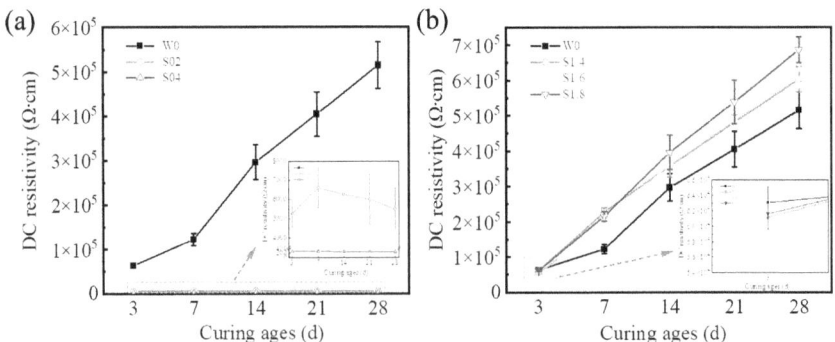

Figure 10.13 DC resistivity of UHPC with SSWs and SFs alone. (a) UHPC with SSWs alone; (b) UHPC with SFs alone

decreases by two and three orders of magnitude, respectively, compared with that of composites without SSWs, manifesting that 0.2 vol% and 0.4 vol% SSWs are in the percolation threshold zone. Particularly, the incorporation of 0.4 vol% SSWs provides UHPC with more stable and outstanding electrical behavior; thus, in the following section, the electrical and sensing performances of UHPC with 0.4 vol% SSWs are investigated.

Figure 10.13(b) demonstrates that the DC resistivity of UHPC containing SFs increases drastically as the curing age increases from 3 days to 28 days; meanwhile, the resistivity increases with increasing content of SFs in general. Especially, at the curing age of 3 days, the DC resistivity of UHPC increases first and then decreases as the content of SFs increases. With 1.6 vol% SFs, 9.4% decrements for maximum resistivity can be obtained by the UHPC. This can be attributed to the conductivity in the early hydration of UHPC being mainly determined by ions or gel water of pore solution and electrons of unreacted cement particles. Furthermore, as the curing age increases, the conductivity of UHPC generally reduces due to the decrease of conductive ions and the increase of contact resistance in the process of hydration [32].

Figure 10.14 shows that the increment in DC resistivity of UHPC with hybrid SSWs and SFs is generally reduced as the curing age increases. The incorporation of SSWs can significantly improve the electrical performance of UHPC containing SFs, and the DC resistivity of hybrid SSWs and SFs reinforced UHPC decreases by one or two orders of magnitude compared with that of UHPC with SFs only. This can be attributed to the fact that SSWs play a key role in the conductive system, and the hybrid SSWs with SFs can form an effective three-dimensional conductive network in UHPC, indicating a positive synergistic effect. In addition, the conductivity of hybrid SFs and SSWs-engineered multifunctional HPC is negatively correlated with the content of SFs, and the optimum conductivity can be realized by UHPC incorporating hybrid 0.4 vol% SSWs and 1.4 vol% SFs, whose

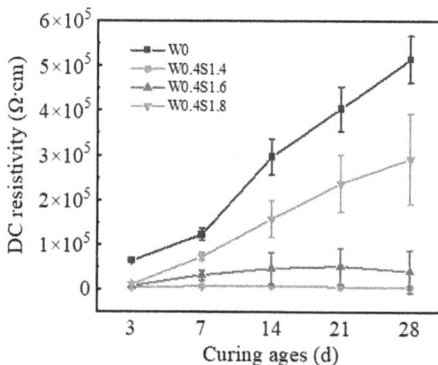

Figure 10.14 DC resistivity of SSWs-engineered multifunctional UHPC incorporating SFs

resistivity is reduced by 99% compared with that of composites without fillers.

The sinusoidal variation of alternating current (AC) voltage can avoid the polarization effect and is favorable to obtain relatively stable resistance. In order to make the results more realistic and accurate, the AC resistivity measured at the frequency of 100 Hz is selected to analyze the electrical performance of UHPC. Figure 10.15(a) depicts the AC resistivity of UHPC, which presents an ascending trend as the curing age increases, and the increment gradually stabilizes when the curing age is within the range of 14 days to 28 days, which can be attributed to the completion of the hydration reaction and the formation of hydration products. Moreover, the variation in the AC resistivity of UHPC with SSWs or hybrid SSWs and SFs is similar to the DC resistivity, whereas the results of the UHPC containing SFs are different from the DC resistivity because of the polarization effect and interface contact resistance in the DC conductive system. The AC resistivity of UHPC containing SFs decreases with increasing content of SFs; the resistivity of UHPC with 1.4 vol%, 1.6 vol%, and 1.8 vol% SFs is 82.9%, 83.9%, and 87.0% lower than that of composites without fillers, respectively. Furthermore, the AC resistivity of hybrid SFs and SSWs-engineered UHPC reduces with decreasing content of SFs; the minimum resistivity of UHPC incorporating hybrid 0.4 vol% SSWs and 1.4 vol% SFs can be reduced to 1310 $\Omega \cdot cm$, which is 98.3%, 22.8%, and 48.2% lower than that of composites without fillers, UHPC containing hybrid SSWs and 1.6 vol% SFs, and UHPC containing hybrid SSWs and 1.8 vol% SFs, respectively. Consequently, the electrical performance of UHPC can be significantly improved by hybrid SSWs and SFs, as the conductive path is enhanced by their hybrid effect. Figure 10.15(b) demonstrates that the normalized AC resistivity of UHPC decreases as the frequency increases from

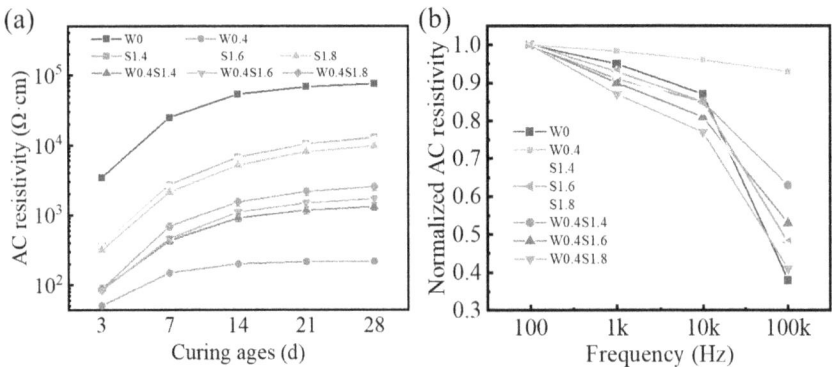

Figure 10.15 AC resistivity of SSWs-engineered multifunctional UHPC incorporating SFs. (a) AC resistivity at the frequency of 100 Hz; (b) Normalized AC resistivity at different frequencies

100 Hz to 100 kHz at the curing age of 28 days. It can be seen that the normalized resistivity of UHPC with 0.4 vol% SSWs is almost unaffected by the frequency; thus, its conductive system is close to pure electrical conductivity, while the UHPC with SFs is remarkably exposed to the frequency on account of the accessory circuit elements formed by the hydration products in the conductive system [34]. Moreover, hybrid SSWs and SFs improve the conductive circuits and electrical systems of UHPC with their exceptional synergistic effect.

10.3.2 Self-sensing properties

In order to evaluate the self-sensing performance of UHPC with SSWs and SFs in single and hybrid forms, the fractional change in electrical resistivity (FCR)–displacement curves under flexural loading are investigated, and the linear slopes of these curves in different stages are analyzed, as shown in Figure 10.16 and Table 10.3. Meanwhile, the values of FCR/stress sensitivity corresponding to peak displacement (point B in Figure 10.16) are measured, as demonstrated in Figure 10.16(h). The FCR of UHPC containing SSWs is negative at the beginning and then becomes positive, meaning that the specimen is dominated first by compression and then by tensile stress in the test. Notably, the reorganization of the conductive path by overlapped SSWs contributes to the conductive system. However, as initial cracking of the specimens occurs, the FCR rises dramatically in a straight line, while the load-carrying capacity of UHPC rapidly decreases. This can be explained by the fact that the conductive network of UHPC containing SSWs is destroyed as SSWs are pulled off with the emergence and expansion of cracks [33].

As presented in Figure 10.16(b)–(d), in the pre-peak zone, the FCR of UHPC with SFs first shows a growth trend with increasing load and displacement, and then, as the load increases from the cracking A to the peak B, the increment tends to be even steadier with the increase of SFs. In the post-peak zone, the FCR of UHPC with SFs quickly increases with reducing load-carrying capacity. It should be noted that the steady variation of FCR in the stage of AB is caused by the phenomenon of crack localization, and the conductive pathway is almost unaffected by the increase of loading [35]. As shown in Table 10.3, the linear slopes of FCR–displacement curves in the stages of OA and BC increase with increasing content of SFs, representing the improvement of sensing ability.

As shown in Figure 10.16(e)–(g), the FCR of hybrid SFs and SSWs-engineered multifunctional UHPC increases linearly until the load reaches the initial cracking point A. The FCRs of UHPC reinforced by hybrid 0.4 vol% SSWs and 1.4 vol%, 1.6 vol%, and 1.8 vol% SFs increase to 6.20%, 7.21%, and 9.08%, respectively. Table 10.3 shows that the linear slope of FCR–displacement curves in stage OA for hybrid SFs and SSWs-engineered UHPC is lower than that for UHPC with SFs alone, benefitting from the

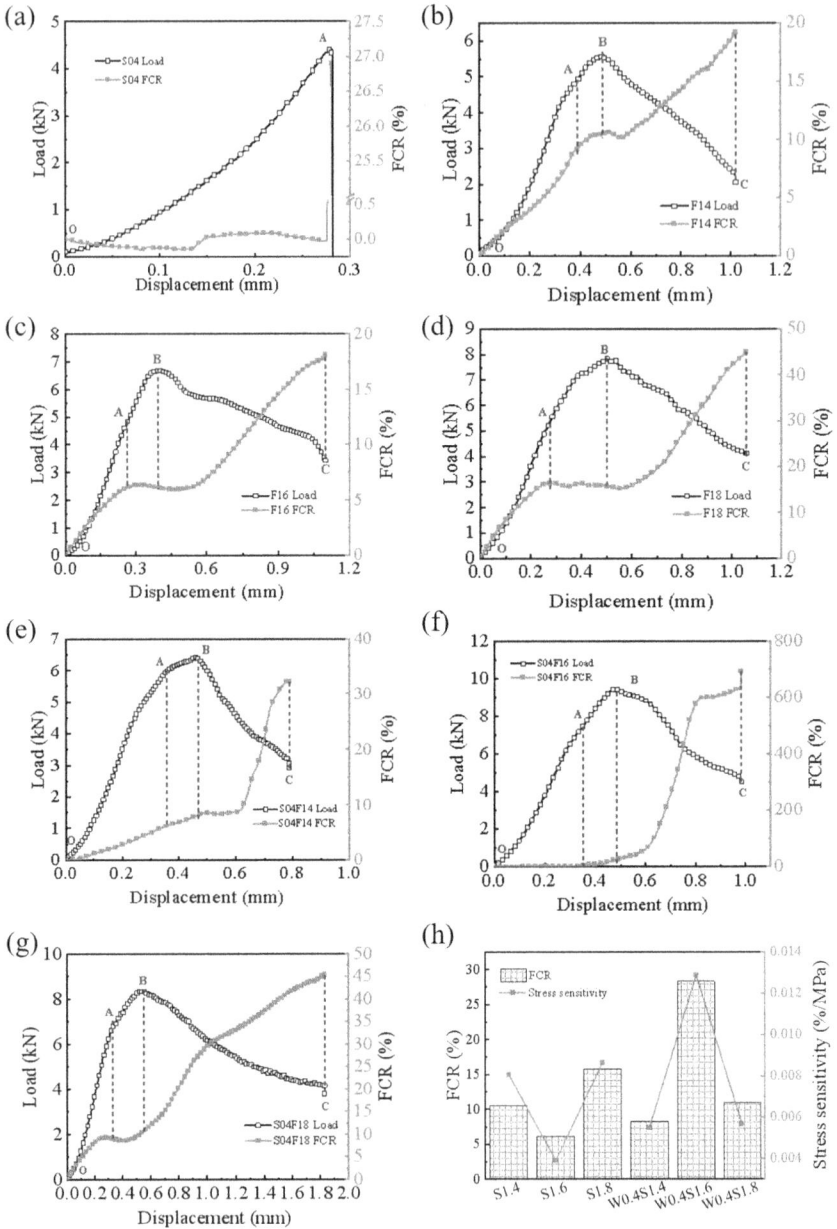

Figure 10.16 Load–displacement and FCR–displacement curves of UHPC. (a) UHPC with 0.4 vol% SSWs alone; (b) UHPC with 1.4 vol% SFs alone; (c) UHPC with 1.6 vol% SFs alone; (d) UHPC with 1.8 vol% SFs alone; (e) UHPC with hybrid 0.4 vol% SSWs and 1.4 vol% SFs; (f) UHPC with hybrid 0.4 vol% SSWs and 1.6 vol% SFs; (g) UHPC with hybrid 0.4 vol% SSWs and 1.8 vol% SFs; (h) FCR and stress sensitivity corresponding to peak displacement

Table 10.3 Linear slopes in different stages for FCR–displacement curves of UHPC with SSWs and SFs in both single and hybrid forms

Materials	Slope in different stages of FCR-displacement curves	
	OA	BC
W0.4	96.11	–
S1.4	21.98	17.35
S1.6	23.26	20.31
S1.8	59.30	61.32
W0.4S1.4	18.79	88.61
W0.4S1.6	18.76	1611.47
SW0.4S1.8	29.64	26.93

inhibiting effect of SSWs on the initiation and convergence of microcracks. This indicates that adding SSWs is conducive to improving the sensing stability of UHPC with SFs at the linear ascending stage of flexural loading and displacement. With the crack emerging and expanding, the load is gradually borne by SSWs and SFs [36], and the synergistic controlling effect of SSWs and SFs on cracking development leads to the increase rate of FCR showing a downward tendency in the AB section. Moreover, in the post-peak zone, the FCR presents a rapid growth trend with the decrease of the bearing capacity of UHPC, which mainly results from the disconnection of conductive fibers [37]. Table 10.3 also shows that the increase rate of the FCR with the increase of displacement in the BC stage for UHPC with hybrid 0.4 vol% SSWs and 1.4 vol%/1.6 vol% SFs is obviously larger than that for UHPC with SFs alone. This can be attributed to the extensively conductive network formed by both SSWs and SFs and the SSWs being pulled off with the expansion of cracks. Especially, the FCR and stress sensitivity (i.e. FCR per unit stress) of UHPC with hybrid 0.4 vol% SSWs and 1.6 vol% SFs increase by one or two orders of magnitude under peak flexural displacement (Figure 10.16(h)) compared with UHPC with 1.6 vol% SFs alone, achieving values of 28.31% and 0.01288%/MPa, respectively. It can be seen from the comparative analysis of Figsure 10.16(d) and (g) that the addition of 0.4 vol% SSWs reduces the slope of FCR–displacement curves in the BC stage and the FCR/stress sensitivity values corresponding to peak displacement for UHPC with 1.8 vol% SFs (Fig. 10.16(h)). This is because 1.8 vol% SFs has an effective bridging and inhibiting role in crack development, thus reducing the influence of cracking on the conductive network.

Hence, using hybrid SSWs and SFs can effectively improve the sensing stability of FCR in the linear ascending stage of flexural load, and can also increase the sensibility of UHPC composites to the load descending process, indicating a positive synergistic effect of SSWs and SFs on enhancing the sensing performance. On one hand, the enhancement of sensing capacity is caused by the hybrid effect between SSWs and SFs. On the other hand, this can be attributed to the fact that the high-magnitude FCR can be affected by some external factors, i.e. the number and width of cracks, and the spacing and orientation of fibers [37].

The previous research of Qiu et al. [35] and Han et al. [38] reported that the pre-peak and post-peak mechanical behaviors can be simulated based on the relationship between measured FCR and loading by non-linear regression analysis, and the relationship of load and FCR under flexural loading can be expressed by the polynomial Eq. (10.12). It should be noted that the fitted parameters in this equation vary with the loading regimes and materials.

$$Y = AX^2 + BX \qquad\qquad (10.12)$$

where Y is FCR, X is load, and A and B are the fitted parameters. Figure 10.17 displays the experimental and fitted FCR–load curves of hybrid SFs and SSWs-engineered UHPC under flexural load. The detailed correlation coefficients and fitting degrees of polynomial regression curves in different stages are shown in Table 10.4.

It can be seen from Figure 10.17(a) that the FCR of UHPC containing SSWs can be divided into two parts under flexural loading. When the applied load is lower than the peak, the structural deformation of UHPC is not enough to affect the conductive system due to the fact that the deformations of SSWs and UHPC are in the elastic regime. However, once the loading reaches the peak, SSWs are gradually pulled off and broken, with cracking appearing and expanding [39]; meanwhile, the FCR shows a dramatic growth trend. Therefore, the failure of UHPC containing SSWs can be perfectly studied by measuring the FCR under flexural loading. As demonstrated in Figure 10.17(b)–(g), the FCR–load curves of UHPC reinforced with SFs alone or hybrid SSWs and SFs can be divided into three stages: stage I, stage II, and stage III. In the case of stage I, the FCRs of both UHPCs increase significantly with increasing content of SFs. The increasing process of FCR in stage I for FCR–load curves of hybrid SFs and SSWs-engineered UHPC is more stable than that of UHPC with SFs alone, and the corresponding value of the polynomial regression fitting degree (R^2) is improved due to the addition of 0.4 vol% SSWs (as shown in Table 10.4), representing the increase of sensing stability to flexural loading. In the case of stage II, the bridging effect of SFs on macrocracks contributes to the sensing performance, and the improvement of FCR in UHPC reinforced by mono SFs alone or hybrid SSWs and SFs is sluggish, as is the increase of loading and SFs content. In the case of stage III, the load–FCR curves present substantial growth as flexural loading decreases, meaning that the conductive paths overlapped by SFs are gradually failing. The polynomial regression fitting degree (R^2) values in stage III for FCR–load curves of UHPC with hybrid 0.4 vol% SSWs and 1.6 vol%/1.8 vol% SFs is reduced

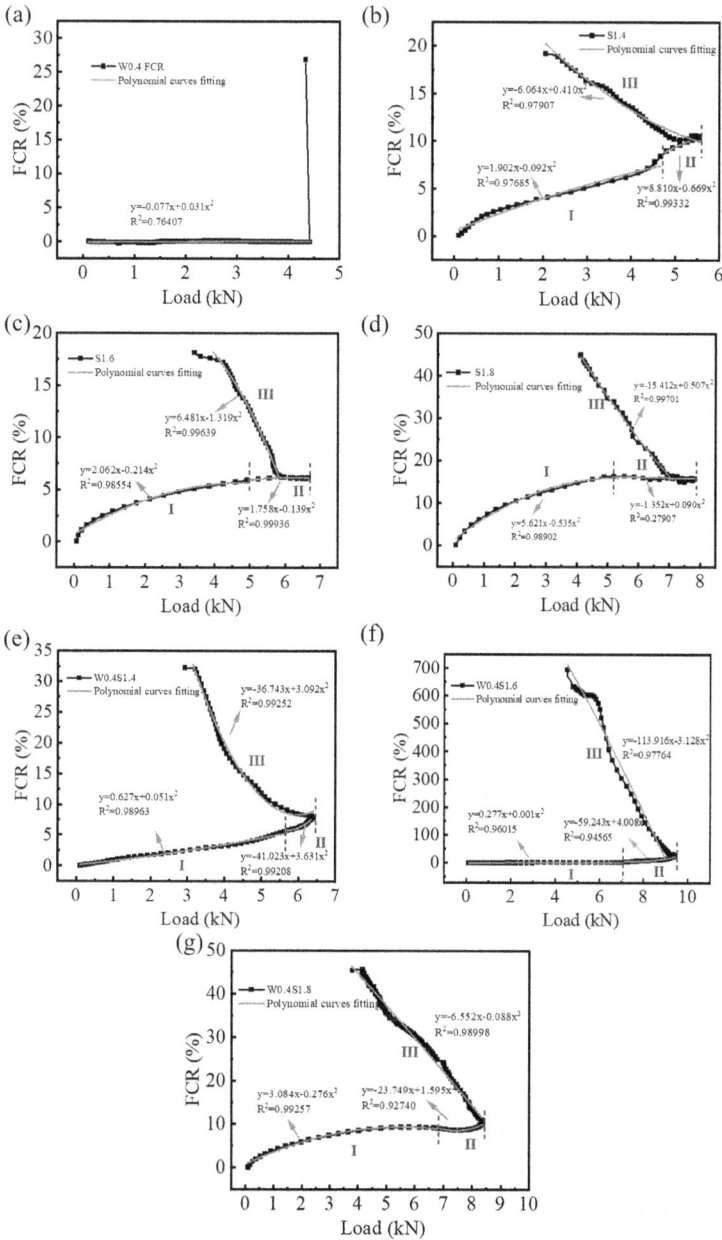

Figure 10.17 Experimental and fitted load–displacement and FCR–displacement curves for UHPC. (a) UHPC with 0.4 vol% SSWs alone; (b) UHPC with 1.4 vol% SFs alone; (c) UHPC with 1.6 vol% SFs alone; (d) UHPC with 1.8 vol% SFs alone; (e) UHPC with hybrid 0.4 vol% SSWs and 1.4 vol% SFs; (f) UHPC with hybrid 0.4 vol% SSWs and 1.6 vol% SFs; (g) UHPC with hybrid 0.4 vol% SSWs and 1.8 vol% SFs

Table 10.4 Correlation coefficient and fitting degree of the polynomial regression curves

Materials	Stage I			Stage II			Stage III		
	A	B	R²	A	B	R²	A	B	R²
W0.4	−0.077	0.031	0.76407	—	—	—	—	—	—
S1.4	1.902	0.092	0.97695	8.81	−0.669	0.99332	6.064	0.41	0.97907
S1.6	2.062	0.214	0.98554	1.758	−0.139	0.99936	6.481	1.319	0.99639
S1.8	5.621	0.535	0.98902	1.352	−0.09	0.27907	15.412	0.507	0.99701
W0.4S1.4	0.627	0.051	0.98963	41.023	3.631	0.99208	−36.743	3.092	0.99252
W0.4S1.6	0.247	0.009	0.98768	−45.130	3.192	0.97609	202.483	4.189	0.99596
W0.4S1.8	3.084	0.276	0.99257	−23.749	1.595	0.9274	−6.552	−0.088	0.98998

compared with that for UHPC with 1.6 vol%/1.8 vol% SFs alone, indicating that the conductive pathway in UHPC formed by both SSWs and SFs is susceptible to flexural loading.

Consequently, UHPC containing SSWs can be regarded as a smart material to estimate the brittle failure of specimens under flexural loading. Meanwhile, UHPC containing SFs or hybrid SSWs and SFs is intelligent in monitoring the propagation of macrocracks and the failure of the specimens. Furthermore, the hybrid SFs and SSWs-engineered multifunctional UHPC has been proved to be more stable and accurate in monitoring the initial cracking, and more sensitive in monitoring residual flexural loading and cracking development, compared with UHPC with SFs alone. This can be attributed to the inhibiting effect of SSWs on microcracks and the extensive conductive network formed by both SSWs and SFs, illustrating the brilliant synergistic effect of SSWs and SFs.

10.3.3 Conductive and sensing mechanisms

Figures 10.18 and 10.19 depict the microstructures and the equivalent circuits of hybrid SFs and SSWs-engineered UHPC. The equivalent circuit of UHPC without fillers is composed of three parts, as shown in Figure 10.19(a): the pore solution, the cement matrix, and the electrodes. It is dominated by ion conduction. Moreover, it can be seen from Figure 10.18(a) that the SSWs overlap each other by crossing, forming a three-dimensional conductive network in the cement matrix. Thus, the conductive compositions of SSWs are added to the equivalent circuit based on the composites without fillers, as shown in Figure 10.19(b), and this conductive system is dominated by the pore solution and SSWs. Furthermore, it can be seen from Figure 10.18(b) that the SFs are randomly distributed in the matrix but are not fully overlapped. Therefore, the conductive system of UHPC containing SFs alone is mainly determined by ion conduction, while SFs make up only a small fraction. Its equivalent circuit may be composed of the pore solution, the cement matrix, the electrodes, and the circuit elements caused by the SFs in series, as shown in Figure 10.19(c). In addition, Figure 10.18(c) shows that the fiber spacing can be obviously reduced by the incorporation of hybrid SSWs and SFs. While the fractional conductive pathway formed by the three-dimensional disordered SSWs and SFs

Figure 10.18 The microstructure of UHPC. (a) SSWs; (b) SFs; (c) Hybrid SSWs and SFs

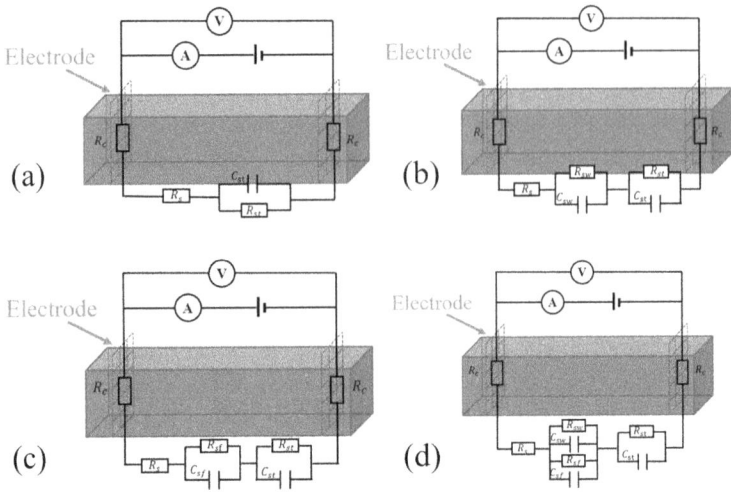

Figure 10.19 Equivalent circuit diagrams of UHPC. (a) UHPC without fillers; (b) UHPC with SSWs; (c) UHPC with SFs; (d) UHPC with hybrid SSWs and SFs R_s, R_{st}, R_{sw}, R_{sf}, and R_c represent the resistance of the pore solution, the cement matrix, SSWs, SFs and the electrodes; C_{st}, C_{sw}, and C_{sf} represent the capacitance of the cement matrix, SSWs, and SFs

facilitates the circuit integrity and effectiveness, it is worth mentioning that the connectivity between the SSWs and SFs is a key factor affecting the electrical performance [40]. Therefore, it can be concluded that the equivalent circuit of hybrid SFs and SSWs-engineered UHPC is assembled by the series circuit of the pore solution, the cement matrix, the electrodes, as well as the conductive element composed by SSWs and SFs in parallel, as shown in Figure 10.19(d).

Figure 10.20 is a schematic diagram of the sensing mechanisms of hybrid SFs and SSWs-engineered UHPC under flexural load. It should be noted that compressive stress works in the upper part of the specimen and tensile stress works in the lower part. Because SSWs can reduce the original flaws in the cement matrix, bridge the microcracks. and then form efficient conductive pathways on flexural behavior, the FCR of UHPC containing SSWs presents small fluctuation during the test. But, the FCR increases dramatically as the SSWs are gradually pulled off, and the conductive pathways are destroyed with the emerging macrocrack.

Moreover, the SFs scattered in the cement matrix fill the macroscopic pores and overlap the conductive pathway, as shown in Figure 10.20(b). Particularly, the connection of the conductive pathway is almost impervious to the increase of loading as the microcrack occurs. This can be attributed to the essentially higher intensity of UHPC containing SFs [41]. However, the conductive system reorganizes itself with the microcrack generation

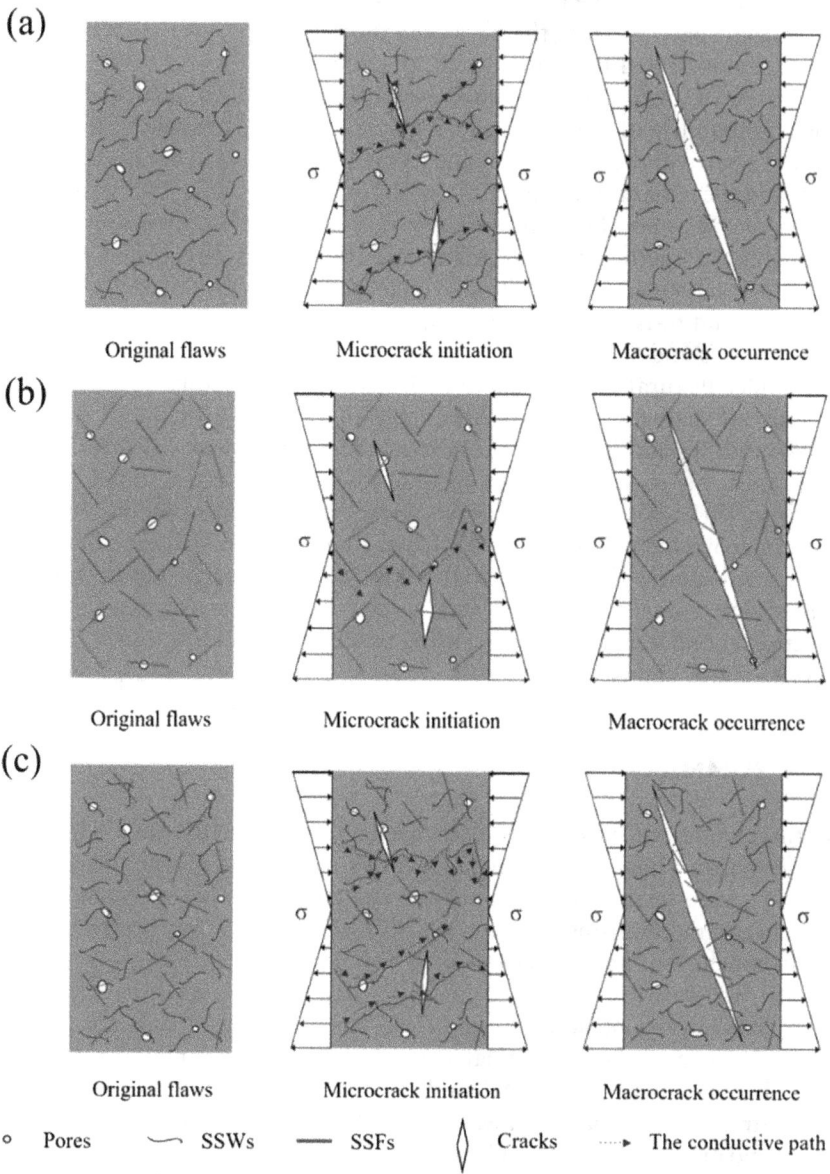

Figure 10.20 Schematic diagram of sensing mechanism of UHPC

and expansion, and the resistivity increases rapidly due to the formation of interfacial space between the fibers and the matrix.

As demonstrated in Figure 10.20, the addition of hybrid SSWs and SFs decreases the distance between these conductive fillers, which improves their probability of connecting in the local region. Meanwhile, the variation of resistivity becomes slow due to the increase of the conductive pathway and the enhancement of crack inhibition caused by the hybrid bridging effect of SWs and SFs. However, as the load increases further, the SSWs bridging the microcracks are pulled off, and the SFs bridging the macrocrack are debonded out, resulting in increased destruction of the conductive network and resistivity. Therefore, it can be concluded that the synergistic effect of SFs and SSWs causes UHPC to have excellent self-sensing properties under flexural loading for the following three reasons. (1) Due to the microscale diameter and high aspect ratio of SSWs, a low content of hybrid SSWs and SFs can form an extensive conductive network in UHPC, making it possible to sense the initiation of microcracks under initial flexural loading. (2) SSWs may not only be widely distributed in UHPC but may also improve the efficiency of SFs. This is beneficial to enhance the monitoring sensitivity of SFs fabricated UHPC to macrocrack propagation. (3) The stainless steel profile of SSWs is conducive to improving the stability of electrical conductivity and self-sensing properties of UHPC under different service status and environments.

10.4 SUMMARY

In this chapter, the flexural properties, electrical resistivity, and self-sensing performance under flexural load of SSWs-engineered UHPC incorporating SFs were introduced. Meanwhile, the corresponding toughening and conductive mechanisms for using hybrid SSWs and SFs were presented. The main conclusions are as follows:

1) SSWs at a content as low as 0.2 vol% can enhance the flexural-tensile elastic modulus, the initial macrocracking stress, and the interfacial bond strength between SFs and the UHPC matrix by modifying microstructural compactness. Meanwhile, SSWs can optimize the dispersion and orientation of SFs and form a widely distributed reinforcing network together with SFs, effectively inhibiting the initiation and convergence of microcracks to optimize the flexural cracking process of UHPC.

2) The flexural load–displacement curves of hybrid 1.8 vol% SFs and 0.2 vol% SSWs reinforced UHPC possess a multiple-cracking stage, thus increasing the flexural strength and toughness index by 87% and 86%, respectively. Meanwhile, the calculated first cracking energy is enhanced by 48%, and the additional energy of new crack is reduced

by 40%, verifying the occurrence of multiple-cracking failure. Adding SSWs has been proved to be an effective way to coordinate the performance of the UHPC matrix and SFs, making it possible to prepare highly ductile UHPC by using high–tensile strength SFs.

3) Incorporating SSWs can significantly improve the conductivity of UHPC containing SFs. SSWs-engineered multifunctional UHPC incorporating SFs can sense its crack initiation and propagation by the measured FCR, and shows more stable and accurate performance in monitoring the initial cracking and greater sensitivity in monitoring the residual flexural load and cracking development. The FCR and stress sensitivity of UHPC with hybrid 0.4 vol% SSWs and 1.6 vol% SFs achieve values of 28.31% and 0.01288%/MPa, respectively, under peak flexural displacement.

4) The conductive mechanisms of SSWs-engineered multifunctional UHPC incorporating SFs can be interpreted as an equivalent circuit composed of four parts, i.e. the series circuit of the pore solution, the cement matrix, and the electrodes, as well as the conductive element connected by SWs and SFs in parallel. The excellent sensing performances of composites mainly result from the inhibiting effect of SSWs on microcracks and the extensive conductive pathway formed by both SSWs and SFs.

REFERENCES

1. S. Ding, X. Wang, L. Qiu, Y. Ni, X. Dong, Y. Cui, A. Ashour, B. Han, J. Ou. Self-sensing cementitious composites with hierarchical carbon fiber-carbon nanotube composite fillers for crack development monitoring of a maglev girder, *Small*. 19(9) (2023) 2206258.
2. D. Wang, S. Dong, X. Wang, N. Maimaitituersun, S. Shao, W. Yang, B. Han. Sensing performances of hybrid steel wires and fibers reinforced ultra-high performance concrete for in-situ monitoring of infrastructures, *Journal of Building Engineering*. 58 (2022) 105022.
3. B. Han, X. Yu, J. Ou. *Self-Sensing Concrete in Smart Structures*, Elsevier, 2014.
4. B. Han, S. Dong, L. Zhang, S. Ding, S. Sun, Y. Wang. R&D of China's strategic new industries-functional materials, chapter 6: Functional civil engineering materials, China Machine Press (2016) 195–298. (In Chinese).
5. B. Han, L. Zhang, J. Ou. *Smart and Multifunctional Concrete Toward Sustainable Infrastructures*, Springer, 2017.
6. W. Meng, K.H. Khayat. Effect of hybrid fibers on fresh properties, mechanical properties, and autogenous shrinkage of cost-effective UHPC, *ASCE Journal of Materials in Civil Engineering*. 30(4) (2018) 04018030.
7. J.P. Romualdi, J.A. Mandel. Tensile strength of concrete affected by uniformly distributed and closely spaced short lengths of wire reinforcement, *ACI Structural Journal*. 61(6) (1964) 27–37.

8. A.E. Naaman, H.W. Reinhardt. *Strain Hardening and Deflection Hardening Fiber Reinforced Cement Composites, HPFRCC-4*, RILEM Publications, 30 (2003) 95–113.

9. S.A.S. Akers, J.B. Studinka. Ageing behavior of cellulose fibre cement composites in natural weathering and accelerated tests, *International Journal of Cement Composites and Lightweight Concrete*. 11(2) (1989) 93–97.

10. A. Bentur, S. Mindess. *Fibre Reinforced Cementitious Composites*, Elsevier, 2007.

11. J. Qi, Y. Yao, J. Wang, F. Han. Effect of sand grain size and fiber size on macro-micro interfacial bond behavior of steel fibers and UHPC mortars, *Magazine of Concrete Research*. 73(5) (2021) 228–239.

12. B. Han, S. Dong, J. Ou, C. Zhang, Y. Wang, X. Yu, S. Ding. Microstructure related mechanical behaviors of short-cut super-fine stainless wire reinforced reactive powder concrete, *Materials and Design*. 96 (2016) 16–26.

13. S. Dong, B. Han, X. Yu, J. Ou. Constitutive model and reinforcing mechanisms of uniaxial compressive property for reactive powder concrete with super-fine stainless wire, *Composites Part B: Engineering*. 166 (2019) 298–309.

14. S. Dong, D. Zhou, A. Ashour, B. Han, J. Ou. Flexural toughness and calculation model of super-fine stainless wire reinforced reactive powder concrete, *Cement and Concrete Composites*. 104 (2019) 103367.

15. S. Dong, D. Wang, X. Wang, A. D'Alessandro, S. Ding, B. Han, J. Ou. Optimizing flexural cracking process of ultra-high performance concrete via incorporating microscale steel wires, *Cement and Concrete Composites*. 134 (2022) 104830.

16. Y. Niu, J. Wei, C. Jiao. Crack propagation behavior of ultra-high-performance concrete (UHPC) reinforced with hybrid steel fibers under flexural loading, *Construction and Building Materials*. 294(10) (2021) 123510.

17. D. Yoo, S. Kim, J. Park. Comparative flexural behavior of ultra-high-performance concrete reinforced with hybrid straight steel fibers, *Construction and Building Materials*. 132 (2017) 219–229.

18. J. Liu, C. Li, Z. Du, G. Cui. Characterization of fiber distribution in steel fiber reinforced cementitious composites with low water-binder ratio, *Indian Journal of Engineering and Materials Sciences*. 18(6) (2011) 449–457.

19. W. Meng, K.H. Khayat. Improving flexural performance of ultra-high-performance concrete by rheology control of suspending mortar, *Composites Part B: Engineering*. 117 (2017) 26–34.

20. S. Lee, J.Youn, J. Hyun. Prediction of fiber orientation structure for injection molded short fiber composites, *Materials Research Innovations*. 6(4) (2002) 189–197.

21. G. Zak, M. Haberer, C.B. Park, B. Benhabib. Estimation of average fibre length in short-fibre composites by a two-section method, *Composites Science and Technology*. 60(9) (2000) 1763–1772.

22. M.J. Shannag, R. Brincker, W. Hansen. Pullout behavior of steel fibers from cement-based composites, *Cement and Concrete Research*. 27(6) (1997) 925–936.

23. S. Ding, Y. Xiang, Y. Ni, V.K. Thakur, X. Wang, B. Han, J. Ou. In-situ synthesizing carbon nanotubes on cement to develop self-sensing cementitious composites for smart high-speed rail infrastructures, *Nano Today*. 43 (2022) 101438.

24. D. Fan, Y. Rui, K. Liu, J. Tan, Z. Shui, C. Wu, S. Wang, Z. Guan, Z. Hu, Q. Su. Optimized design of steel fibres reinforced ultra-high performance concrete (UHPC) composites: Towards to dense structure and efficient fibre application, *Construction and Building Materials*. 273 (2021) 121698.

25. A.O. Al-Mwanes, R. Agayari. Studying the effect of hybrid fibers and silica fumes on mechanical properties of ultra-high-performance concrete, *IOP Conference Series: Materials Science and Engineering*. 1076(1) (2021) 012128.

26. V.C. Li. From micromechanics to structural engineering-the design of cementitious composites for civil engineering applications, *Structural Engineering/ Earthquake Engineering*. 10 (1993) 37–48.

27. I. Markovic, J. Walraven, J. Mier. Development of high-performance hybrid fibre concrete // *High Performance Fiber Reinforced Cement Composites (HPFRCC4)*, RILEM Publications. (2003) 277–300.

28. P. Tjiptobroto, W. Hansen. Tensile strain hardening and multiple cracking in high performance cement-based composites containing discontinuous fibers, *ACI Materials Journal*. 90 (1993) 16025.

29. H. Wei, T. Liu, A. Zhou, D. Zou, Y. Liu. Multiscale insights on enhancing tensile properties of ultra-high performance cementitious composite with hybrid steel and polymeric fibers, *Journal of Materials Research and Technology*. 14 (2021) 743–753.

30. M. Xu, J. Yu, J. Zhou, Y. Bao, V.C. Li. Effect of curing relative humidity on mechanical properties of engineered cementitious composites at multiple scales, *Construction and Building Materials*. 284 (2021) 122834.

31. S. Dong, D. Zhou, Z. Li, X. Yu, B. Han. Super-fine stainless wires enabled multifunctional and smart reactive powder concrete, *Smart Materials and Structures*. 28(12) (2019) 125009.

32. S. Dong, B. Han, J. Ou, Z. Li, L. Han, X. Yu. Electrically conductive behaviors and mechanisms of short-cut super-fine stainless wire reinforced reactive powder concrete, *Cement and Concrete Composites*. 72 (2016) 48–65.

33. S. Dong, W. Zhang, D. Wang, X. Wang, B. Han. Modifying self-sensing cement-based composites through multiscale composition, *Measurement Science and Technology*. 32(7) (2021) 074002.

34. L. Qiu, S. Dong, X. Yu, B. Han. Self-sensing ultra-high performance concrete for in-situ monitoring, *Sensors and Actuators A: Physical*. 331 (2021) 113049.

35. A. Dinesh, D. Suji, M. Pichumani. Electro-mechanical investigations of steel fiber reinforced self-sensing cement composite and their implications for real-time structural health monitoring, *Journal of Building Engineering*. 51 (2022) 104343.

36. I. You, D.Y. Yoo, S. Kim, M.J. Kim, G. Zi. Electrical and self-sensing properties of ultra-high-performance fiber-reinforced concrete with carbon nanotubes, *Sensors*. 17(11) (2017) 2481.

37. B. Han, L. Zhang, S. Sun, X. Yu, X. Dong, T. Wu, J. Ou. Electrostatic self-assembled carbon nanotube/nano carbon black composite fillers reinforced cement-based materials with multifunctionality, *Composites Part A: Applied Science and Manufacturing*. 79 (2015) 103–115.

38. S. Dong, X. Dong, A. Ashour, B. Han, J. Ou. Fracture and self-sensing characteristics of super-fine stainless wire reinforced reactive powder concrete, *Cement and Concrete Composites*. 105 (2020) 103427.

39. D.Y. Yoo, I. You, S.J. Lee. Electrical and piezoresistive sensing capacities of cement paste with multi-walled carbon nanotubes, *Archives of Civil and Mechanical Engineering*. 18(2) (2018) 371–384.

40. S.J. Lee, I. You, S. Kim, H.O. Shin, D.Y. Yoo. Self-sensing capacity of ultra-high-performance fiber-reinforced concrete containing conductive powders in tension, *Cement and Concrete Composites*. 125 (2022) 104331.

41. S. Ding, C. Xu, Y. Ni, B. Han. Extracting piezoresistive response of self-sensing cementitious composites under temperature effect via Bayesian blind source separation method, *Smart Materials and Structures*. 30(6) (2021) 065010.

Chapter 11

Stainless steel wires-engineered multifunctional ultra-high performance concrete fabricated with sea water and sea sand

11.1 INTRODUCTION

The increasing depletion of river sand and freshwater resources has raised a series of challenges to the development of coastal and marine infrastructure, such as ecological damage and high project costs [1–3]. Furthermore, the earth can provide almost infinite seawater and sea sand resources [4]. Consequently, utilizing seawater and sea sand in place of freshwater and river sand in the production of concrete is beneficial for mitigating natural resource depletion, and the costs of materials and transportation can be substantially decreased [5, 6]. However, the endogenous chloride ions inside concrete made with seawater and sea sand lead to a significant corrosion risk to steel fibers or steel bars used for toughening concrete [7]. The rusted steel fibers/steel bars can accelerate the degradation of the mechanical properties of concrete structures [8] as well as decrease the durability and aesthetic performance of buildings [9].

Using stainless steel fibers and improving the microstructural compactness and chloride binding capacity of concrete makes it possible to prepare concrete with seawater and sea sand. Ultra-high performance concrete (UHPC) is a kind of concrete with compact packing structure and excellent strength, toughness, and durability [10, 11]. Meanwhile, one of the mixing ratio features of UHPC is the use of large amounts of mineral admixtures, resulting in a high content of calcium silicate hydrate gels in its hydration products and a corresponding high chloride binding capacity [12]. Meanwhile, stainless steel wires (SSWs) with high corrosion resistance and micro diameter can further improve the structural compactness of UHPC, reduce the weak interface in UHPC, and then endow UHPC with excellent mechanical and multifunctional/smart properties, as verified by Chapters 2–8. Based on this analysis, seawater and sea sand have the potential to be used in place of freshwater and river sand to fabricate SSWs-engineered multifunctional UHPC.

Hence, SSWs-engineered multifunctional UHPC fabricated with seawater and sea sand (UHPSSC) is introduced in this chapter, and its mechanical properties and chloride binding capacity are also introduced. Furthermore,

DOI: 10.1201/9781003276357-11

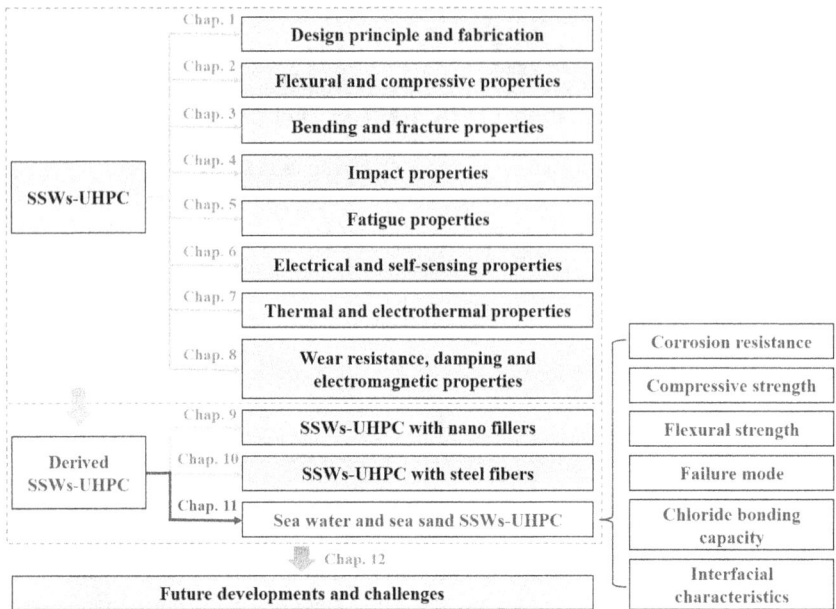

Figure 11.1 Main contents of Chapter 11

analyses of the microscopic appearance and chemical composition of SSWs after immersion in seawater for 2 years are presented using scanning electron microscope (SEM) and energy dispersive spectrometry (EDS) observation, together with the interface of SSWs–matrix and sea sand–matrix. The main contents of this chapter are shown in Figure 11.1.

11.2 SEAWATER CORROSION RESISTANCE OF SSWS

Seawater is the primary source of chloride in SSWs-engineered multifunctional UHPSSC, imposing a high requirement for the seawater corrosion resistance of reinforcing fibers in UHPSSC. SSWs were immersed in natural seawater for 2 years at room temperature (20 °C). Then, the microstructures and chemical composition of SSWs before and after immersion in seawater for 2 years were analyzed using SEM and EDS, as shown in Figure 11.2. It can be seen that the diameter of the SSWs did not change significantly, and no obvious pitting corrosion was observed on the surface of the SSWs after immersion in seawater for 2 years. Then, EDS point scanning analysis was carried out on the surface of the SSWs. Cr, Ni, Fe, and Mo are the main elements of 316L stainless steel. Figure 11.3 illustrates that the Cr, Ni, Fe, and Mo element contents of the two samples before and after immersion in seawater for 2 years were the same, indicating that the SSWs did not

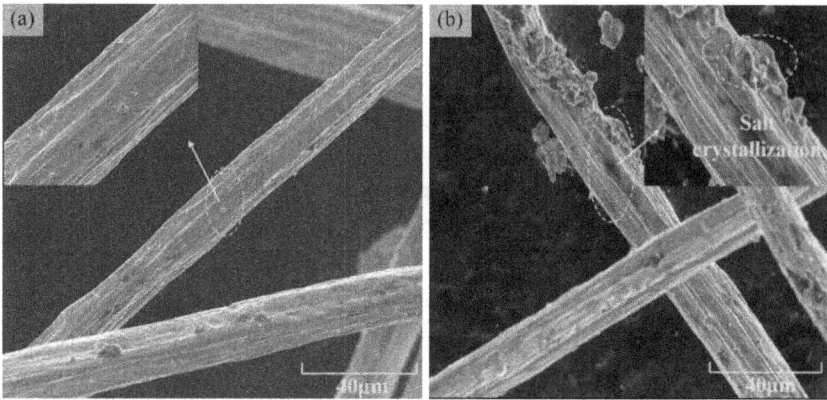

Figure 11.2 Micromorphology of SSWs before and after immersion in seawater for 2 years. (a) Before immersion in seawater; (b) After immersion in seawater for 2 years

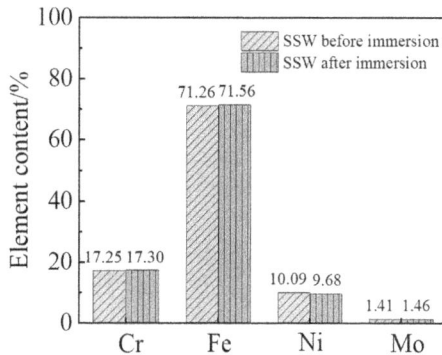

Figure 11.3 Element contents of SSWs before and after immersion in seawater for 2 years

rust in seawater. It is worth noting that there were some crystals on the surface of the SSWs after immersion in seawater. EDS results showed that the main elements in this area were Cl and Na, indicating that the crystals were crystalline salts in seawater rather than rust. The SSWs are made from 316L stainless steel by strong drawing. There is a dense passive film mainly composed of Cr_2O_3 located at the surface of ordinary stainless steel, but the passive film can be destroyed in the chloride-rich environment, resulting in the transformation of stainless steel from a passive state to an active state [13]. Unlike ordinary stainless steel, 316L stainless steel contains a certain amount of Mo element, which provides a beneficial effect on pitting resistance in the test solution, because the molybdate is an anodic inhibitor [14]. The synergistic effect of Mo and Cr can improve the passivation ability of stainless steel and the stability of the passivation film [15]. Therefore, SSWs are expected to endow UHPSSC with long-term retention of mechanical properties due to the high resistance to corrosion in seawater.

11.3 MECHANICAL PROPERTIES OF SSWS-ENGINEERED MULTIFUNCTIONAL UHPC FABRICATED WITH SEAWATER AND SEA SAND

The flexural and compressive strengths of SSWs-engineered multifunctional UHPSSC are shown in Figure 11.4. Because the compressive strength of UHPSSC reinforced with 0.5 vol%/1.0 vol%/1.5 vol% SSWs (marked W201005, W201010, and W201015, respectively) at the curing age of 28 days is greater than 120 MPa, the composites prepared in this study all reach the mechanical requirement of UHPC. Besides, the flexural and compressive strengths of UHPSSC increase as the SSWs content increase from 0 vol% (marked W0) to 1.5 vol%. Due to the addition of 0.5 vol%, 1.0 vol%, and 1.5 vol% SSWs, the flexural strength of UHPSSC is improved by 2.14 MPa/24.3%, 3.88 MPa/44.1%, and 5.00 MPa/56.8%, respectively, whereas the compressive strength is improved by 6.61 MPa/5.7%, 17.76 MPa/15.3%, and 22.48 MPa/19.4%, respectively. The extensive three-dimensional network formed by SSWs with micro diameters and high aspect ratios exerts a crack-blocking and toughening effect, thus contributing to the increase in flexural strength. In addition, porosity is closely related to compressive strength [16]. Incorporating SSWs reduces the porosity of UHPSSC and refines the pore size distribution, resulting in a better level of compactness in the matrix. This could be a significant factor contributing to the enhancement effect of SSWs.

The mechanical properties of SSWs-engineered multifunctional UHPSSC are compared with SSWs-engineered multifunctional UHPC in Chapter 2

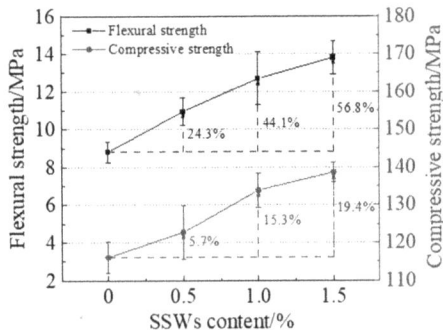

Figure 11.4 Flexural and compressive strengths of SSWs-engineered multifunctional UHPSSC

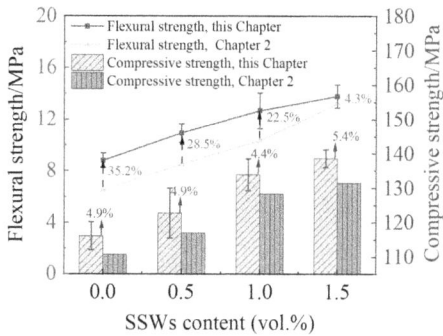

Figure 11.5 Flexural and compressive strengths of SSWs-engineered multifunctional UHPSSC compared with SSWs-engineered multifunctional UHPC

with freshwater and quartz sand under the same mix proportion. Figure 11.5 demonstrates that the flexural strengths of UHPSSC reinforced with 0 vol%, 0.5 vol%, 1.0 vol%, and 1.5 vol% SSWs are 2.29 MPa/35.2%, 2.43 MPa/28.5%, 2.33 MPa/22.5%, and 0.57 MPa/4.3%, respectively, higher than those of UHPC, and the compressive strengths are 5.43 MPa/4.9%, 5.74 MPa/4.9%, 5.59 MPa/4.4%, and 7.11 MPa/5.4% higher, respectively, than those of UHPC. Interestingly, the flexural strength of SSWs-engineered multifunctional UHPSSC exceeds that of SSWs-engineered multifunctional UHPC, but as SSWs content increases, the gap between them gradually narrows. However, the compressive strength of UHPSSC slightly exceeds that of UHPC with the same mix proportion, and the maximum increase range is only 5.4%. It has been shown that at the age of more than 28 days, there is almost no variance between the mechanical properties of freshwater-mixed and seawater-mixed composites [2, 5], indicating that UHPSSC has excellent mechanical properties equivalent to those of UHPC, and the use of seawater and desalinated sea sand instead of freshwater and quartz sand has no adverse effect on the early mechanical properties of UHPSSC. Considering that chlorides are mainly from natural seawater in this study, the difference in strength between UHPSSC and UHPC is mainly due to the presence of seawater. The effects of seawater on the early strength gain of UHPSSC can be summarized as follows. Firstly, the abundant ions inside seawater can accelerate the hydration of cementitious materials and increase the degree of polymerization of the hydration products [17], particularly enhancing the early hydration reaction of silica fume and thus improving the early strength [18]. Additionally, seawater has the ability to improve the indentation modulus and indentation hardness of the cement paste and interfacial transition zone (ITZ) [19]. Furthermore, seawater can also reduce porosity and refine the pore size distribution by enhancing the hydration reaction [19], and the chloride ion and AFm-like phase can form Friedel's salt, thus promoting the densification of the microstructure [20].

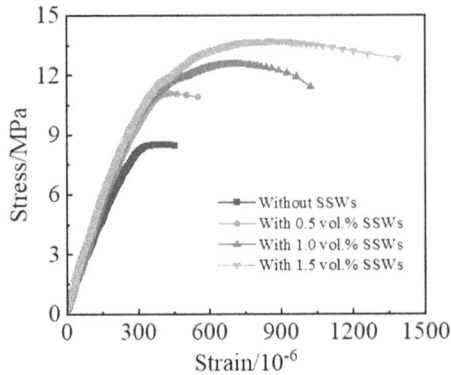

Figure 11.6 Flexural stress–strain curves of SSWs-engineered multifunctional UHPSSC

Table 11.1 Flexural toughness and peak strain of SSWs-engineered multifunctional UHPSSC

SSWs content (vol%)	Flexural toughness (J/m³)	Increasing range (%)	Peak strain (×10⁻⁶)	Increasing range (%)
0	2682.57	–	392.08	–
0.5	4304.34	60.4	436.59	11.4
1.0	10307.81	286.6	694.61	77.2
1.5	14187.12	428.9	859.78	119.3

The stress–strain curves of SSWs-engineered multifunctional UHPSSC are shown in Figure 11.6. As the SSWs content increases, the elastic stage of the stress–strain curves steadily grows longer, and incorporating SSWs slightly increases the slope of the linear segment representing the modulus of elasticity, as illustrated in Figure 11.6. The proportional limit gradually increases; meanwhile, the non-linear strain-hardening stage gradually becomes noticeable. The falling branch of the stress–strain curves continues to gradually extend at a lower rate, and the absence of a drop section in the stress–strain curve is because the primary fracture occurs at the bottom of the specimens upon achieving the peak load, leading to the breakage of the strain gauge at this position. The flexural toughness and peak strain of SSWs-engineered UHPSSC are shown in Table 11.1. As a consequence of adding 0.5 vol%, 1.0 vol%, and 1.5 vol% SSWs, the flexural toughness based on the stress–strain curves is improved by 60.4%, 286.6%, and 428.9%, respectively. Moreover, the peak strain is improved by 11.4%, 77.2%, and 119.3%, respectively. Overall, the incorporation of SSWs greatly improves the flexural toughness and ductility of UHPSSC, making it a crucial aspect for the construction of highly resilient coastal and marine infrastructure.

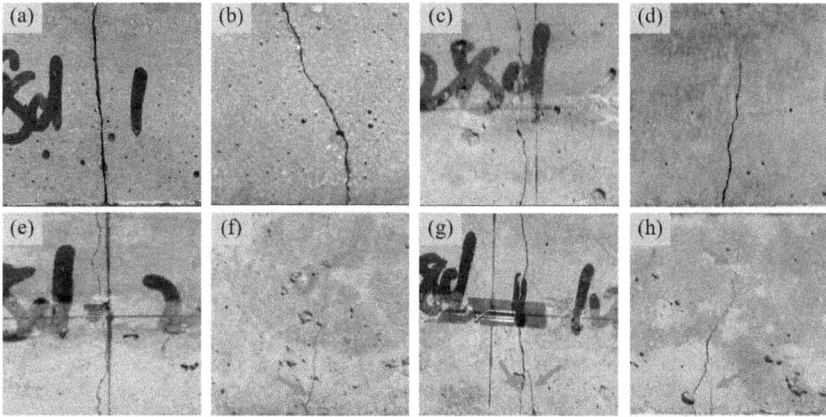

Figure 11.7 Crack images of the failure specimens after flexural load. (a) W0 flexural-tensile bottom surface; (b) W0 flexural-tensile side surface; (c) W201005 flexural-tensile bottom surface; (d) W05 flexural-tensile side surface; (e) W201010 flexural-tensile bottom surface; (f) W201010 flexural-tensile side surface; (g) W201015 flexural-tensile bottom surface; (h) W201015 flexural-tensile side surface

Figure 11.7 illustrates the crack images of the fracturing specimens after flexural load. The flexural-tensile bottom surface refers to the surface subjected to tension under bending, and it is adjacent to the flexural-tensile side surface. As shown in Figure 11.7(a)–(f), as the SSWs content increases from 0 to 1.0 vol%, the crack widths on the flexural-tensile bottom surface gradually decrease, and the bending degree of cracks increases. Meanwhile, the crack lengths on the flexural-tensile side surface gradually decrease as the SSWs content increases, showing that SSWs can effectively hinder the extension of large cracks. Figure 11.7(g) and (h) also show that a fine micro-crack forms near the main crack when the SSWs content is 1.5 vol%, suggesting a transition to a more ductile pseudo-strain-hardening behavior of UHPSSC with 1.5 vol% SSWs. It can be concluded that the flexural toughness of UHPSSC is significantly improved due to the inhibition of crack propagation and the enhanced strain-hardening behavior of SSWs.

The flexural stress–strain curves, flexural toughness, and peak strain of SSWs-engineered multifunctional UHPSSC are compared with those of SSWs-engineered UHPC in Chapter 2 as well as steel fibers reinforced UHPC in Chapter 10. As shown in Figure 11.8(a), the addition of seawater and sea sand increases the proportional limit, peak stress, and peak strain, and there is no significant change in the modulus of elasticity and strain-hardening stage, confirming that UHPSSC has excellent mechanical properties equivalent to UHPC composed of freshwater and quartz sand. Figure 11.8(a) also shows that the flexural strength of UHPC with 1.4 vol% and 1.6 vol% steel fibers is 9.33 MPa and 14.3 MPa, respectively, while

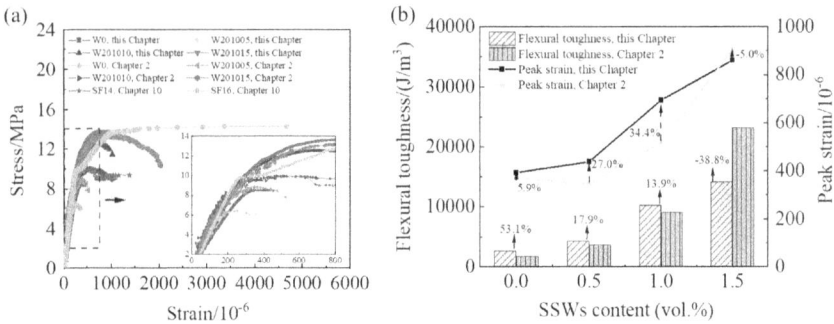

Figure 11.8 Stress–strain curves, flexural toughness, and peak strain of SSWs-engineered multifunctional UHPSSC compared with SSWs-engineered multifunctional UHPC and steel fibers reinforced UHPC. (a) Stress–strain curves; (b) Flexural toughness and peak strain

the flexural strength of UHPSSC with 1.5 vol% SSWs in this study reaches 13.8 MPa. Moreover, there is no significant difference between the strain-hardening stage caused by the addition of SSWs and that resulting from the addition of steel fibers, illustrating the great potential of SSWs as an alternative material to steel fiber as the reinforcing fiber. It should be noted that the diameter of SSWs is finer than that of steel fibers, and fibrous rupture may occur when SSWs-engineered multifunctional UHPSSC is subjected to flexural load, resulting in a loss of flexural toughness. The higher flexural toughness of steel fibers reinforced UHPC can be explained by the pull-out effect of steel fiber [21]. Figure 11.8(b) shows that the flexural toughness of UHPSSC reinforced with 0 vol%, 0.5 vol%, and 1.0 vol% SSWs is 53.1%, 17.9%, and 13.9% higher, respectively, than that of UHPC composed of freshwater and quartz sand, whereas the peak strain is 5.9%, 27.0%, and 34.4% higher, respectively, than that of UHPC composed of freshwater and quartz sand. Interestingly, the flexural toughness and peak strain of UHPSSC with 1.5 vol% SSW are 38.8% and 5.0% lower, respectively, than those of UHPC composed of freshwater and quartz sand. It can be concluded that the flexural toughness of UHPSSC can be improved due to the use of seawater.

11.4 CHLORIDE BINDING CAPACITY OF SSWS-ENGINEERED MULTIFUNCTIONAL UHPC FABRICATED WITH SEAWATER AND SEA SAND

The chloride content of SSWs-engineered UHPC fabricated with seawater and sea sand (UHPSSC) is exhibited in Figure 11.9, indicating that the bound chloride content of UHPSSC with different SSWs content is

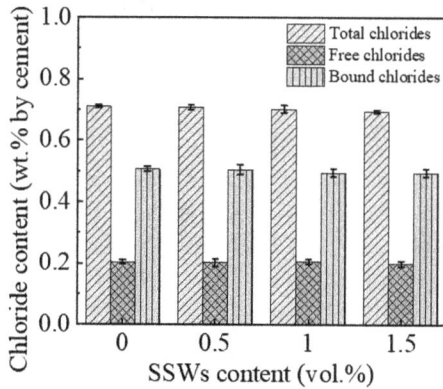

Figure 11.9 Chloride content of SSWs-engineered multifunctional UHPSSC

approximately the same. In this study, the chlorides in UHPSSC are mainly introduced by seawater, which falls into the category of endogenous chlorides. The bound chloride content of UHPSSC reinforced with 0 vol%, 0.5 vol%, 1.0 vol%, and 1.5 vol% SSWs is 0.51%, 0.51%, 0.50%, and 0.50% by weight of cement, respectively, which is consistent with the range of 0.38% to 0.52% of seawater paste previously reported by Wang et al. [22]. The chloride binding capacity of SSWs-engineered multifunctional UHPSSC is mainly related to the UHPSSC matrix. Because of the high strength of the UHPSSC matrix, chlorides introduced by seawater are physically adsorbed by the large amount of C-S-H gels in the UHPSSC matrix due to the formation of an ionic bond between chloride ions and unionized SiOH [23]. Moreover, fly ash and silica fume can undergo secondary hydration reactions with CH crystals to generate more C-S-H gels due to high pozzolanic activity, further immobilizing more chlorides inside the UHPSSC matrix. Hence, SSWs inside UHPSSC have a low risk of rusting due to the excellent corrosion resistance of SSWs to chloride in seawater and the high chloride binding capacity of the UHPSSC matrix.

Figure 11.9 shows that the free chloride content of SSWs-engineered multifunctional UHPSSC is approximately 0.20% by weight of cement and was approximately 0.13% by weight of cementitious materials. According to ACI 318-19 [24], the maximum limit of water-soluble chloride content in non-prestressed concrete exposed to seawater (C2 exposure class) is 0.15% by weight of cementitious materials, indicating that the free chloride content of SSWs-engineered multifunctional UHPSSC meets the requirements of the specification, and the use of seawater and desalinated sea sand has great potential for the production of UHPC with low chloride corrosion risk.

11.5 INTERFACIAL CHARACTERISTICS OF SSWS-ENGINEERED MULTIFUNCTIONAL UHPC FABRICATED WITH SEAWATER AND SEA SAND

11.5.1 Fiber–matrix interface

Figure 11.10 shows the C-S-H gels of the reference mixture and the C-S-H gels and fiber–matrix interface of 1.5 vol% SSWs-engineered multifunctional UHPSSC, while the results for the molar ratios of CaO to SiO_2 (Ca/Si ratios) analyzed by EDS are demonstrated in Figure 11.11(a). The average Ca/Si ratios at the locations of C-S-H gels of the reference mixture without SSWs and C-S-H gels of 1.5 vol% SSWs reinforced UHPSSC, the surface of SSWs, and the fiber–matrix interface are 1.127, 1.143, 0.637, and 0.737, respectively. SSWs have no influence on the Ca/Si ratios of C-S-H gels. The Ca/Si ratios of the C-S-H gels are 1.79 and 1.55 times those of the surface of SSWs and the concrete matrix, respectively. Silica fume with high pozzolanic activity has the tendency to become enriched on the high specific surface area of SSWs during stirring, resulting in the reduction of water/binder ratios and Ca/Si ratios during hydration in this region [25, 26]. Consequently, a large amount of C-S-H gel has the potential to form on the surface of SSWs, thus achieving a strong interfacial bond between SSWs and concrete. Therefore, no weak ITZ between SSWs and the concrete matrix is observed, explaining the underlying mechanisms by which SSWs reinforce UHPSSC, leading to significant improvement of its toughness without reducing its compressive strength. Therefore, the application of SSWs to UHPSSC has great benefits due to their corrosion-resistant properties, strong bonding with the matrix, significant toughness enhancement, and no adverse effect on strength.

Because of the large quantity of endogenous chlorides inside UHPSSC, it is essential to understand the chloride distribution in SSWs-engineered multifunctional UHPSSC. Wang et al. [27] used Cl/(Ca + Al) ratios to evaluate

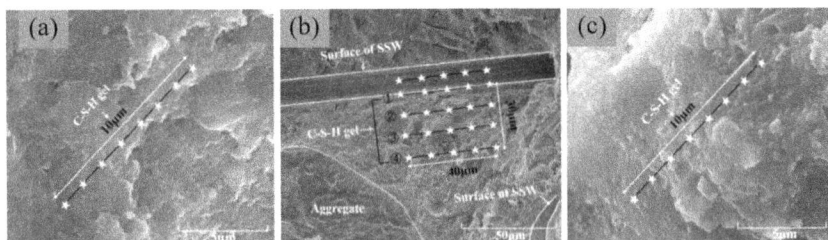

Figure 11.10 C-S-H gel of reference mixture, C-S-H gel, and fiber–matrix interface of 1.5 vol% SSWs -engineered multifunctional UHPSSC. (a) C-S-H gel of reference mixture; (b) C-S-H gel of 1.5 vol% SSWs-engineered multifunctional UHPSSC; (c) Fiber–matrix interface of 1.5 vol% SSWs-engineered multifunctional UHPSSC

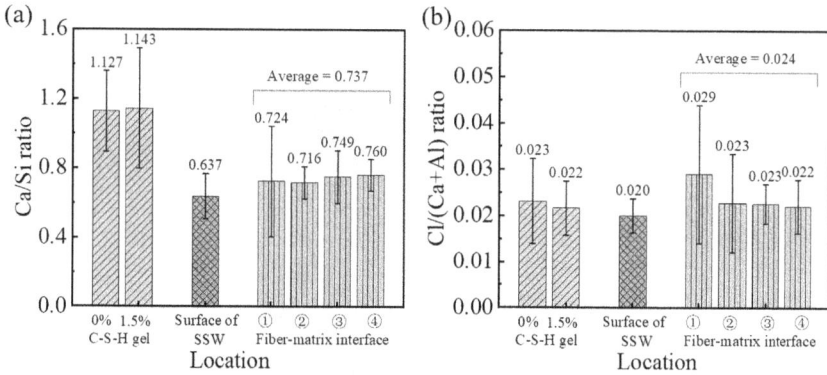

Figure 11.11 Ca/Si ratios and Cl/(Ca + Al) ratios of C-S-H gel, the surface of SSWs, and the fiber–matrix interface. (a) Ca/Si ratios; (b) Cl/(Ca + Al) ratios

the chloride binding capacity, because it is mainly correlated with the calcium phase and aluminum phase. The results for Cl/(Ca + Al) ratios at different locations analyzed by EDS are demonstrated in Figure 11.11(b). The average Cl/(Ca + Al) ratios at the locations of C-S-H gels of the reference mixture and C-S-H gels of 1.5 vol% SSWs-engineered multifunctional UHPSSC, the surface of SSWs, and the fiber–matrix interface are 0.023, 0.022, 0.020, and 0.024, respectively. Therefore, the addition of SSWs has no influence on the chloride content of C-S-H gels, consistent with the findings obtained earlier. Interestingly, a relatively high Cl/(Ca + Al) ratio is observed near SSWs. But, chlorides are evenly distributed at the SSWs–matrix interface, while no obvious chloride enrichment has been observed in this area.

Figure 11.12 shows the microstructures of SSWs-engineered multifunctional UHPSSC. According to the preceding analysis, the following categories can be used to classify the reinforcing effect of SSWs:

1) Pull out and pull off. As shown in Figure 11.12(a), when the composites are loaded to cracking, the matrix transmits the load to SSWs through the interface, consequently blocking any further crack growth, because pulling out or pulling off involves absorbing a lot of energy.

2) Bridging effect. As shown in Figure 11.12(b), SSWs can transfer stress from the crack tip and bridge the crack, resulting in stress redistribution and thus delaying the further expansion of cracks. Furthermore, after the flexural loading process is completed, cracks are only observed in the tensile area at the bottom of the specimen. No overall fracture occurs due to the bridging effect of SSWs.

3) Overlapping effect. Figure 11.12(c) illustrates that a great number of SSWs overlap each other to form a three-dimensional network. These

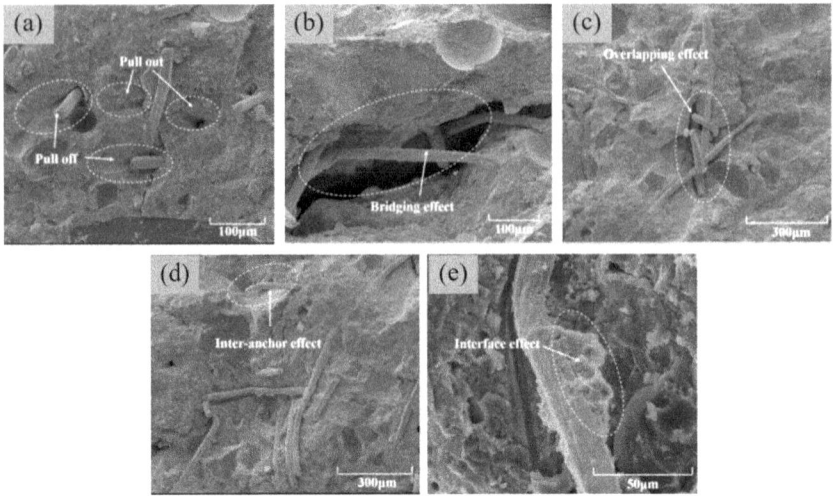

Figure 11.12 The reinforcing effect of SSWs on UHPSSC. (a) Pull out and pull off; (b) Bridging effect; (c) Overlapping effect; (d) Inter-anchor effect; (e) Interface effect

networks markedly hinder the creation and propagation of cracks as well as prolonging the propagation path.

4) Inter-anchor effect. Figure 11.12(d) illustrates many hydration products wrapping a single SSW to form an inner-anchor interface. The interface resistance overcome by the crack propagation increases greatly when the SSWs are pulled off or pulled out.

5) Interface effect. Figure 11.12(e) also reveals that there is a thick layer of hydration products on the surface of SSWs. The high specific surface area of SSWs and enrichment of silica fume on its surface enhance the interfacial bond between SSWs and concrete, thereby giving full play to the above mechanisms and causing SSWs to consume more energy when they are pulled off or pulled out.

11.5.2 Interfacial transition zone between sea sand and cement paste

Figures 11.13 and 11.14 show the ITZ between sea sand and cement paste of 1.5 vol% SSWs-engineered multifunctional UHPSSC and element contents along the perpendicular direction of the aggregate surface, respectively. The thickness of the ITZ can be estimated from the variable tendency of element contents. From Figure 11.14(a), it can be inferred that the interfacial thickness of SSWs-engineered multifunctional UHPSSC is about 20 μm according to the changing trend of Al and Mg. The Ca/Si ratio increases gradually and then decreases to a stable value when the distance from the aggregate surface increases, in agreement with previous research results [28]. As illustrated in Figure 11.14(b), the Cl/(Ca + Al) ratio at a distance of

Figure 11.13 ITZ of 1.5 vol% SSWs-engineered multifunctional UHPSSC

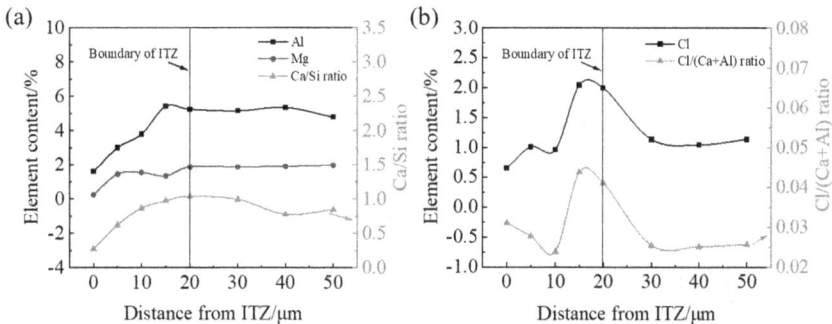

Figure 11.14 Element contents along the perpendicular direction of the aggregate surface. (a) Ca/Si ratio; (b) Cl/(Ca + Al) ratio

15 μm reaches 0.044; then, the Cl/(Ca + Al) ratio gradually decreases and stabilizes at 0.025, which agrees with the finding that the Cl/(Ca + Al) ratio of the fiber–matrix interface is 0.024. It can be concluded that chlorides are evenly distributed in the matrix and fiber–matrix interface of UHPSSC, but chlorides are enriched in the ITZ, where the maximum chloride content reaches 1.76 times that of the matrix.

Figure 11.15 shows the hydration products in the ITZ of 1.5 vol% SSWs-engineered multifunctional UHPSSC. As shown in Figure 11.15(a), CH crystals with large size and disordered arrangement are observed in the scope of aggregate ITZ, consistent with the characteristics of low hydration degree and high CH content in ITZ. As depicted in Figure 11.15(b), EDS mapping analysis is conducted in this location. Chlorides are enhanced in the region surrounding CH crystals, proving the existence of Friedel's salt because the main elements are mainly O, Ca, Al, and Cl; thus, more chlorides are

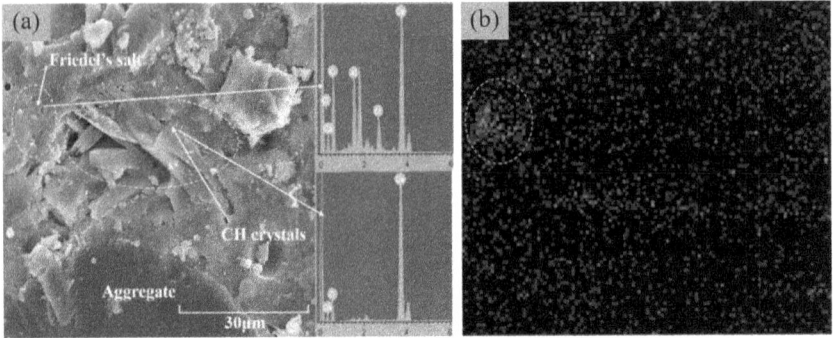

Figure 11.15 The hydration products in ITZ. (a) CH crystals and Friedel's salt; (b) Distribution of chlorides in ITZ by EDS mapping

immobilized in ITZ. Additionally, Friedel's salt and the C-S-H gels with a high Ca/Si ratio in the ITZ are beneficial for immobilizing chlorides. The following causes can be used to explain the chloride enrichment in ITZ. Firstly, the high Ca/Si ratio influences the surface electricity of C-S-H gels in ITZ, thereby requiring more chloride ions to be adsorbed physically to balance the positive charge [29], consistently with the findings in Figure 11.14(a). As the distance from the aggregate surface increases to 10–20 μm, the Ca/Si ratio in this area is larger than that in the matrix, thus increasing the physical adsorption of C-S-H gels to chlorides. Secondly, Friedel's salt has been formed in the ITZ through an anion exchange mechanism to chemically bind more chlorides [30, 31].

The chloride immobilization mechanisms of the ITZ can be summarized as follows. The strong physical adsorption of C-S-H gels with a high Ca/Si ratio makes chloride migration difficult through the cement paste matrix, and Friedel's salt further improves chloride migration resistance through the fine pore structure. Due to the chloride immobilization effect of ITZ, the risk of long-term corrosion of SSWs inside the UHPSSC can be significantly reduced, holding great potential for SSWs-engineered multifunctional UHPSSC to maintain excellent long-term mechanical properties and durability as well as the quality of its appearance.

11.6 SUMMARY

In this chapter, the corrosion resistance of SSWs and the mechanical properties and chloride binding capacity of SSWs-engineered multifunctional UHPC fabricated with seawater and sea sand (UHPSSC) were introduced. The reinforcing mechanisms of SSWs on UHPSSC and the chloride immobilization mechanisms were presented. The main conclusions are as follows:

1) SSWs do not rust after immersion in seawater for 2 years, providing excellent fibers for fabricating UHPSSC with long-term high mechanical properties.

2) The flexural and compressive strength of UHPSSC reinforced with 1.5 vol% SSWs was 13.8 MPa and 138.6 MPa, respectively, and the flexural toughness is increased by 428.9% compared with that of UHPSSC without SSWs, reaching the basic mechanical requirements of UHPC.

3) The main mechanisms by which SSWs reinforce UHPSSC include being pulled out and pulled off, bridging, the interface effect, the inter-anchor effect, and overlapping network effects. Moreover, the high specific surface area of SSWs and enrichment of silica fume on its surface enhance the interfacial bond between SSWs and matrix, thus further giving full play to these mechanisms.

4) C-S-H gels with a high Ca/Si ratio within the ITZ as well as Friedel's salt are conducive to immobilizing chlorides, thereby blocking the migration of chlorides through the cement paste matrix and further mitigating the risk of long-term chloride corrosion of SSWs.

REFERENCES

1. J. Teng, T. Yu, J. Dai, G. Chen. FRP composites in new construction: Current status and opportunities, 7th National Conference on FRP Composition, Beijing, Industrial Construction (2011).

2. J. Xiao, C. Qiang, A. Nanni, K. Zhang. Use of sea-sand and seawater in concrete construction: Current status and future opportunities, *Construction and Building Materials*. 155 (2017) 1101–1111.

3. M. Marvila, P. de Matos, E. Rodriguez, S.N. Monteiro, A.R.G. de Azevedo. Recycled aggregate: A viable solution for sustainable concrete production, *Materials*. 15(15) (2022) 5276.

4. M. Elimelech, W.A. Phillip. The future of seawater desalination: Energy, technology, and the environment, *Science*. 333(6043) (2011) 712–717.

5. Y. Zhao, X. Hu, C. Shi, Z. Zhang, D. Zhu. A review on seawater sea-sand concrete: Mixture proportion, hydration, microstructure and properties, *Construction and Building Materials*. 295 (2021) 123602.

6. Z. Li, S. Ding, L. Kong, X. Wang, A. Ashour, B. Han, J. Ou. Nano TiO_2-engineered anti-corrosion concrete for sewage system, *Journal of Cleaner Production*. 337 (2022) 130508.

7. Y. Chen, R. Yu, X. Wang, J. Chen, Z. Shui. Evaluation and optimization of ultra-high performance concrete (UHPC) subjected to harsh ocean environment: Towards an application of Layered Double Hydroxides (LDHs), *Construction and Building Materials*. 177 (2018) 51–62.

8. J.L. Granju, S.U. Balouch. Corrosion of steel fibre reinforced concrete from the cracks, *Cement and Concrete Research*. 35(3) (2005) 572–577.

9. S.U. Balouch, J.P. Forth, J.L. Granju. Surface corrosion of steel fibre reinforced concrete, *Cement and Concrete Research*. 40(3) (2010) 410–414.

10. B. Han, L. Zhang, J. Ou. *Smart and Multifunctional Concrete Toward Sustainable Infrastructures*, Springer, 2017.
11. S. Dong, Y. Wang, A. Ashour, B. Han, J. Ou. Uniaxial compressive fatigue behavior of ultra-high performance concrete reinforced with super-fine stainless wires, *International Journal of Fatigue*. 142 (2021) 105959.
12. B. Han, S. Dong, L. Zhang, S. Ding, S. Sun, Y. Wang. R&D of China's strategic new industries-functional materials, chapter 6: Functional civil engineering materials, China Machine Press (2016) 195–298. (In Chinese)
13. G.T. Burstein, P.C. Pistorius, S.P. Mattin. The nucleation and growth of corrosion pits on stainless steel, *Corrosion Science*. 35(1–4) (1993) 57–62.
14. G.O. Ilevbare, G.T. Burstein. The role of alloyed molybdenum in the inhibition of pitting corrosion in stainless steels, *Corrosion Science*. 43(3) (2001) 485–513.
15. R. Qvarfort. Some observations regarding the influence of molybdenum on the pitting corrosion resistance of stainless steels, *Corrosion Science*. 40(2–3) (1998) 215–223.
16. M.T. Marvila, A.R.G. de Azevedo, P.R. de Matos, S.N. Monteiro, C.M.F. Vieira. Materials for production of high and ultra-high performance concrete: Review and perspective of possible novel materials, *Materials*. 14(15) (2021) 4304.
17. P. Li, W. Li, T. Yu, F. Qu, V.W.Y. Tam. Investigation on early-age hydration, mechanical properties and microstructure of seawater sea sand cement mortar, *Construction and Building Materials*. 249 (2020) 118776.
18. W.L. Lam, P. Shen, Y. Cai, Y. Sun, Y. Zhang, C.S. Poon. Effects of seawater on UHPC: Macro and microstructure properties, *Construction and Building Materials*. 340 (2022) 127767.
19. M. Sun, R. Yu, C. Jiang, D. Fan, Z. Shui. Quantitative effect of seawater on the hydration kinetics and microstructure development of ultra high performance concrete (UHPC), *Construction and Building Materials*. 340 (2022) 127733.
20. U. Ebead, D. Lau, F. Lollini, A. Nanni, P. Suraneni, T. Yu. A review of recent advances in the science and technology of seawater-mixed concrete, *Cement and Concrete Research*. 152 (2022) 106666.
21. J. Gong, Y. Ma, J. Fu, J. Hu, X. Ouyang, Z. Zhang, H. Wang. Utilization of fibers in ultra-high performance concrete: A review, *Composites Part B: Engineering*. 241 (2022) 109995.
22. J. Wang, E. Liu, L. Li. Multiscale investigations on hydration mechanisms in seawater OPC paste, *Construction and Building Materials*. 191 (2018) 891–903.
23. Y. Elakneswaran, T. Nawa, K. Kurumisawa. Electrokinetic potential of hydrated cement in relation to adsorption of chlorides, *Cement and Concrete Research*. 39(4) (2009) 340–344.
24. ACI (American Concrete Institute). *Building Code Requirements for Structural Concrete*, ACI 318-19. ACI, 2019.
25. S. Dong, X. Wang, A. Ashour, B. Han, J. Ou. Enhancement and underlying mechanisms of stainless steel wires to fatigue properties of concrete under flexure, *Cement and Concrete Composites*. 126 (2022) 104372.

26. J.E. Rossen, B. Lothenbach, K.L. Scrivener. Composition of C-S-H in pastes with increasing levels of silica fume addition, *Cement and Concrete Composites.* 75 (2015) 14–22.
27. Y. Wang, Z. Shui, X. Gao, R. Yu, Y. Huang, S. Cheng. Understanding the chloride binding and diffusion behaviors of marine concrete based on Portland limestone cement-alumina enriched pozzolans, *Construction and Building Materials.* 198 (2019) 207–217.
28. X. Wang, Q. Zheng, S. Dong, A. Ashour, B. Han. Interfacial characteristics of nano-engineered concrete composites, *Construction and Building Materials.* 259 (2020) 119803.
29. P. Li, W. Li, Z. Sun, L. Shen, D. Sheng. Development of sustainable concrete incorporating seawater: A critical review on cement hydration, microstructure and mechanical strength, *Cement and Concrete Composites.* 121 (2021) 104100.
30. K. De Weerdt. Chloride binding in concrete: Recent investigations and recognised knowledge gaps: RILEM Robert L'Hermite Medal Paper 2021, *Materials and Structures.* 54(6) (2021) 214.
31. K. De Weerdt, B. Lothenbach, M.R. Geiker. Comparing chloride ingress from seawater and NaCl solution in Portland cement mortar, *Cement and Concrete Research.* 115 (2019) 80–89.

Chapter 12

Future development and challenges of stainless steel wires-engineered multifunctional ultra-high performance concrete

12.1 INTRODUCTION

As the core for developing multifunctional fiber reinforced concrete, high-performance and multifunctional fibers should not only have high mechanical properties and excellent function as well as strong bonding with the concrete matrix, but should also have excellent durability to avoid the degradation or loss of their composite effect due to severe environmental actions, thus ensuring that multifunctional concrete can maintain long-term stability of mechanical properties, durability, and functionality.

Stainless steel wires (SSWs) are a type of high-performance and multi-functional fiber with high tensile strength, excellent electrical and thermal properties, and high corrosion resistance. Meanwhile, SSWs have spatial morphology effects through their micron diameter and high aspect ratio to realize the formation of a wide range of strengthening/toughening/elect rical/thermal conductive networks at a low content level. Combined with the advantages of ultra-high performance concrete (UHPC) (i.e. a dense packing structure, without coarse aggregate, and using a large amount of fine powder), SSWs-engineered multifunctional UHPC can be developed into a new generation of multifunctional concrete in the true sense, with excellent mechanics, durability, functionality, and intelligence.

Therefore, many excellent properties of SSWs-engineered multifunctional UHPC have been verified, such as high static/dynamic strength, good toughness, good electrically/thermally conductive properties, and stable self-sensing and self-heating properties. Meanwhile, three kinds of derived SSWs-engineered multifunctional UHPC are proposed to broaden the application of UHPC. SSWs-engineered multifunctional UHPC has the potential to improve infrastructure reliability, safety, and longevity, and reduce the impact of engineering structures on resources, energy, and the environment. However, there are some issues (e.g. the industrial preparation of SSWs, the performance optimization/expansion and modification mechanisms of SSWs-engineered multifunctional UHPC, and the application of

DOI: 10.1201/9781003276357-12

SSWs-engineered multifunctional UHPC) that remain to be solved. The continuing development and in-depth discovery of SSWs-engineered multifunctional UHPC are needed.

This chapter will discuss current challenges in the research and development of SSWs-engineered multifunctional UHPC, involving aspects from fabrication to applications that should be addressed in the future. Preventive measures and future developments of SSWs-engineered multifunctional UHPC will also be put forward.

12.2 INDUSTRIAL PREPARATION OF SSWS

The development of industrial manufacturing technology and high-efficiency dispersing equipment/methods for SSWs are the premise for the preparation of SSWs-engineered multifunctional UHPC with high cost-effectiveness.

SSWs with micro diameter and excellent electrically conductive properties made from stainless steel by strong drawing have been widely used for producing electromagnetic shielding fabrics [1, 2]. SSWs need to be cut into a certain length before being added to UHPC. Research into high-efficiency drawing and cutting apparatus is very necessary; even advanced manufacturing technology such as intelligent construction, 3D printing, and digital manufacturing can be applied to produce SSWs. This will contribute to the industrial production of SSWs and reduce the production cost as well as energy consumption. This will further help maximize the composite efficiency of SSWs at low cost increments of UHPC and achieve the mass production of SSWs-engineered multifunctional UHPC.

Improving the dispersion of SSWs in UHPC is also an urgent problem, because good dispersion can reduce the content of SSWs and control the cost of SSWs-engineered multifunctional UHPC under the equivalent performance improvement effect. The commonly used dispersion methods include adjusting the addition sequence of SSWs and controlling the mixing rate, mixing form (vibrating mixing or forced mixing), and mixing time of concrete. The optimal mixing process considering these factors can be obtained through orthogonal or uniform experimental design. Meanwhile, blowing SSWs into UHPC using air-blasting equipment or sifting SSWs into UHPC using a steel screen of suitable particle size is expected to be an effective way to improve the dispersion of SSWs in UHPC. It should be noted that SSWs with micro diameter adhere easily to the surface of objects in the process of being blown or dispersed, and anti-static measures should be considered during the process of microfilament dispersion. Besides, applying an induced electric or magnetic field to SSWs-engineered multifunctional UHPC also has potential to improve the dispersion of SSWs in UHPC, but attention should be paid to the effect of the temperature increase caused by the electric field or magnetic field on hydration products.

In addition, nanotechnology, advanced composite technology, and bio-technology provide feasible technical means for fabricating new types of SSWs with high dispersion. For example, combining SSWs and nanomaterials using in-situ growth, chemical deposition, or grafting technology [3–6] to produce a new type of easily dispersed, high-performance, multifunctional SSWs will bring unexpected modification effects for UHPC.

12.3 PERFORMANCE OPTIMIZATION AND EXPANSION OF SSWS-ENGINEERED MULTIFUNCTIONAL UHPC

To meet the resources, energy, and environment challenges, concrete needs to develop in the direction of high performance (i.e., excellent mechanics, durability, machining properties, etc.) and multifunctionality (including electrical, thermal, electrothermal, electromagnetic, self-sensing, etc.) to maintain its sustainability [7, 8]. SSWs-engineered multifunctional UHPC has demonstrated excellent strength, toughness, electrical, thermal, self-sensing, and self-heating properties, and thus shows great application prospects in ultra-high and large-span structures. Nonetheless, the performance of SSWs-engineered multifunctional UHPC still needs to be continuously improved and expanded in order to meet the requirements of a complex service environment, long-life structures and special protective structure design, and low carbon emissions.

For example, the effect of SSWs on the workability and rheological behavior of UHPC should be clarified in order to obtain hardened composites with dense microstructure and guide on-site construction. The feasibility of using digital construction technology, i.e. 3D printing technology, to fabricate SSWs-engineered multifunctional UHPC specimens or components is well worth exploring. The performance stability of composites should be further investigated under complex service environments such as marine, saline, and extreme cold environments. Durability testing of SSWs-engineered multifunctional UHPC can be carried out in two forms: an accelerated corrosion environment and the same environment with infrastructures. In view of the special requirements of certain structures, in-depth research is also needed on the high temperature resistance, electromagnetic wave absorption/shielding characteristics, and bullet/artillery fire resistance properties of SSWs-engineered multifunctional UHPC. In addition, it is necessary to carry out a performance test on full-size components of SSWs-engineered multifunctional UHPC before application.

Besides, the three types of derived SSWs-engineered multifunctional UHPC demonstrated in Chapters 9–11 have been preliminarily investigated, but further research should be carried out on their comprehensive performance. Furthermore, the diameter and length of SSWs used for fabricating multifunctional UHPC in this book are constant. In future research, SSWs with different aspect ratios can be incorporated or combined with other types of functional fillers such as nanofillers, bio-based fillers, and shape memory alloy fillers to enhance the performance of UHPC.

12.4 MECHANISMS AND MODELS OF SSWS-ENGINEERED MULTIFUNCTIONAL UHPC

The acquisition of modification mechanisms (SSWs dispersion state, SSWs failure mode, interface feature between SSWs and concrete) of SSWs and the establishment of a constitutive model for SSWs-engineered multifunctional UHPC have an important guiding function on its application.

The mechanisms by which SSWs modify UHPC that have been explained mainly include overlapping, bridging, deflection, and being pulled off. However, these mechanisms have been obtained by planar image analysis of composites, and the 3D random distribution of SSWs in UHPC remains unclear. The acquisition of the 3D random distribution state of SSWs using micro computer tomography (CT) technology is very helpful to determine the conductive threshold of SSWs and regulate the self-sensing and self-heating performance of SSWs-engineered multifunctional UHPC. Meanwhile, finite element models (e.g. Abaqus, Ansys) can be used to obtain the 3D random dispersion parameters (orientation coefficient, distribution coefficient, and SSWs spacing) of SSWs. Furthermore, the probability of encounter between widely distributed SSWs and random cracks generated under load obtained by digital image correlation (DIC) technology can be used to explain the strengthening and toughening mechanisms of SSWs more scientifically.

Due to the high specific surface area of SSWs, fine powder can be adsorbed on the surface of SSWs and continue to hydrate. This is conducive to the enhancement of the interface between SSWs and the UHPC matrix. The bond strength and failure mode between SSWs and the UHPC matrix can be obtained by the single SSW drawing test. The microstructure, pore characteristics, hydration product morphology, and nanomechanical characteristics in the interfacial transition zone between SSWs and the UHPC matrix can be demonstrated to verify the low rate of defects in SSWs-engineered multifunctional UHPC. The relationships between SSWs' failure state (being pulled out or being pulled off) and interface bonding strength, as well as geometric parameters of SSWs, can be obtained by designing different anchoring lengths during the single SSW drawing test and numerical simulation calculation. And then, the critical pulling-out strength of SSWs can be calculated to provide guidance for the selection of geometric parameters of SSWs used in multifunctional UHPC.

The uniaxial compressive constitutive model based on damage theory, the electro-mechanical constitutive model based on percolation threshold theory, and the electrothermal constitutive model based on energy conservation have been established for SSWs-engineered multifunctional UHPC. However, coupling factors such as environment and load should be considered to modify these models and improve the fitness between the constitutive model and actual structures. Meanwhile, the influence of specimen size needs to be excluded in the constitutive model.

12.5 NUMERICAL CALCULATION OF SSWS-ENGINEERED MULTIFUNCTIONAL UHPC

Numerical calculation can run through the whole process of design, fabrication, performance, mechanism analysis, and application of SSWs-engineered multifunctional UHPC to achieve more precise design and reduce experimental workload, as shown in Figure 12.1.

Materials genomics [9–12] can be used to reveal the relationships between the composition of raw materials and the properties of composites. By establishing basic units such as SSWs-centered micro-enhanced regions and using high-throughput characterization techniques, it is possible to achieve the purposeful design of SSWs-engineered multifunctional UHPC. Molecular dynamics and the first principle method can be used to explain the source of good interfaces between SSWs and the UHPC matrix. Based on probabilistic numerical simulation methods such as the Monte Carlo method [13–15], the SSWs content with the optimal modification effect on the mechanical properties of UHPC and the electrically/thermally conductive threshold content of SSWs can be obtained.

The multifunctional properties of SSWs-engineered multifunctional UHPC can be analyzed based on multiscale physical field simulation. Meanwhile, the mechanical and durability properties of composites at full size under the actual load are simulated according to finite element numerical calculation. Machine learning models [16, 17, 18] and digital

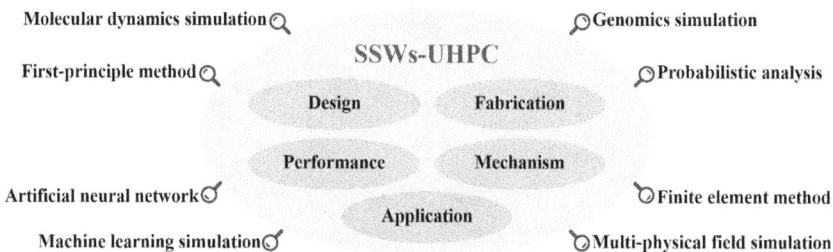

Figure 12.1 Numerical calculation methods

twin technology [14], such as neural networks, support vector machines, decision trees, and random forests, can be used individually or in combination to predict the mechanical and functional properties of specimens and components of SSWs-engineered multifunctional UHPC.

12.6 APPLICATION OF SSWS-ENGINEERED MULTIFUNCTIONAL UHPC

Due to its excellent mechanical and multifunctional properties, SSWs-engineered multifunctional UHPC is expected to be used in post-pouring nodes of prefabricated structures, key parts of transportation infrastructures (e.g. high-speed railways, highways, long tunnels, and airport runways), long-span bridge structures, and nuclear industrial facilities, as shown in Figure 12.2.

It is worth noting that the life-cycle assessment should be considered during the application of SSWs-engineered multifunctional UHPC. This comprises the drawing and cutting of SSWs, the mixing, forming, and curing of composites, and the maintenance, service life, and recycling of composite structures. Prior to application, relevant technical guidelines for SSWs-engineered multifunctional UHPC should be specified. The following three aspects in the relevant application technical procedures are suggested for inclusion. The first part may involve the mechanical properties, mixing/pouring procedure, and property test methods of SSWs-engineered

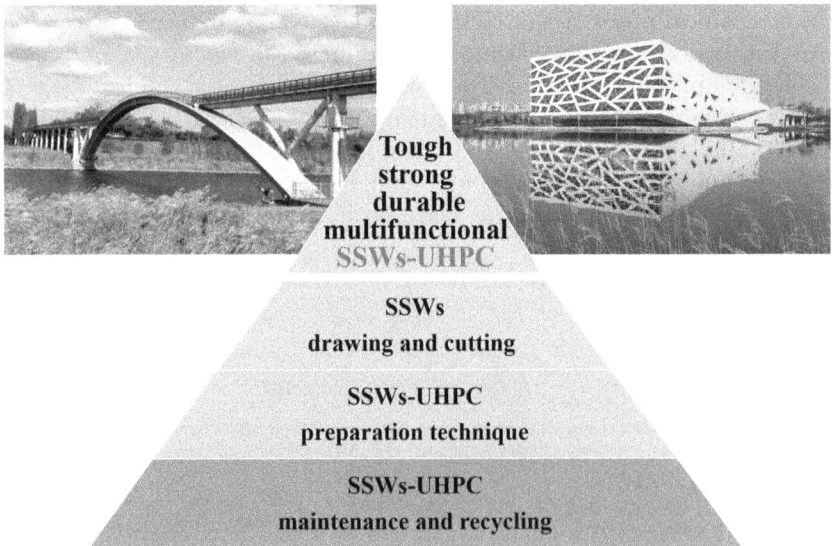

Figure 12.2 Application of SSWs-engineered multifunctional UHPC

multifunctional UHPC, including compressive strength, bending strength, tensile strength, elastic modulus, Poisson's ratio, thermal expansion coefficient, creep shrinkage characteristics, impact resistance characteristics, etc. The design and analysis methods, including the flexural, shearing, and torsional resistance of components under the normal service limit state and the bearing capacity limit state of SSWs-engineered multifunctional UHPC structures may be described in the second part. The third part may focus on durability design of SSWs-engineered multifunctional UHPC structures, involving porosity, chloride ion diffusion, calcium hydroxide content, stability of admixture, hydration delay, corrosion resistance of SSWs and steel bar, etc.

12.7 SUMMARY

Due to its strong, durable, resilient, and smart/multifunctional nature, ease of fabrication, and low environmental footprint characteristics, SSWs-engineered multifunctional UHPC has been proved to be an excellent multifunctional material with wide application prospects. However, basic theories, including properties, mechanisms, and the quantitative

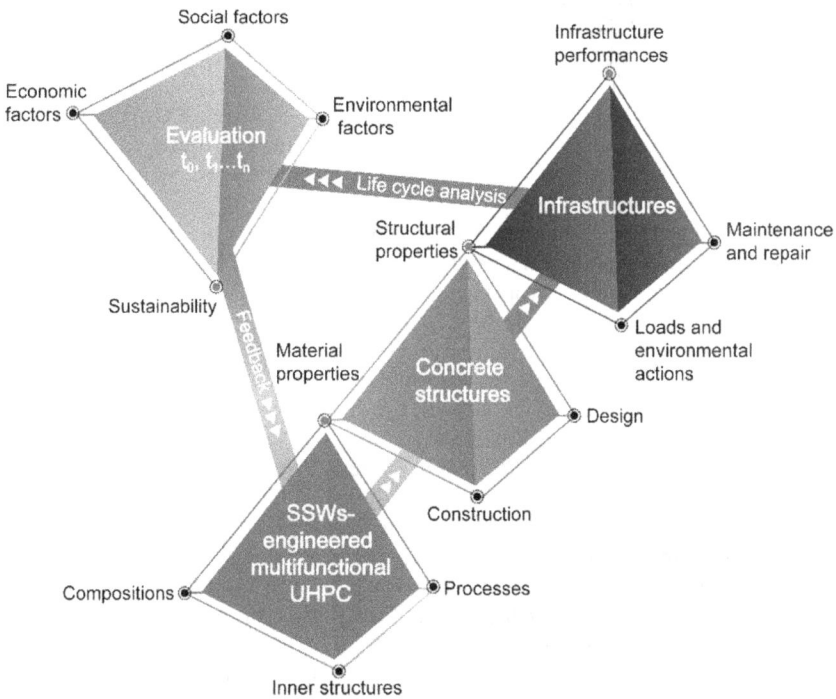

Figure 12.3 Life-cycle assessment of SSWs-engineered multifunctional UHPC

relationship between composition–structure–process–performance; life-cycle analysis, including economic, social, and environmental aspects; and the development of standards/specifications for engineering applications of SSWs-engineered multifunctional UHPC urgently need further investigation to provide guidance for its performance control and application, as shown in Figure 12.3.

SSWs-engineered multifunctional UHPC with fine design can be used in large-span and high-rise civil infrastructures, important transportation infrastructures (e.g. high-speed railways, highways, long tunnels, and airport runways), nuclear industrial facilities, and infrastructures in complex and extreme environments, presenting the unlimited possibility for energizing the sustainable development of concrete structures.

REFERENCES

1. K.F. Casey. Electromagnetic shielding behavior of wire-mesh screens, *IEEE Transactions on Electromagnetic Compatibility.* 30(3) (1988) 298–306.
2. A. Khodadadi, M.H. Nazari, S.H. Hosseinian. Designing an optimal lightning protection scheme for substations using shielding wires, *Engineering, Technology and Applied Science Research.* 7(3) (2017) 1595–1599.
3. S.Q. Ding, X.Y. Wang, L.S. Qiu, Y.Q. Ni, X.F. Dong, Y.B. Cui, A. Ashour, B.G. Han, J.P. Ou. Self-sensing cementitious composites with hierarchical carbon fiber-carbon nanotube composite fillers for crack development monitoring of a maglev girder, *Small.* 19(9) (2023) 2206258.
4. S.Q. Ding, Y. Xiang, Y.Q.Ni, V.K. Thakur, X.Y. Wang, B.G. Han, J.P. Ou. In-situ synthesizing carbon nanotubes on cement to develop self-sensing cementitious composites for smart high-speed rail infrastructures, *Nano Today.* 43 (2022) 101438.
5. S.F. Dong, D.N. Wang, A. Ashour, B.G. Han, J.P. Ou. Nickel plated carbon nanotubes reinforcing concrete composites: From Nano/micro structrues to macro mechanical properties, *Composites Part A: Applied Science and Manufacturing.* 141 (2021) 106228.
6. S.F. Dong, Y.L. Wang, A. Ashour, B.G. Han, J.P. Ou. Nano/micro-structures and mechanical properties of ultra-high performance concrete incorporating graphene with different lateral sizes, *Composites Part A: Applied Science and Manufacturing.* 137 (2020) 106011.
7. B. Han, X. Yu, J. Ou. *Self-Sensing Concrete in Smart Structures*, Elsevier, 2014.
8. B. Han, L. Zhang, J. Ou. *Smart and Multifunctional Concrete Toward Sustainable Infrastructures*, Springer, 2017.
9. A. Agrawal, A. Choudhary. Perspective: Materials informatics and big data: Realization of the "fourth paradigm" of science in materials science, *APL Materials.* 4(5) (2016) 053208.
10. J.J. de Pablo, N.E. Jackson, M.A. Webb, L. Chen, J.E. Moore, D. Morgan, R. Jacobs, T. Pollock, D.G. Schlom, E.S. Toberer, J. Analytis, I. Dabo, D.M. DeLongchamp, G.A. Fiete, G.M. Grason, G. Hautier, Y. Mo, K. Rajan, E.J.

Reed, E. Rodriguez, V. Stevanovic, J. Suntivich, K. Thornton, J. Zhao. New frontiers for the materials genome initiative, *NPJ Computational Materials.* 5(1) (2019) 41.

11. A. Jain, S.P. Ong, G. Hautier, W. Chen, W.D. Richards, S. Dacek, S. Cholia, D. Gunter, D. Skinner, G. Ceder, K.A. Persson. Commentary: The materials project: A materials genome approach to accelerating materials innovation, *APL Materials.* 1(1) (2013) 011002.

12. Y. Wang, Y. Liu, S. Song, Z. Yang, X. Qi, K. Wang, Y. Liu, Q. Zhang, Y. Tian. Accelerating the discovery of insensitive high-energy-density materials by a materials genome approach, *Nature Communications.* 9(1) (2018) 2444.

13. H. Diaz, A.P. Teixeira, C.G. Soares. Application of Monte Carlo and fuzzy analytic hierarchy processes for ranking floating wind farm locations, *Ocean Engineering.* 245(1) (2022) 110453.

14. J. LeBlanc, A.E. Antipov, F. Becca, I.W. Bulik, G. Chan, C.M. Chung, Y. Deng, M. Ferrero, T.M. Henderson, C.A. Jimenez-Hoyos, E. Kozik, X. Liu, A.J. Millis, N.V. Prokof'Ev, M. Qin, G.E. Scuseria, H. Shi, B.V. Svistunov, L.F. Tocchio, I.S. Tupitsyn, S.R. White, S.W. Zhang, B. Zheng, Z. Zhu, E. Gull, C.M. Simons. Solutions of the two-dimensional Hubbard model: Benchmarks and results from a Wide Range of numerical algorithms, *Physical Review X.* 5(4) (2015) 041041.

15. H. Janssen. Monte-Carlo based uncertainty analysis: Sampling efficiency and sampling convergence, *Reliability Engineering and System Safety.* 109 (2013) 123–132.

16. M. Shariati, M.S. Mafipour, B. Ghahremani, F. Azarhomayun, M. Ahmadi, N.T. Trung, A. Shariati. A novel hybrid extreme learning machine-grey wolf optimizer (ELM-GWO) model to predict compressive strength of concrete with partial replacements for cement, *Engineering with Computers.* 38(1) (2022) 757–779.

17. W.B. Chaabene, M. Flah, M.L. Nehdi. Machine learning prediction of mechanical properties of concrete: Critical review, *Construction and Building Materials.* 260 (2020) 119889.

18. Q. Min, Y. Lu, Z. Liu, C. Su, B. Wang. Machine learning based digital twin framework for production optimization in petrochemical industry, *International Journal of Information Management.* 49 (2019) 502–519.

Index

For Product Safety Concerns and Information please contact our EU
representative GPSR@taylorandfrancis.com
Taylor & Francis Verlag GmbH, Kaufingerstraße 24, 80331 München, Germany